AF443734

Animals live in a wide range of habitats, and many often face a shortage of oxygen, an essential requirement for life. This book, written by experts in their fields, explains the problems animals face, from annelid worms living in oxygen-deficient muds in the sea, to the active bird whose requirement for oxygen during flight makes great demands on its respiratory system.

The approach taken is essentially comparative, with individual authors chosen so that their own areas of expertise complement the whole subject. Each chapter starts with a short review and this is then supplemented by the latest research results. The resulting book provides a wider synthesis and integration of data with existing ideas and theories than is possible in traditional journals. Researchers studying respiratory physiology as well as students studying zoology and physiology will find this book of great interest.

SOCIETY FOR EXPERIMENTAL BIOLOGY
SEMINAR SERIES · 41

PHYSIOLOGICAL STRATEGIES FOR GAS EXCHANGE AND METABOLISM

SOCIETY FOR EXPERIMENTAL BIOLOGY
SEMINAR SERIES

A series of multi-author volumes developed from seminars held by the Society for Experimental Biology. Each volume serves not only as an introductory review of a specific topic, but also introduces the reader to experimental evidence to support the theories and principles discussed, and points the way to new research.

1. Effects of air pollution on plants. *Edited by T.A. Mansfield*
2. Effects of pollutants on aquatic organisms. *Edited by A.P.M. Lockwood*
3. Analytical and quantitative methods. *Edited by J.A. Meek and H.Y. Elder*
4. Isolation of plant growth substances. *Edited by J.R. Hillman*
5. Aspects of animal movement. *Edited by H.Y. Elder and E.R. Trueman*
6. Neurones without impulses: their significance for vertebrate and invertebrate systems. *Edited by A. Roberts and B.M.H. Bush*
7. Development and specialisation of skeletal muscle. *Edited by D.F. Goldspink*
8. Stomatal physiology. *Edited by P.G. Jarvis and T.A. Mansfield*
9. Brain mechanisms of behaviour in lower vertebrates. *Edited by P.R. Laming*
10. The cell cycle. *Edited by P.C.L. John*
11. Effects of disease on the physiology of the growing plant. *Edited by P.G. Ayres*
12. Biology of the chemotactic response. *Edited by J.M. Lackie and P.C. Williamson*
13. Animal migration. *Edited by D.J. Aidley*
14. Biological timekeeping. *Edited by J. Brady*
15. The nucleolus. *Edited by E.G. Jordan and C.A. Cullis*
16. Gills. *Edited by D.F. Houlihan, J.C. Rankin and T.J. Shuttleworth*
17. Cellular acclimatisation to environmental change. *Edited by A.R. Cossins and P. Sheterline*
18. Plant biotechnology. *Edited by S.H. Mantell and H. Smith*
19. Storage carbohydrates in vascular plants. *Edited by D.H. Lewis*
20. The physiology and biochemistry of plant respiration. *Edited by J.M. Palmer*
21. Chloroplast biogenesis. *Edited by R.J. Ellis*
22. Instrumentation for environmental physiology. *Edited by B. Marshall and F.J. Woodward*
23. The biosynthesis and metabolism of plant hormones. *Edited by A. Crozier and J.R. Hillman*
24. Coordination of motor behaviour. *Edited by B.M.H. Bush and F. Clarac*
25. Cell ageing and cell death. *Edited by I. Davies and D.C. Sigee*
26. The cell division cycle in plants. *Edited by J.A. Bryant and D. Francis*
27. Control of leaf growth. *Edited by N.R. Baker, W.J. Davies and C. Ong*
28. Biochemistry of plant cell walls. *Edited by C.T. Brett and J.R. Hillman*
29. Immunology in plant science. *Edited by T.L. Wang*
30. Root development and function. *Edited by P.J. Gregory, J.V. Lake and D.A. Rose*
31. Plant canopies: their growth, form and function. *Edited by G. Russell, B. Marshall and P.G. Jarvis*
32. Developmental mutants in higher plants. *Edited by H. Thomas and D. Grierson*
33. Neurohormones in invertebrates. *Edited by M. Thorndyke and G. Goldsworthy*
34. Acid toxicity and aquatic animals. *Edited by R. Morris, E.W. Taylor, D.J.A. Brown and J.A. Brown*
35. Division and segregation of organelles. *Edited by S.A. Boffey and D. Lloyd*
36. Biomechanics in evolution. *Edited by J.M.V. Rayner*
37. Techniques in comparative respiratory physiology: A comparative approach. *Edited by C.R. Bridges and P.J. Butler*
38. Herbicides and plant metabolism. *Edited by A.D. Dodge*
39. Plants under stress. *Edited by H.G. Jones, T.J. Flowers and M.B. Jones*
40. *In situ* hybridisation: application to developmental biology and medicine. *Edited by M. Harris and D.G. Wilkinson*

PHYSIOLOGICAL STRATEGIES FOR GAS EXCHANGE AND METABOLISM

Edited by

A.J. Woakes
School of Biological Sciences
University of Birmingham, UK

M.K. Grieshaber
Department of Zoology
University of Düsseldorf, FRG

and

C.R. Bridges
Department of Zoology
University of Düsseldorf, FRG

CAMBRIDGE UNIVERSITY PRESS
Cambridge
New York   Port Chester
Melbourne   Sydney

Published by the Press Syndicate of the University of Cambridge
The Pitt Building, Trumpington Street, Cambridge CB2 1RP
40 West 20th Street, New York, NY 10011–4211, USA
10 Stamford Road, Oakleigh, Melbourne 3166, Australia

First published 1991

Printed in Great Britain at the University Press, Cambridge

British Library cataloguing in publication data
Physiological strategies for gas exchange and metabolism.
1. Animals, Respiration
I. Woakes, A.J.   II. Grieshaber, M.K.   III. Bridges, C.R.   IV. Series
501.12

Library of Congress cataloguing in publication data
Physiological strategies for gas exchange and metabolism / edited by
    A.J. Woakes, M.K. Grieshaber, and C.R. Bridges.
      p.   cm. — (Seminar series / Society for Environmental Biology
    ; 41)
      1. Respiration.   2. Energy metabolism.   3. Physiology,
Comparative.   I. Woakes, A.J.   II. Grieshaber, M.K.
III. Bridges, C.R.   IV. Series: Seminar Series (Society for
Experimental Biology (Great Britain)) ; 41.
QP121.P54   1991
591.1'2   dc20   90-1766   CIP

ISBN 0 521 36602 X hardback

CONTENTS

CONTRIBUTORS

Atkinson, R.J.A.
University Marine Biological Station, Millport, Isle of Cumbrae, Scotland, UK

Brackenbury, J.H.
Subdepartment of Veterinary Anatomy, University of Cambridge, Tennis Court Road, Cambridge CB2 1QS, UK

Bridges, C.R.
Department of Zoology IV, University of Düsseldorf, D-4000 Düsseldorf, FRG

Butler, P.J.
School of Biological Sciences, University of Birmingham, PO Box 363, Birmingham B15 2TT, UK

Burggren, W.W.
Department of Zoology, University of Massachusetts, Amherst, MA 01003-0027, USA

Köhler, P.
Institute of Parasitology, University of Zurich, Winterthurerstrasse 266a, Zurich, Switzerland

Glass, M.L.
Max Planck Institut für experimentelle Medizin, D-3400 Göttingen, FRG (present address: Departamento de Fisiologia, Universidade de São Paulo, Rebeirao Preto, Brazil)

Gnaiger, E.
Department of Zoology, University of Innsbruck, A-6020 Innsbruck, Austria

Grieshaber, M.K.
Department of Zoology IV, University of Düsseldorf, D-4000
Düsseldorf, FRG

Gros, G.
Zentrum Physiologie, Medizinische-Hochschule Hannover,
Konstanty-Gutschowstrasse 8, 3000 Hannover 61, FRG

Jensen, F.B.
Institute of Biology, Odense University, 5230 Odense M,
Denmark

Taylor, A.C.
Department of Zoology, University of Glasgow, Glasgow G12
8QQ, UK

Taylor, E.W.
School of Biological Sciences, University of Birmingham, PO
Box 363, Birmingham B15 2TT, UK

Toulmond, A.
Laboratoire de Zoologie, Université Pierre-et-Marie-Curie,
Bâtiment A, 4 Place Jussieu, 75230 Paris, France

Van den Thillart, G.
Department of Biology (Animal Physiology), Gorlaeus
Laboratories, Einsteinweg 5, PO Box 9502, 2300 RA Leiden,
The Netherlands

Van Waarde, A.
Department of Biology (Animal Physiology), Gorlaeus
Laboratories, Einsteinweg 5, PO Box 9502, 2300 RA Leiden,
The Netherlands

Wheatly, M.G.
Department of Zoology, University of Florida, Gainesville,
FLA 32611, USA

Whiteley, N.M.
School of Biological Sciences, University of Birmingham, PO
Box 363, Birmingham B15 2TT, UK

Woakes, A.J.
School of Biological Sciences, University of Birmingham, PO Box 363, Birmingham B15 2TT, UK

Wood, S.C.
Lovelace Medical Foundation, Albuquerque, New Mexico 87108, USA

PREFACE

In the course of evolution animals have adopted a variety of strategies in their 'fight for survival'. The aim of this volume is therefore to present a comparative view of some of the physiological strategies evolved in both invertebrates and vertebrates in their successful colonization of a particular environment. Comparisons are drawn from both aquatic and terrestrial physiology and range from the cellular or biochemical level to that of the whole organism. It is hoped that this comparative viewpoint will make the understanding of such processes clearer and also stimulate research in these fields.

Initially this volume arose out of two symposia organized by the Animal Respiration Group of the Society of Experimental Biology and hosted by the second International Congress of Comparative Physiology and Biochemistry held at Louisiana State University, Baton Rouge, Louisiana, USA, on August 1–5, 1988. Review lectures were presented along the lines of the two major themes of this book: (i) strategies in oxygen and carbon dioxide transport and (ii) respiratory adaptations to limited oxygen supply. These, together with two invited review papers, then formed a basis for the review chapters that appear in the present volume.

The success of this venture, culminating in the publication of this volume in the Seminar Series of the SEB, could not have been achieved without the help of many of our colleagues. We are therefore grateful to the contributors to this volume and for the efforts of a number of anonymous referees who helped in the reviewing of the manuscripts. We would also like to thank T. H. Dietz and W. B. Stickle of LSU for the local organization of the meeting, and extend our thanks to the SEB and the Royal Society for providing travel support for SEB members.

A. J. Woakes, M. K. Grieshaber and C. R. Bridges
Editors for the Society for Experimental Biology

WARREN W. BURGGREN

Does comparative respiratory physiology have a role in evolutionary biology (and vice versa)?

Rationale for this chapter

This comprehensive volume draws together timely contributions from comparative respiratory physiologists studying diverse organisms by using a variety of techniques and methodologies. Yet all have addressed a central issue in comparative respiratory physiology: adaptations of animals that allow them to survive, if not thrive, in a wide range of environments that threaten respiratory homeostasis. Most authors have taken a truly comparative approach, in which no single organism is isolated for study, but rather broad taxonomic groups are investigated. By taking this approach, each author has successfully revealed not only respiratory adaptations unique to particular taxa, but also adaptations common to a diverse range of organisms. Thus, we are slowly but surely identifying 'universal' versus 'exceptional' respiratory adaptations.

In listening to the symposium presentations, and in reading the resulting papers, I quickly discovered that the word '*adaptation*' has been used numerous times by each author. When we discuss respiratory adaptations we are, of course, looking at the most recent product of a sequence of evolutionary changes from ancestral animals. I suspect that virtually every author in this volume (and I venture most authors of journal articles on comparative respiratory physiology) has, during the conception, execution or analysis of their experiments, thought about the evolutionary processes leading to the respiratory adaptations they study. Indeed, as will be emphasized below, understanding the evolution of physiological processes is one of the chief goals of comparative physiology. Yet the word '*evolution*' is but rarely mentioned in this volume, and there are few explicit references to actual evolutionary changes in physiological systems.

The purpose of this essay is to encourage comparative respiratory physiologists to: (i) continue to expand the scope of their studies to examine the actual evolutionary processes leading to respiratory adap-

1

tations (in addition to examining the adaptations in a strictly physiological context); and (ii) approach questions of physiological evolution by using the numerous tools of contemporary evolutionary biology that have proven so useful to comparative anatomists and systematists.

Studying physiological evolution

To observe that most authors in this volume have not explored the evolutionary implications of their findings is certainly not to offer criticism. To the contrary, because we know so little about many of the fascinating physiological systems that are being discussed in this volume, a detailed discussion of evolutionary processes would be highly speculative at best, and the authors are to be complimented on successfully restricting the potentially broad scope of their subjects. Yet, as the field of respiratory physiology matures, and as we come to know more of physiological mechanisms and the respiratory adaptations that they represent, we will turn increasingly to examining *how these respiratory mechanisms and adaptations evolved*. That is, we shall begin to integrate comparative respiratory physiology and evolutionary biology.

The degree of success with which we shall achieve this integration will depend to some extent upon how thoroughly we adopt the tools of modern evolutionary biology. As physiologists, we all know instances in which anatomists have incorrectly predicted physiology on the basis of anatomy, of how ecologists have incorrectly implied physiological mechanisms on the basis of whole animal energetics, etc. The intent has always been sincere, but in these instances the anatomist or ecologist has been unwilling to learn about physiology before venturing, however fleetingly, into a new field. The same lesson can be directed at us. Accordingly, we as physiologists must be willing to learn the tools of evolutionary biology if we are to analyse physiological evolution.

What are these tools? It is beyond the scope of this essay to define them in detail; such tools deserve more than the passing mention that could be afforded here and have been discussed repeatedly (mostly outside the physiological literature) (Lauder, 1982; Cracraft, 1983; Felsenstein, 1985; Futuyma, 1987; Huey & Bennett, 1987; Patterson, 1987; Huey, 1987; Coddington, 1988; Herring, 1988). Let it suffice to present two examples of useful tools of modern evolutionary biology. First, there are new *statistical procedures designed for interspecific comparisons*. One might reasonably ask 'what is different about interspecific comparisons?' As Huey (1987) has recently discussed in detail, species are not statistically independent sampling units, but rather are linked together by often strong phylogenetic affinities. If closely related species are automatically assumed to be statistically independent units, as almost invariably is the case in physiological studies, we risk an exaggerated Type I statistical error by using an artificially inflated degree of freedom in our calculations. Statistical tools to deal with lack of independence between species (e.g. nested analysis of variance; variance partitioning) have been proposed (Crook, 1965; Ridley, 1983; Clutton-Brock & Harvey, 1984; Mace *et al.*, 1981; Stearns, 1983, 1984; Cheverud *et al.*, 1985; Dunham & Miles, 1985; Felsenstein, 1985). Unfortunately, this potential problem in comparative studies is rarely recognized by comparative physiologists (but see Huey, 1987; Huey & Bennett, 1987).

A second example of a useful tool to be borrowed from evolutionary biologists is *cladistic analysis*. This approach uses the presence of shared, derived characters to establish systematic relationships between species (Hennig, 1979; Wiley, 1981). Traditional cladistic analysis has relied extensively upon anatomical characters that are qualitative in nature (e.g. presence or absence of a structure, rather than its size). While many cladists would argue that quantitative physiological characters do not lend themselves well to cladistic analyses, systematic inferences drawn from qualitative physiological characters (e.g. endothermy, formation of hypertonic urine, haemocyanin modulation by lactate) are just as viable for cladistic analysis as are skeletal articulations or protein patterns in gel electrophoresis (see Burggren & Bemis, 1990). In any event, most comparative physiologists will use the concepts inherent in cladistic methods to predict patterns of physiological evolution within already established phylogenies, rather than to construct new phylogenies (though we may be able to contribute to them).

I wish to emphasize that, while I advocate that physiologists study the evolution of respiratory adaptations using the tools of modern evolutionary biology, I am not simply airing my personal opinion that we should do so. Rather, we must realize that many contemporary evolutionary biologists[1] appear to have a general and broad disrespect for how comparative physiologists have regarded and continue to regard evolutionary processes, and for the utility of any resulting information in understanding the evolution of a species (see discussion in Burggren & Bemis, 1990). To overstate the case bluntly, *no one outside of comparative physiology is listening to what we say about evolution*!

---

1 By 'evolutionary biologist' I mean the great number of those biologists who consider ecology, systematics, behaviour, physiology or other disciplines as tools for studying evolution, rather than as an end in themselves.

To help understand the current nature of the interaction (or lack of interaction) between comparative physiology (including comparative *respiratory* physiology) and evolutionary biology, let us examine the 'evolution' of the goals of comparative respiratory physiology.

Goals of comparative respiratory physiology

During the late 1800s and early 1900s, concomitant with major advances and refinements in theories and techniques, many innovative new approaches to comparative respiratory physiology were being established, led by Paul Bert and then August Krogh. Measurement of whole animals' metabolic rates remained a staple of respiratory physiology during the first half of this century (and well beyond), but experimental techniques were also being developed to permit the investigation of blood O_2 and CO_2 transport and acid–base balance, as just two examples. Attention began to be turned towards respiratory adaptations of unusual animals in unusual environments, an approach championed by August Krogh (see Krebs, 1975) and later used to great effectiveness by Lawrence Irving and Per Scholander, among many others.

These early 'modern' studies were guided by two related goals of comparative respiratory physiology: (i) to understand how respiratory systems function; and (ii) to understand the physiological processes allowing animals to exchange and transport gases in extreme environments. The first goal was a common one shared by comparative physiologists and more clinically related physiologists alike. The second goal essentially involved the study of respiratory adaptations. Of course, these goals remain at the centre of most contemporary studies in comparative respiratory physiology.

The goals of comparative physiology (including comparative *respiratory* physiology) have been explicitly stated and re-examined numerous times in recent years (see chapters in Greenberg *et al.*, 1975; see also Ross, 1981; Greenberg, 1985; Randall, 1986; Downer, 1986; Feder *et al.*, 1987). In fact, no biological field of which I am aware seems to subject itself to such intense, regular (and healthy!) self-scrutiny. Less common, however, have been pragmatic statements of how to achieve the goals of comparative physiology most effectively (Huey, 1987). A survey of the early literature in respiratory physiology indicates that the prevailing attitude of the first half of this century was that the goal of understanding respiratory adaptations could best be achieved by making metabolic and respiratory measurements on a wide variety of animals, often exotic species from exotic places. As a result of this approach, comparative respiratory physiologists can base new experiments on an extremely large and diverse collection of data on respiratory adaptations in animals. Yet, to understand completely respiratory adaptations to the environment, it is an implicit but fundamental requirement that we understand *how such adaptations evolved*, not just how they currently operate in living animals. Presumably, most comparative physiologists regard this as a truism. Yet, with the tremendous advantage of hindsight, the 'encyclopaedic' categorization of the myriad respiratory adaptations of animals, only infrequently guided by testable hypotheses constructed *a priori*, has largely failed to achieve the fundamental goal of understanding how physiological processes actually evolved, as distinct from how extant animals are adapted to their environment. This criticism applies not just to comparative respiratory physiology, of course, but to all of comparative physiology (see Greenberg *et al.*, 1975; Ross, 1981; numerous chapters in Feder *et al.*, 1987).

The view that our in-depth studies have not revealed to us in equal depth how physiological processes have evolved is somewhat controversial, and will not be endorsed by comparative respiratory physiologists following highly traditional paths. It therefore is illuminating to ask 'what has been the overall contribution of comparative respiratory physiologists to evolutionary biology?' To answer this question, we should see how biologists outside of our discipline regard our studies on respiratory adaptations. After all, our studies will be entirely self-serving if we can't make generalizations useful to other biologists. Up until the past few years, I had naively assumed that comparative physiologists were indeed studying evolutionary problems by using reasonable paradigms applied to reasonable theory. Thus, I was disturbed to discover recently that many (if not most) contemporary evolutionary biologists hold the view that there is little or no merit in studying physiology if one's goal is to learn how animals evolve (for detailed discussion see Burggren & Bemis, 1990). In fact, recent books on evolutionary biology (Mayre & Provine, 1980; Futuyma, 1987) typically do not even list the word 'physiology' in otherwise extensive indexes. As C. Ladd Prosser (1986) (under)states, 'evolutionists pay little attention to physiology'.

Why are 'evolutionary' studies in comparative physiology (respiratory physiology included) so broadly and readily ignored by evolutionary biologists outside of our field? A complete answer is beyond the scope of this introductory chapter, because it involves a critique of how physiologists have traditionally studied animal adaptations and evolutionary processes (see Huey, 1987; Bennett, 1987; Burggren & Bemis, 1990, for reviews of this subject). Suffice it to say that the views of our colleagues in evolutionary biology have been shaped by the sad fact that traditional comparative physiology is performed outside of a framework of modern evolutionary biology, often embracing theories, positions or approaches that contemporary morphologists, evolutionary biologists and geneticists have long abandoned (cf. Mayr & Provine, 1980; Bennett, 1987; Huey, 1987). Too many comparative investigations that purport to investigate the evolution of respiratory adaptations do not test a specific hypothesis that has been formulated on the basis of rigorous evolutionary theory and sound systematics. While Carl Gans (1970) warned us of this problem in his classic paper *'Respiration in early tetrapods – the frog is a red herring'*, it would seem that as a group comparative physiologists have not been overly receptive to the suggestion that they make forays into evolutionary biology and systematics. Again, to paraphrase C. Ladd Prosser (1986), 'most physiologists have a only a superficial knowledge of evolution.'

To develop a specific example of how comparative physiologists can err in studies with evolutionary overtones, the present author singles out one of his own early papers for criticism (primarily because this study is a good example of a bad evolutionary study, but also to disarm those who would suggest that he is quick to criticize others). Burggren (1976) conducted a study examining the relationship between ventilation patterns and heart rate in two chelonians: the freshwater turtle *Pseudemys scripta* and the terrestrial tortoise *Testudo graeca*. The intent was to contrast the effects of terrestrial and aquatic life on cardiorespiratory function. Sidestepping the issue of the physiological merits of the study, *from the perspective of a study in evolutionary biology* the author committed significant trangressions. First, Burggren (1976) made a rather popular, simple assumption: that the physiological differences between an aquatic and a terrestrial species would reflect only adaptations to the aquatic and terrestrial habitat. Unfortunately, this common assumption ignores the fact that *Testudo* and *Pseudemys* are phylogenetically distinct members of lineages which have been evolving separately for millions of years. Many of the observed differences may have little or nothing to do with aquatic or terrestrial habits. No attempt was made to determine whether the observed breathing patterns, ventilation tachycardia, atropine sensitivity of heart rate, beta-adrenergic excitation of the heart, etc. were (i) shared primitive features common to other Chelonia, (ii) shared derived features common to only these Chelonia, or (iii) unique to the genus *Testudo* or *Pseudemys*. As a result, the evolutionary meaning (as distinct from the physiological meaning) of any of the differences Burggren (1976) found is ambiguous. In fact, the physiological differences between these two chelonian genera might be due solely to *the phylogenetic distance between them* and not to physiological adaptations to different environments, a possibility that cannot be refuted owing to the design of Burggren's study (or most other comparative respiratory studies purporting to differentiate between aquatic and terrestrial species).

A far more appropriate way to have tested the hypothesis that habitat affects physiology would have been to compare three or more closely related species that occupy different habitats. The more closely related the species (preferably within the same genus), the more confidence we can have that the adaptations we find are related to specific environmental differences. The power of this approach, which really is just the standard comparative method applied rigorously in the context of an animal's phylogenetic history, has been beautifully demonstrated by Huey (1987) in his discussion of the comparative method in physiology. Figure 1, modified and expanded from Huey's (1987) paper, indicates that ignoring phylogeny when comparing physiological data from just two species (especially distantly related species) can lead to quite erroneous conclusions about the effect of environment on a physiological response or adaptation.

Returning to Burggren's (1976) study of cardiorespiratory adaptations in chelonians, the results from chelonians were compared with other 'reptiles' such as alligators. The 'reptiles' are a paraphyletic taxonomic grouping: i.e. a taxonomic group, more convenient than accurate, that does not include all of the descendants of a single common ancestor and whose members do not uniquely share derived characters (Figure 2). This contrasts with a monophyletic group (e.g. the Lepidosaurs: sphenodonts, lizards and snakes), which all share derived characters that are not found in any other tetrapod taxa. By placing birds and crocodilians in separate taxonomic groups, even though crocodilians are much more closely related to birds than to turtles, we emphasize differences between birds and crocodilians rather than similarities. From a strictly phylogenetic perspective, there is no reason to expect greater similarities in physiology between turtles and alligators than between turtles and birds. Now, we know after decades of *empirical* experimentation that reasonable and interesting comparisons can be made between alligators and turtles. These comparisons are valid because both groups, for example, are ectotherms, exhibit intermittent breathing, and have central cardiovascular shunts, but *not* because of a spurious close phylogenetic relationship suggested by the word 'reptile'!

The appropriateness of using the word 'reptile' (or 'great apes' or 'annelids') may seem like a simple semantic argument. Yet, when we begin

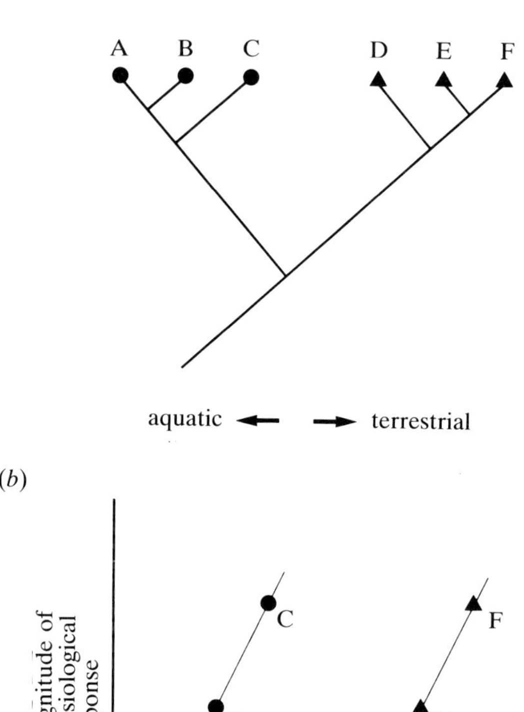

Figure 1. (*a*) A cladogram of six hypothetical animal species. Superimposed on this phylogenetic analysis is an indication of the environment these animals live in, ranging from completely aquatic to fully terrestrial. Parts (*b*), (*c*) and (*d*) show possible regressions of physiological characteristics of the animals against environment. Part (*b*) shows that correlations between environment and physiology are clearly evident when phylogeny is taken into account and several species are examined. Separate analysis of the two taxonomic groups indicates a very pro-

to investigate new taxa or make new comparisons where we don't have the benefit of empirically derived knowledge, we should be particularly careful to avoid structuring our comparisons around paraphyletic groups. If we fail to plan our experiments according to sound phylogenetics (i.e. if we do no studies on monophyletic taxa), comparative physiologists may be led to

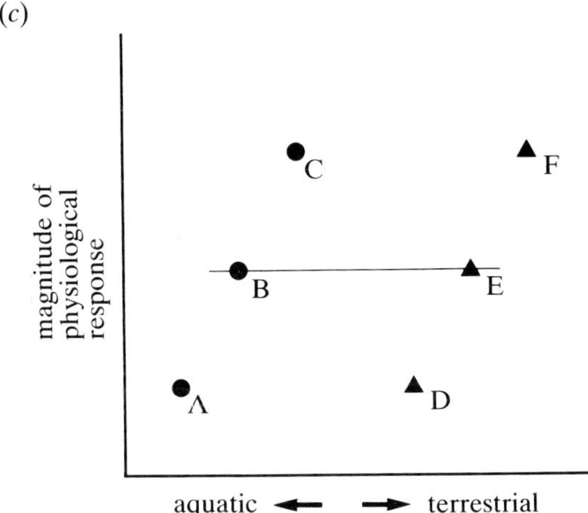

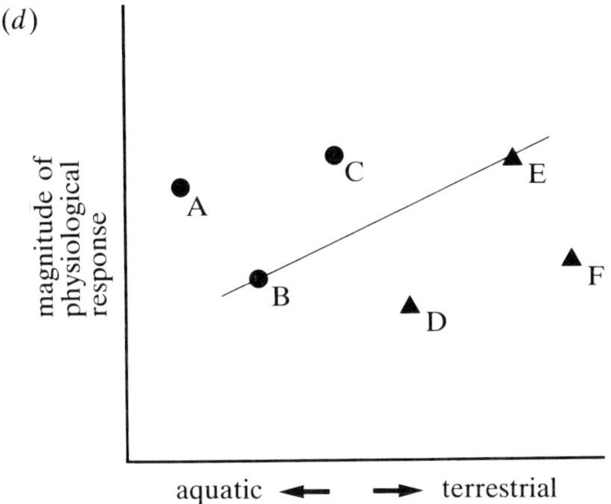

Caption to fig. 1. (cont.)
nounced effect of environment. (*c*) The regression when comparing only species B and E (a common approach in comparative physiology). The erroneous conclusion from this regression is that environment has no effect on physiological characteristics. Part (*d*) shows an alternative situation in which there is no effect of environment, but analysis of only two distantly related species erroneously suggests that there is. (Modified and expanded from Huey, 1987.)

erroneous interpretations of the evolution of physiological processes. Just as unfortunately, we may miss the most interesting of the *predictive* possibilities of our work, because predictions about evolutionary processes can only make sense in the light of phylogeny.

To give one brief example of the power of using rigorous phylogenetic

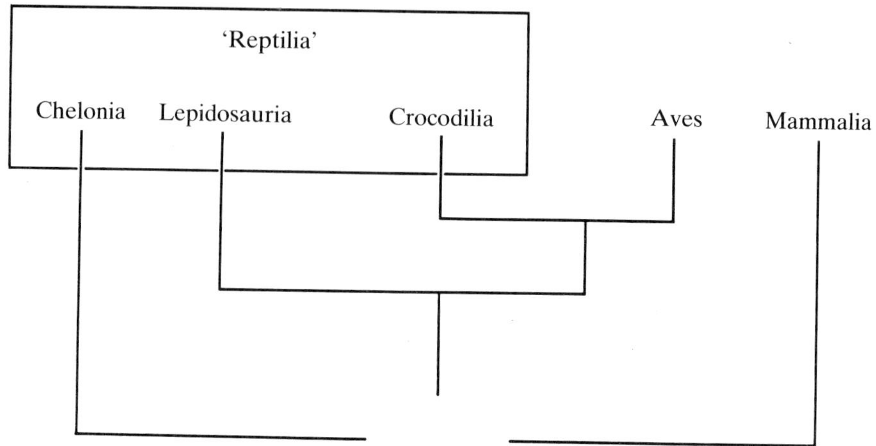

Figure 2. As shown in this conventional phylogeny of the living amniote vertebrates, the 'Reptilia' are a paraphyletic taxonomic group. For example, the Crocodilia are included with the Chelonia, even though these two taxa are very distantly related. Note that for clarity numerous fossil groups have been excluded from the Reptilia. Also, since it is not the purpose of this figure to debate the origins of the amniotes, the relationship between stem reptiles, thecodonts and therapsids has purposely been left undefined.

approaches to physiological studies, consider the unpublished findings of R. Huey, F. van Berkum, A.F. Bennett & P.E. Hertz (discussed in Huey, 1987) on the thermal dependence of sprint speed in lizards. The authors' operating hypothesis was that lizards that are normally active over a broad range of body temperature should also sprint effectively through a broad range of body temperatures: that is, they should be 'thermal generalists'. This hypothesis predicts that species with a high standard deviation of field body temperatures should also be able to sprint well over a wide range of body temperatures. Huey and co-workers carefully considered the phylogenetic relationships of the species they used and, quite appropriately, they initially analysed patterns within groups of closely related lizards (genera and families) rather than lumping both close and distant relatives in a single grouping. Taking this approach, dictated by sound systematics, they found that, indeed, thermal generalists (i.e. those that sprinted well over a wide range of body temperatures) were typically active over a wide range of body temperatures in nature. However, to illustrate the pitfalls of ignoring phylogeny, the authors purposely then used an approach all too common to comparative physiological studies, and lumped together for comparison data on sprint speeds from twenty distantly related lizard genera. This re-analysis led to quite the opposite conclusion: that is, an increasing body temperature range was associated with a *reduced* rather than increased temperature range in which rapid sprinting could occur! This is perhaps an unusually graphic example, but as Huey (1987) comments 'By reducing phylogenetic artifacts, studies of closely related species reduce the risk of obscuring a real pattern or of even forcing a spurious one.'

A challenge for comparative respiratory physiologists

One of the original goals of comparative physiology – to understand how physiological adaptations have evolved and currently operate is as challenging and exciting now as it was at the turn of the century. The last few years have seen David Randall (1986), C. Ladd Prosser (1986), Martin Feder (1987) and many others predict that comparative physiology will play an important role in evolutionary biology in the coming decades. If this prophecy is to be fulfilled, more comparative physiologists will need to make a concerted effort not just to understand the tools and approaches of evolutionary biologists, but actually to integrate evolutionary theory and methodologies into our comparative physiological studies. If we simultaneously embrace contemporary theories and practices in both physiology *and* evolutionary biology, comparative physiology will not only achieve its self-established goals of understanding physiological adaptation, but will also make broader contributions to evolutionary biology (this despite the current scepticism of evolutionary biologists viewing our field from outside). Respiratory physiology, in particular, is one of the most active and thriving fields in contemporary comparative physiology, and respiratory physiologists can play a leading role in moving comparative physiology into the mainstream of evolutionary biology.

The author is grateful to William Bemis, Robert Infantino, Peter Kimmel and, in particular, Raymond Huey, for criticizing drafts of this manuscript.

References

Bennett, A.F. (1987). The accomplishments of ecological physiology. In *New Directions in Ecological Physiology* (ed. M.E. Feder, A.F. Bennett, W.W. Burggren & R.B. Huey), pp. 147–65. Cambridge University Press, New York.

Burggren, W.W. (1976). A quantitative analysis of ventilation tachycardia and its control in two chelonians, *Pseudemys scripta* and *Testudo graeca*. *Journal of Experimental Biology* **63**, 367–80.

Burggren, W.W. & Bemis, W.E. (1990). Studying physiological evolution: paradigms and pitfalls. In *Evolutionary Innovations: Patterns and*

Processes (ed. M. Nitecki & D. Nitecki). University of Chicago Press.

Cheverud, J.M., Dow, M.M. & Leutenegger, W. (1985). The quantitative assessment of phylogenetic constraints in comparative analyses: sexual dimorphism in body weight in primates. *Evolution* **29**, 1335–51.

Clutton-Brock, T.H. & Harvey, P.H. (1984). Comparative approaches to investigating adaptation. In *Behavioural Ecology: an Evolutionary Approach* (ed. J.R. Krebs & N.B. Davies), pp. 7–29. Sinauer, Sunderland, Massachusetts.

Coddington, J.A. (1988). Cladistic tests of adaptational hypotheses. *Cladistics* **4**(1), 3–22.

Cracraft, J. (1983). The significance of phylogenetic classifications for systematic and evolutionary biology. In: *Numerical Taxonomy* (ed. J. Felsenstein), pp. 1–17. Springer-Verlag, Berlin.

Crook, J.H. (1965). The adaptive significance of avian social organization. *Symposia of the Zoological Society of London* **14**, 181–218.

Downer, R.G.H. (1986). Projected research trends in invertebrate physiology and biochemistry. *Canadian Journal of Zoology* **65**, 797–802.

Dunham, A.E. & Miles, D.B. (1985). Patterns of covariation in life-history traits of squamate reptiles: the effects of size. *American Naturalist* **126**, 56–72.

Feder, M.E. (1987). New directions and conclusion. In *New Directions in Ecological Physiology* (ed. M.E. Feder, A.F. Bennett, W.W. Burggren & R.B. Huey), pp. 38–70. Cambridge University Press, New York.

Feder, M.E., Bennett, A.F., Burggren, W.W. & Huey, R.B. (eds) (1987). *New Directions in Ecological Physiology*. Cambridge University Press, New York.

Felsenstein, J. (1985). Phylogenies and the comparative method. *American Naturalist* **125**, 1–15.

Futuyma, D.J. (1987). *Evolutionary Biology*. Sinauer Associates, Sunderland, Massachusetts.

Gans, C. (1970). Respiration in early tetrapods – the frog is a red herring. *Evolution* **24**, 723–34.

Greenberg, M.J. (1985). Ex bouillabaisse lux: the charm of comparative physiology and biochemistry. *American Zoologist* **25**, 737–49.

Greenberg, M.J., Hochachka, P.W. & Mangum, C.P. (eds) (1975). New directions in comparative physiology and biochemistry. *Journal of Experimental Zoology* **194**, 1–347.

Hennig, W. (1979). *Phylogenetic Systematics*. University of Illinois Press, Urbana, Illinois.

Herring, S. (1988). Introduction: How to do functional morphology. *American Zoologist* **28**, 189–92.

Huey, R.B. (1987). Phylogeny, history and the comparative method. In *New Directions in Ecological Physiology* (ed. M.E. Feder, A.F. Bennett, W.W. Burggren & R.B. Huey), pp. 76–98. Cambridge University Press, New York.

Huey, R.B. & Bennett, A.F. (1987). Phylogenetic studies of coadaptation: preferred temperatures versus optimal performance temperatures of lizards. *Evolution* **41**, 1098–115.

Krebs, H.A. (1975). The August Krogh principle: 'For many problems there is an animal on which it can be most conveniently studied'. *Journal of Experimental Zoology* **194**, 309–44.

Lauder, G.V. (1982). Historical biology and the problem of design. *Journal of Theoretical Biology* **97**, 57–67.

Mace, G.M., Harvey, P.H. & Clutton-Brock, T.H. (1981). Brain size and ecology in small mammals. *Journal of Zoology (London)* **193**, 333–54.

Mayr, E. & Provine, W.B. (eds) (1980). *The Evolutionary Synthesis: Perspectives on the Unification of Biology*. Harvard University Press, Cambridge, Massachusetts.

Patterson, C. (ed.) (1987). *Molecules and Morphology in Evolution: Conflict or Compromise?* Cambridge University Press, London.

Randall, D.J. (1986). The next 25 years: vertebrate physiology and biochemistry. *Canadian Journal of Zoology* **65**, 794–6.

Ridley, M. (1983). *The Explanation of Organic Diversity: The Comparative Method and Adaptations for Mating*. Oxford University Press.

Ross, D.M. (1981). Illusion and reality in comparative physiology. *Canadian Journal of Zoology* **59**, 2151–8.

Stearns, S.C. (1983). The influence of size and phylogeny on patterns of covariation among life-history traits in the mammals. *Oikos* **41**, 173–87.

Stearns, S.C. (1984). The effects of size and phylogeny on patterns of covariation in the life history traits of lizards and snakes. *American Naturalist* **123**, 56–72.

Wiley, E. O. (1981). *Phylogenetics: The Theory and Practice of Phylogenetic Systematics*. John Wiley, New York.

PETER KÖHLER

Energy metabolism in helminths

Introduction

Helminths are an extremely heterogeneous group of organisms which have adapted to a specialized parasitic mode of life. The vast majority of these worm parasites are representatives of two particular phyla, the Platyhelminthes and Nematoda. These organisms possess a greater diversity of biochemical energy-yielding pathways (reactions that can be coupled to the phosphorylation of ADP to form ATP) than is found in higher animals. The generation of this chemical energy is achieved through the fundamental principles of either substrate-level phosphorylation (SLP) or electron transport (respiration)-associated phosphorylation (ETP). A further striking feature of helminth metabolism is the prevalence of anaerobic energy-yielding routes that function even if oxygen is available. Indeed, the contribution of oxygen-dependent processes is in many cases limited; this limitation is indicated by the lack, or low activity, of the tricarboxylic acid cycle, fatty acid oxidation sequence and a cytochrome-oxidase-linked mitochondrial respiratory chain. Consequently, many species are only capable of limited oxidation of carbon substrates to CO_2 and water, and are thus unable to make use of the concomitant large decrease in free energy for ATP synthesis. Marked oxidative capacities are restricted to some small helminths and to larval and other intermediate stages.

In adult helminths, carbohydrates are the predominant energy-yielding substrates, of which glycogen is the major storage polysaccharide. Trehalose, the blood sugar of insects, can also serve as a carbohydrate reserve in some worm species and be used as an energy source. After cleavage of this disaccharide to two molecules of glucose, its metabolism is identical to that undergone by glucose. Until recently it was believed that amino acids were of only limited importance for energy generation in helminths. There is now, however, increasing evidence to suggest that some types of amino acids may play a significant role as energy-rich fuels in these organisms (see

15

below). In accordance with the limited oxidative capacities of helminth metabolism, fatty acids cannot be utilized as energy-yielding substrates by most adult worms, and the catabolism of carbohydrates and amino acids beyond the acetate step is often only barely feasible.

Possibly because of the great differences in their morphology and the environmental conditions of their habitats, helminths show a great diversity in the pattern, efficiency and power output of their energy metabolism. In spite of this variability, two main trends in the metabolic strategies of anaerobic substrate catabolism are evident. In one group, carbohydrate breakdown is essentially restricted to glycolysis with lactate as the end-product (lactate fermentation). The other, larger group of helminths is characterized by a type of energy metabolism commonly found in free-living invertebrates, in particular aquatic species (e.g. annelids and molluscs), during periods of environmental hypoxia (Grieshaber, 1986). This metabolism is rather complex and shows a high adaptive potential towards various intrinsic and environmental factors. Characteristically, a variety of fermentations and anaerobic electron transport processes are major constituent routes of this metabolism, resulting in the formation of several organic end products, the most common of which are volatile fatty acids, succinate, alcohols, lactate and CO_2. Besides these anaerobic processes, oxygen-dependent pathways are also used for energy generation in helminths, but the contribution of oxidative processes to ATP production can vary considerably between different helminth species and developmental stages.

Within the past decade, energy metabolism in helminths has been repeatedly reviewed. The present work is intended to update this earlier literature (Saz, 1981; Köhler, 1982, 1985, 1988; Ward, 1982; Barrett, 1984). Particular emphasis will be placed on areas in which major advances have recently been achieved and which have provided answers to a number of important questions in helminth bioenergetics. These involve some aspects of the regulation of carbohydrate metabolism, the properties of the fumarate-reducing mitochondrial electron transport system, the relevance of oxygen-dependent pathways for energy generation, and the role of amino acids as energy-rich fuels. The new findings obtained on the metabolic alterations associated with the developmental transitions within the life cycle of helminths will be also briefly discussed. As detailed investigations within these fields have largely been confined to a few 'model systems' (i.e. the nematode *Ascaris suum*, some filarial species, *Schistosoma mansoni*, and the liver fluke *Fasciola hepatica*), the discussion will primarily deal with the data obtained from studies of these species.

Anaerobic carbohydrate metabolism in adult helminths

Carbohydrates play the most important role in furnishing energy in helminths. These nutrients can be stored in very large amounts (10–20% of the worm's dry mass) in tissues from which it can rapidly be mobilized during sudden demands for energy (Ward, 1982; Tielens & van den Bergh, 1987). As in most other cells, glycolysis serves as the central pathway of carbohydrate catabolism in all helminths, so far studied (Figure 1). This sequence of reactions differs between helminths and other animal species only in how the rate is regulated and in the subsequent metabolic fate of the products formed. A comparison of the mass action ratios of the glycolytic enzymes from a number of helminths with their apparent equilibrium constants showed that hexokinase, phosphofructokinase and pyruvate kinase are, as in most other systems, non-equilibrium regulatory enzymes (Barrett *et al.*, 1986*a,b*). There is also some evidence to suggest that under certain conditions aldolase may become rate-limiting and thus could be considered as pseudoregulatory. One of the regulatory enzymes, phosphofructokinase (PFK), has been purified from *Ascaris* muscle tissue to homogeneity and its kinetic and regulatory properties have been extensively investigated (Hofer *et al.*, 1982; Starling *et al.*, 1982). These studies demonstrate that the ascarid enzyme appears to be the primary site for control of glycolytic flux, as it is in the cells of most other organisms. In addition, the PFK from the nematode was found to be similar to mammalian PFKs, in that it was inhibited by ATP and stimulated by a number of other effectors. The structure of the parasite enzyme also resembles that of the mammalian form, although the subunits of the ascarid PFK are slightly larger and contain noticeable differences in amino acid composition. A more significant difference between the parasite and mammalian PFKs has been demonstrated for the influence of phosphorylation on the regulatory properties of these enzymes (Kulkarni *et al.*, 1987; Thalhofer *et al.*, 1988). Although phosphorylation of the ascarid PFK by the catalytic subunit of mammalian cyclic AMP-dependent protein kinase has resulted in significant stimulation of activity, this treatment has failed to demonstrate such an effect on the mammalian PFK. Recent work by Daum *et al.* (1986) has shown that the degree of activation of the ascarid enzyme was greater when PFK phosphorylation was catalysed by a protein kinase purified from *Ascaris* muscle rather than by the corresponding mammalian enzyme. Also recently, an undecapeptide carrying a phosphorylated serine has been isolated from proteolytic digests of the nematode phosphoenzyme (Kulkarni *et al.*, 1987). The sequence of this peptide was found to be

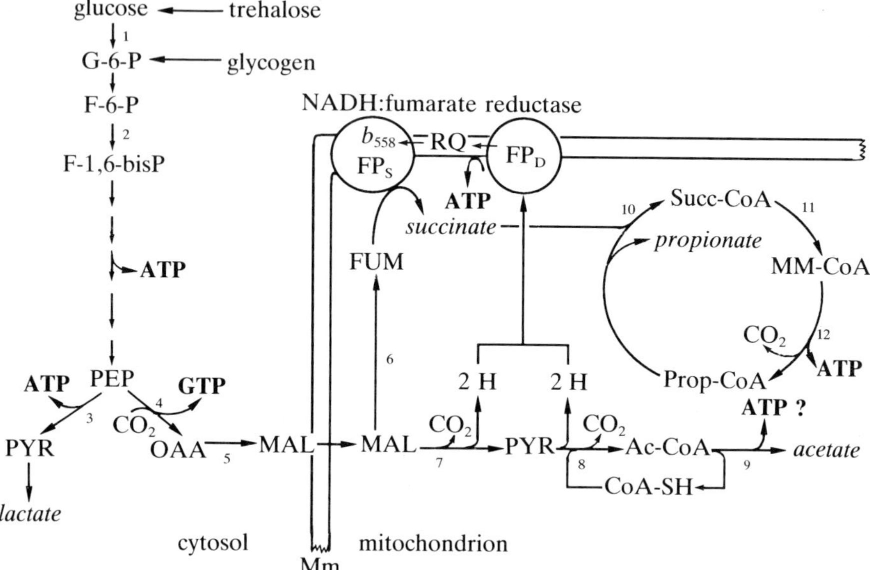

Figure 1. The major pathways of anaerobic energy metabolism and their associated sites of ATP formation in adult helminths. The numbers represent regulatory and other important enzymes of helminth anaerobic carbohydrate catabolism: 1, hexokinase; 2, phosphofructokinase; 3, pyruvate kinase; 4, PEP carboxykinase; 5, malate dehydrogenase; 6, fumarate hydratase; 7, malate dehydrogenase, decarboxylating; 8, pyruvate dehydrogenase complex; 9, acetyl CoA synthetase (present, but relevance to be determined); 10, acyl CoA transferase; 11, methylmalonyl CoA mutase; 12, propionyl CoA carboxylase. Abbreviations: Ac-CoA, acetyl CoA; CoA-SH, coenzyme A; F-1,6-bisP, fructose 1,6-bisphosphate; F-6-P, fructose 6-phosphate; G-6-P, glucose 6-phosphate; FD$_D$, flavoprotein dehydrogenase of respiratory chain complex I; FP$_S$, flavoprotein dehydogenase of respiratory chain complex II; FUM, fumarate; MAL, malate; Mm, mitochondrial membrane; MM-CoA, methylmalonyl CoA; OAA, oxaloacetate; PEP, phosphoenolpyruvate; Prop-CoA, propionyl CoA; PYR, pyruvate; Succ-CoA, succinyl CoA. End-products are shown in italics.

entirely different from the phosphoserine-containing peptide sequence derived from rabbit muscle PFK. The observation that phosphorylation of the ascarid PFK can be reversed by a protein phosphatase, also obtained from the nematode's muscle, indicates that the complete interconversion system for PFK regulation is present in the ascarid muscle tissue (Daum *et al.*, 1986). It is the first time that a reversible phosphorylation system has been demonstrated for a PFK, and it is suggested that this system provides a highly sensitive mechanism for regulation of PFK and, thus, of entire glycolysis.

In many helminth species the sequence of reactions in glycolysis differs from that operating in most other animals in that it proceeds only up to the step of phosphoenolpyruvate (PEP) formation (see Saz, 1981; Köhler, 1985). Only in a minor group of helminths (schistosomes and some filarial worms) does the glycolytic flux continue one step further to form pyruvate through the action of pyruvate kinase, which can be subsequently converted into lactate. In the alternative, more frequently occurring pathway, PEP undergoes carboxylation to yield oxaloacetate (Figure 1). This reaction is catalysed by PEP carboxykinase and is comparable to the pyruvate-kinase-catalysed step in that it serves to regenerate energy-rich phosphate (in the form of GTP) utilized in upper glycolysis. The subsequent step, catalysed by malate dehydrogenase, is the reduction of oxaloacetate to malate. The factors determining the carbon flow at the PEP branch-point are manifold and involve (i) the activities of the two enzymes competing for PEP, pyruvate kinase and PEP carboxykinase, (ii) the different affinities of these two enzymes for PEP, (iii) the concentrations of substrates, in particular CO_2, within the parasite tissues, and (iv) the activities of the dehydrogenases that catalyse the subsequent reductive reactions (see Saz, 1981; Köhler, 1985). In a large variety of helminths (intestinal nematodes and cestodes, such as *Ascaris*, trichostrongylids, *Trichuris*, *Moniezia*, *Hymenolepis*; liver flukes and lungworms) the presence of high activities of PEP carboxykinase, together with an extremely active cytoplasmic malate dehydrogenase, but low activities of the enzymes of the competitive, lactate-producing branch, favour the conversion of PEP into malate. In addition, the high levels of CO_2 prevailing in most parasite habitats may also support carbon flow in the direction of CO_2-fixation: it has been shown that the activity of helminths' PEP carboxykinase is stimulated, but the activity of the pyruvate kinase suppressed, as the concentrations of bicarbonate ions are increased (Podesta *et al.*, 1976). Both, glucose derived malate and lactate formation fulfil identical metabolic functions, i.e. to generate 'energy-rich' phosphate by kinase enzymes and re-oxidize the pyridine nucleotide coenzyme, which is reduced during the middle phase of glycolysis.

The advantage of forming malate instead of lactate is that the structure of the dicarboxylate is much more suitable for further anaerobic enzymatic conversions. After passing from the cytosol to the mitochondrial matrix, this substrate can undergo a coupled oxidation–reduction process

(Figure 1), even in the complete absence of molecular oxygen (see Köhler, 1980). One route involves the sequential catalysis by the malic enzyme and the pyruvate dehydrogenase complex (PDC) and results in the formation of acetyl CoA and CO_2 after two consecutive steps of oxidative decarboxylation. The malic enzyme (malate dehydrogenase decarboxylating) from *Ascaris* muscle, which has been purified and the reaction mechanism of which has been thoroughly studied, reacts preferentially with NAD^+ rather than $NADP^+$ as electron acceptor (Saz & Hubbard, 1957; Allen & Harris, 1981). In some other helminths, such as *Hymenolepis diminuta*, malate oxidation is catalysed by an $NADP^+$-specific malic enzyme, thereby producing intramitochondrial NADPH rather than NADH. However, since the subsequent reductive enzymatic reactions display a preference for NADH (see below), for further utilization transhydrogenation of the reducing equivalents from NADPH to NAD^+ is required (Fioravanti, 1981). The malic enzyme reaction is followed by a second step of NAD^+-mediated oxidative decarboxylation, in which pyruvate is converted to acetyl CoA and CO_2. This process is catalysed by the lipoamide-dependent mitochondrial PDC, which has recently been purified and has had its regulatory functions investigated (Komuniecki *et al.*, 1979; Thissen *et al.*, 1986; Ruff *et al.*, 1988). The ascarid enzyme is regulated by reversible phosphorylation, mediated by a distinct kinase and phosphatase (Komuniecki *et al.*, 1983). On the other hand, a number of differences have emerged, which appear to reflect the function of the parasite PDC in its primarily anaerobic and largely reduced mitochondrial environment. Under these unique conditions high PDC activities are required to provide both NADH and acetyl CoA for the subsequent reductive energy-yielding processes, which result in the formation of succinate and volatile fatty acids. The recent work by Thissen *et al.* (1986) suggests that a continuously high activity of the PDC may be achieved through the much higher affinities of the ascarid PDC for NAD^+ and CoASH, and a lower sensitivity to inhibition by elevated $NADH:NAD^+$ ratios as compared to the corresponding mammalian enzyme. In addition, pyruvate and propionate, two intermediates of the nematode's mitochondrial metabolism, were found to considerably inhibit pyruvate dehydrogenase kinase, indicating that their accumulation could serve to prevent inactivation of the PDC. In the mitochondria of *Ascaris* and a variety of other helminths, the exergonic nature of the two consecutive reactions of the malic enzyme and PDC reactions, and the associated production of reduced pyridine nucleotides provide the driving force to reduce another portion of malate, via fumarate, to succinate and to form higher volatile fatty acids.

As shown in Figure 1, in helminths the NADH-linked reduction of fumarate to succinate is brought about by mitochondrial electron transport by employing respiratory chain complexes I and II (Köhler, 1985). The properties of this enzyme system are quite unique for its high activity of fumarate reductase. The activity ratio of the ascarid fumarate reductase to succinate dehydrogenase was found to be about 400 times higher than that of its mammalian counterpart (Hata-Tanaka *et al.*, 1988). The components of the system required for NADH-linked fumarate reduction are all membrane-bound. Complex I has been isolated from *Ascaris* muscle mitochondria as a complex I–III fragment of the respiratory chain; some of the properties of this enzyme system have been determined (Takamiya *et al.*, 1984). Recent extensive studies have also provided new insights into the structural organization and functional role of mitochondrial respiratory chain complex II of *Ascaris* (Takamiya *et al.*, 1986; Kita *et al.*, 1988*a,b*; Hata-Tanaka *et al.*, 1988). This enzyme system is one of the major components of adult *Ascaris* mitochondria (8% of total mitochondrial protein) and was found to be composed of four polypeptides with molecular masses of 68, 26, 15 and 13.5 kDa. The largest subunit contains covalently bound flavin and is very similar in its properties to the mammalian succinate dehydrogenase. The 26 kDa polypeptide contains three types of iron–sulphur clusters (Hata-Tanaka *et al.*, 1988). In addition, *Ascaris* mitochondrial respiratory complex II contains the major constituent cytochrome of the nematode mitochondrion, designated cytochrome b_{558} (Kita *et al.*, 1988*a,b*). This cytochrome was separated from the flavoprotein and iron–sulphur protein subunits and was shown to be reducible by succinate and, in the presence of *Ascaris* respiratory chain complexes I–III, also by NADH; it was re-oxidizable by fumarate. The properties of the cytochrome were found to differ in amino acid sequence, antigenicity, molecular mass and physicochemical properties from the low-potential cytochrome b_{560} fractionating into beef–heart respiratory chain complex II. In addition, in contrast to the ascarid cytochrome b_{558}, reduction of the heart cytochrome by succinate was reported to be hardly, if at all, feasible. These results, together with previous enzymatic studies, suggest that NADH-linked fumarate reduction in *Ascaris*, and possibly in other helminths, is mediated by electron transport via the respiratory chain complex I (NADH–coenzyme Q reductase), quinone, cytochrome b_{558} and the succinate dehydrogenase operating in reverse (Figure 1). Very recently, NADH–fumarate reductase activity has been reconstituted by employing bovine heart mitochondrial complex I, mitochondrial complex II from *Ascaris* muscle, and phospholipids (Kita *et al.*, 1990). It was demonstrated that the two enzyme systems, phospholipid and rhodoquinone are the only components required to establish electron transport from NADH to fumarate.

Since in typical aerobic cells hydrogen transfer from NADH to fumarate occurs at only low rates, specific requirements not available in aerobic mitochondria must be essential for helminth metabolism to favour electron transport in the direction of succinate formation. It has been suggested that the structural properties of the subunit polypeptides of complex II, in particular those of the *b*-type cytochrome, are likely to be important factors determining the pathway in its reverse direction (Köhler, 1985; Kita *et al.*, 1990). The deficiency of a mitochondrial cytochrome system associated with a very low activity of cytochrome *c* oxidase frequently observed in the mitochondria of adult helminths may also serve to direct electron flow to the fumarate-linked pathway. An additional major factor responsible for electron flow from NADH to fumarate is obviously the nature of the quinone species, which functions as a connecting link between different mitochondrial electron transfer complexes. Except for a few species, all helminths capable of fumarate reduction were found to contain rhodoquinone, either as a sole quinone or in combination with the more frequently occurring ubiquinone (Fioravanti & Kim, 1988). Sato *et al.* (1972) and recent reconstitution experiments by Kita *et al.* (1990) have clearly demonstrated that rhodoquinone-9 from *Ascaris* is required for mitochondrial electron transport from NADH to fumarate and that this function cannot be fulfilled by ubiquinone. Fioravanti & Kim (1988) have recently shown the occurrence of rhodoquinone as a mitochondrial membrane component in the cestode *Hymenolepis diminuta*, whereas ubiquinone was found to be absent. The rhodoquinone specificity of the fumarate-reducing electron transport system in helminths may be closely related to the physicochemical properties of this compound, in particular the relatively low redox potential ($E_0 = -63$ mV). Since this value is more negative than that of the fumarate–succinate couple, the use of this quinone species should favor electron transport in the direction of fumarate. Ubiquinone, the electron carrier of the conventional mitochondrial respiratory chain, has a highly positive redox potential ($E_0 = +113$ mV) and, thus, is a less suitable carrier of electrons to fumarate, but instead donates them preferentially to the cytochrome chain.

Several research groups have previously demonstrated that electron transport from NADH to fumarate in helminth mitochondria is associated with an uncoupler- and oligomycin-sensitive phosphorylation of ADP (Saz, 1971; Köhler & Bachmann, 1980). In this process, one molecule of

ATP is formed per pair of electrons flowing to fumarate at a site comparable to phosphorylation site 1 of the conventional respiratory chain (Figure 1).

In many helminths succinate, a product of fumarate reduction, is a major excretory product of carbohydrate catabolism. In the oxidative branch of the malate dismutation reaction, acetyl CoA is formed as a product of pyruvate oxidation (see above). In typical aerobic organisms, this 2-carbon thioester either enters into the citric acid cycle, where it is ultimately oxidized to CO_2, or is used as a precursor for various synthetic purposes. However, in helminth mitochondria, acetyl CoA is primarily hydrolysed to form free acetate (Figure 1), which is then, like succinate, excrcted into the surrounding host tissue. In a few helminth species acetyl CoA can also serve as a precursor for the synthesis of higher volatile fatty acids (Komuniecki *et al.*, 1981). The enzymatic mechanism resulting in the liberation of acetate from its CoA thioester is still unclear (Figure 1). There is some evidence to suggest that CoA hydrolysis in helminth mitochondria is limited, and that acetate formation may occur as a result of CoA transfer reactions, which would serve to conserve the free energy of the thioester bond (Komuniecki *et al.*, 1987). Distinct enzymes catalysing such reactions have been shown to be present in helminths (McLaughlin *et al.*, 1986; Saz & deBruyn, 1987). Conservation of the 'energy-rich' thioester bond of acetyl CoA in the form of ATP, as catalysed by a specific thiokinase or the combined action of an acyl CoA transferase and a thiokinase, has not been observed in helminths so far, although the enzymes catalysing these reactions appear to be present in the tissues of these organisms (Barrett *et al.*, 1978).

Short-chain acyl CoA compounds are also constituents of the pathway of propionate formation which occurs in a variety of helminth species (Figure 1). This pathway was shown to proceed essentially by reversal of the reactions required for the conversion of propionate to succinate in mammalian tissues (Pietrzak & Saz, 1981). As expected, each cycle of the enzymatic sequence promotes the synthesis of one molecule of ATP through coupling of ADP phosphorylation to the lysis of the carboxyl function of methylmalonyl CoA. Most importantly, however, a net formation of ATP during propionate formation can only be achieved if this pathway is functionally coupled to an acyl CoA transferase, which allows the re-cycling of the 'energy-rich' thioester bond of the CoA intermediates. Both net ATP formation during succinate decarboxylation, and the presence of the enzyme activities required for this process, have been demonstrated in propionate-forming helminths (Köhler *et al.*, 1978; Saz &

Pietrzak, 1980; Pietrzak & Saz, 1981). These enzymes also include a CoA transferase, which is capable of activating succinate by exchange of the thioester bond in either propionyl CoA or acetyl CoA (Saz & deBruyn, 1987). One of the unique properties of this enzyme is to favour an equilibrium overwhelmingly in the direction of succinyl CoA formation, allowing the decarboxylation, rather than deacylation, of the CoA derivative.

Another pathway of short-chain acyl CoA compounds, which might also have high potential for ATP formation, is the synthesis of 2-methyl branched-chain volatile fatty acids. Such compounds accumulate in a few helminths, for example in *Ascaris*, where 2-methylbutyrate and 2-methylvalerate are the predominant end-products of carbohydrate catabolism. As shown in Figure 2, these acids arise from the condensation of acetyl CoA with propionyl CoA, or of two propionyl CoA units, with subsequent reduction of the condensation products to the saturated acids (Komuniecki *et al.*, 1981). The characteristic feature of this entirely mitochondrial process is that it occurs by reversing the β-oxidation process of fatty acid degradation. The shift in the overall equilibrium of the oxidative pathway to favour the synthetic direction is achieved by the strongly reduced environmental conditions within the ascarid mitochondrion (elevated $NADH:NAD^+$ ratios), and a number of distinct kinetic and regulatory properties of the enzymes involved in this pathway. Of particular significance in the malonyl CoA-independent fatty-acid synthesizing system of *Ascaris*, is the step in which the enoyl CoA compounds are reduced to the saturated acyl CoA esters. Recent studies (Komuniecki *et al.*, 1984a,b) have suggested that the reducing power required for this reduction comes from NADH, with its electrons passing through a sequence of electron-transporting enzymes, including the mitochondrial respiratory chain complex I and a soluble electron-transferring flavoprotein (Figure 2). A 2-methyl branched-chain enoyl CoA reductase then serves as the terminal reductase, converting the enoyl CoA into the saturated acyl CoA compound. Both the observed rotenone-sensitivity of the overall process and the thermodynamic considerations suggest that a site 1 ATP synthesis may be associated with this electron transport process, in a way similar to that coupled to NADH-dependent fumarate reduction. Whether a second energy coupling site may be associated at the end of the fatty acid synthetic pathway with the hydrolysis of the acyl CoA thioester bond, which occurs with a large drop in free energy, remains to be determined. A more likely mechanism seems to be the transfer of the CoA

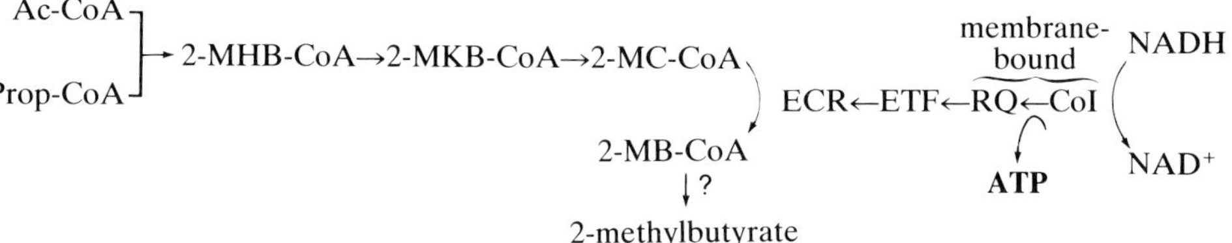

Figure 2. The pathway of branched-chain volatile fatty acid formation in *Ascaris* muscle mitochondria. The example shows the route for 2-methylbutyrate formation from acetyl CoA and propionyl CoA. Ac-CoA, acetyl CoA; CoI, complex I of the mitochondrial respiratory chain; ECR, enoyl CoA reductase; ETF, electron-transferring flavoprotein; 2-MB-CoA, 2-methylbutyryl CoA; 2-MC CoA, 2-methylcrotonyl CoA; 2-MHB-CoA, 2-methyl-3-hydroxybutyryl CoA; 2-MKB-CoA, 2-methyl-3-ketobutyryl CoA; RQ, rhodoquinone.

moiety to suitable CoA acceptors as it is suggested to occur in acetate and propionate formation (see above).

The preceding discussion has shown that anaerobic energy generation in helminths can involve several sites of ATP synthesis. At the substrate level, these are coupled to glycolysis, PEP carboxylation and propionate formation from succinate, whereas NADH-dependent reductions of fumarate and possibly enoyl CoA compounds are associated with electron transport phosphorylation. Whether, and how, a possible coupling of acyl CoA thioester hydrolysis to energy generation is achieved in helminth mitochondria is still unclear. The current evidence suggests that free CoASH is not generated to a significant extent and that the preferred mechanism in acyl CoA metabolism is the exchange of thioester linkages between different CoA acceptor and donor systems. As can be seen in Figure 1, anaerobic energy metabolism in helminths could have a noticeable advantage over simple glycolysis with lactate or ethanol formation. Whereas the latter pathways yield only 2 moles of ATP per mole of glucose catabolized, the propionate-forming route results in approximately 5 moles of ATP per mole of glucose utilized. However, this economic efficiency is still low compared with the amount of energy conserved during complete oxidation of substrates to CO_2 and water.

The existence of such a diverse array of strategies for anaerobic energy generation in helminths may fulfil a variety of functions. For example, they could provide increased flexibility in adapting energetically to the conditions of the parasite's specific environments, in balancing their intracellular redox couples and selecting key intermediates for synthetic purposes.

Komuniecki *et al.* (1987) have recently suggested that the synthesis of branched-chain fatty acids in *Ascaris* mitochondria may have regulatory properties under the reducing conditions present in the host gut by serving as a sink for excess reducing power. A further advantage of volatile fatty acid formation is that these compounds can, because of the relatively low dissociation constants of their carboxyl functions, readily diffuse across biological membranes. They are therefore less damaging to the parasite and host tissue than other acids, such as lactate. It is interesting to note that there seems to be a relationship between the type of bioenergetic strategy evolved by a helminth species and the conditions of the environment in which it lives. The prevalence of multiple fermentations in helminths inhabiting primarily the intestine and bile ducts of vertebrates may provide the parasite with a greater versatility and flexibility to respond to the chemical and nutritional changes often occurring in these habitats. On the other hand, helminths residing in the blood vessels or subcutaneous tissues of vertebrates (schistosomes, filarial parasites) may not require this flexibility and therefore depend primarily on glycolysis, which is not so efficient at energy production.

The relevance of oxygen to energy generation in helminths

Like all free-living invertebrates, helminths utilize molecular oxygen when it is available, and in many species the change from anoxic to aerobic conditions significantly affect the pattern of energy metabolism (see Barrett, 1984; Köhler, 1985). However, in contrast to other animals, in the adult stages of helminths a predominantly anaerobic type of energy metabolism persists if oxygen is available. Aerobiosis usually causes a shift in the proportion of the reductive to oxidative metabolic routes, with the oxidative pathway being increased. However, the balance often lies predominantly on the side of incomplete substrate oxidation (see Barrett, 1984). For example, in *Fasciola hepatica* increasing oxygen pressures result in an enhanced rate of the acetate-forming pathway with a concomitant decrease in propionate production. The filarial worm *Litomosoides carinii*, when maintained under anaerobiosis, degrades carbohydrate solely to lactate, whereas in the presence of oxygen acetate is also formed (Ramp & Köhler, 1984). It is assumed that this pathway increases substantially the overall rate of energy conservation over that obtained from simple lactate fermentation (Ramp *et al.*, 1985). In addition, many helminths seem to be able to use oxygen to oxidize at least part of their energy substrates completely to CO_2 and water (see below).

As in most eukaryotes, helminth respiratory systems are located within the mitochondria. However, a conventional type of electron transport chain, terminating with cytochrome *c* oxidase, often contributes only a minor, or even negligible, portion to the overall rate of respiration. Such is the case with *Ascaris, F. hepatica* and many cestodes. Instead, these respiration systems carry an 'alternative oxidase' to accomplish the transfer of electrons to molecular oxygen. Very little information is available on the nature of this alternative respiration mechanism and the reason remains unclear for its occurrence in preference to, or alongside, the classical respiration chain. The 'alternative oxidase' system present in muscle mitochondria of *Ascaris* and other nematodes has been investigated to some extent (Kmetec & Bueding, 1961; Köhler & Bachmann, 1980; Paget *et al.*, 1987) and has been shown to be insensitive to cyanide and antimycin but susceptible to rotenone inhibition. The membrane-bound enzyme complex involves the NADH–quinone reductase segment (respiratory chain complex I) and an unknown oxidation system passing the NADH-derived electrons to oxygen and reducing it to hydrogen peroxide. Respiration through the cyanide-resistant system is believed to be coupled to a site 1 energy conservation in which, as in fumarate-dependent NADH oxidation, one molecule of ATP is formed for each pair of electrons transferred to oxygen (Köhler & Bachmann, 1980). Such a phosphorylation system is of relatively low efficiency in energy conservation and, if functional in the worm within its natural environment, is probably unable to improve the overall rate of ATP production compared with the anaerobically functioning systems. Recently, the possible role for cytochrome *c* peroxidase, an enzyme previously identified in *Ascaris suum* by Hayashi & Terada (1973), in the nematode's mitochondrial respiration was discussed by Campbell *et al.* (1989).

There are, however, helminth species that can use oxygen to maintain a much more efficiently functioning energy metabolism. Particularly in small worms living in unambiguously aerobic sites in the host body, such as some filarial parasites, trichostrongylids and schistosomes, a functional citric acid cycle, together with a conventional type of phosphorylating respiratory chain, serves to oxidize nutrients completely to CO_2 and water (Ramp *et al.*, 1985; Van Oordt *et al.*, 1985; Paget *et al.*, 1987). However, even in these worms only a small portion of the carbon substrate utilized appears to be capable of entering the pathways of complete oxidation, at least if carbohydrate serves as the sole energy source. Most important, however, is the fact that for each molecule of glucose undergoing complete oxidation considerably more chemical energy is generated (a maximum of 38 molecules of ATP) than is recovered from anaerobic processes (2–6 molecules of ATP). Therefore, even if only a small percentage of total substrate utilization can enter the pathways of complete oxidation, the rates of ATP formation associated with this process will result in a substantial contribution to the parasite's overall energy balance.

Relatively little information is as yet available on the relevance of amino acids for energy generation, although some amino acid species are known to be advantageous for the maintenance of certain helminths in culture. Recent experiments in our laboratory have demonstrated that the filarial worm *Litomosoides carinii* could survive *in vitro* almost as well in the presence of glutamine as the sole carbon substrate as it could in a medium which contained only glucose (Davies & Köhler, 1990). Similar results were also recently obtained for other filarial species, e.g. *Onchocerca*, which preferred media supplemented with glutamine alone to any other carbon source for their *in vitro* maintenance (VandeWaa *et al.*, 1988). More detailed analyses of the effects of glutamine on filarial metabolism have shown that a major portion of the amino acid utilized by the worms was oxidized to CO_2 and that the worms incubated in media containing only glutamine were highly sensitive to the respiratory chain inhibitors antimycin A and cyanide (Rees & Comley, 1988; VandeWaa *et al.*, 1988). Additional work has demonstrated the ability of glutamine to stimulate considerably the rate of glucose oxidation to CO_2 by the filariids (Davies & Köhler, 1990). The precise function of glutamine in this activation process is probably to replenish the pool of intermediates of the tricarboxylic acid cycle (Figure 3). The input of carbon into the cycle would initiate an increased production of acetyl CoA units from glycolytically formed pyruvate for condensation with oxaloacetate, and hence would cause an overall activation of carbon oxidation to CO_2. The most important consequence of this glutamine-dependent coordinated activation of glycolysis and the Krebs cycle would be a significant increase in the economic efficiency of energy production.

The reasons for the generally limited capacity of helminths for complete substrate oxidation are not completely clear but may be manifold. The most plausible explanation for the phenomenon is that the variety of helminth anaerobic energy-yielding systems have evolved in response to the lack of a circulatory system and/or the often hypoxic conditions prevailing in many habitats of these organisms. The body of several helminths is certainly too large for oxygen to diffuse into the deeper tissues, and it has been calculated that the critical thickness to sustain aerobic life throughout the whole organism ranges from 0.4 mm for nematodes to 0.75 mm for cestodes. An analysis carried out by Fry & Jenkins (1984) on the relation-

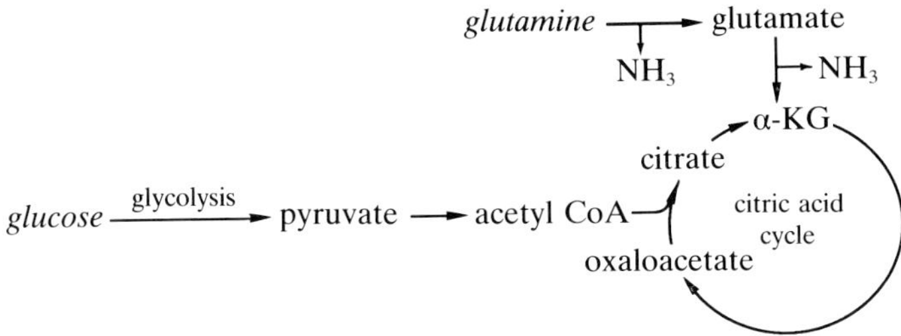

Figure 3. Aerobic substrate catabolism in helminths showing the coupled oxidation of glucose and glutamine; α-KG, α-ketoglutarate.

ship between body diameter of various intestinal nematodes of different sizes and the type of mitochondrial respiratory system has revealed an inverse relationship between the thickness of a worm species and the extent to which a conventional type of respiratory chain can possibly contribute to the overall respiratory process. There are, however, exceptions to this rule; for example, the respiration of the small nematode *Heterakis spumosa*, which resides in the hypoxic environment of the large intestine of the mouse, was found to be resistant to inhibition by antimycin A and low concentrations of cyanide (K.P. Davies & P. Köhler, unpublished observation).

Unlike the adults, the free-living and many of the parasitic developmental stages of helminths are aerobic. These stages usually have a high oxygen requirement, mainly for energy generation, with significant Krebs cycle and mitochondrial respiratory chain activities, and are therefore capable of completely oxidizing the major portion of their substrates to CO_2 and water. The alterations occurring in bioenergetics during the developmental cycles of helminths, and the mechanisms initiating the switch from one pathway to another, have at present been investigated for only a few species (Thompson *et al.*, 1984; Tielens *et al.*, 1984, 1987; Komuniecki & Vanover, 1987). One such species is the liver fluke *Fasciola hepatica*. The early liver parenchymal stages of this species possess a predominantly aerobic metabolism capable of complete substrate oxidation. However, in flukes between 2 and 8 weeks old the activity of the tricarboxylic acid cycle is already largely suppressed, although aerobic reactions still remain functional as characterized by an oxygen-dependent utilization of carbohydrate to acetate. These oxidative capacities then decline gradually until, after arrival of the flukes in their definitive environment (the bile ducts), anaerobic energy-yielding processes become the main strategies for ATP generation (see Figure 1). The more pronounced oxidative capacities of helminth developmental stages compared with the adults may be due to their smaller size, which causes fewer oxygen diffusion problems, along with the fact that sufficient oxygen is often available in the habitats of larval, in particular free-living, stages. Interestingly, most aerobic developmental stages of helminths can also survive under anoxic conditions by using anaerobic energy-conserving pathways, whereas in the adult the latter pathways usually cannot be reversibly replaced by aerobic processes (Tielens *et al.*, 1984). The observation that larval and juvenile stages often possess a typical aerobic metabolism clearly indicates that the genomes of these organisms must contain the complete message for all the enzymes required for the establishment of their terminal respiration. However, the precise mechanism responsible for the initiation and control of the metabolic transitions occurring during the developmental cycle of helminths remain to be elucidated.

References

Allen, B.L. & Harris, B.G. (1981). Purification of malic enzyme from *Ascaris suum* using NAD$^+$-Agarose. *Molecular and Biochemical Parasitology* **2**, 367–72.

Barrett, J. (1984). The anaerobic end-products of helminths. *Parasitology* **88**, 179–98.

Barrett, J., Coles, G.C. & Simpkin, K.G. (1978). Pathways of acetate and propionate production in adult *Fasciola hepatica*. *International Journal for Parasitology* **8**, 117–23.

Barrett, J., Dunn, T. & Butterworth, P.E. (1986*a*). Carbohydrate catabolism in *Onchocerca gutturosa* and *Onchocerca lienalis*. *Tropical Medicine and Parasitology* **37**, 69.

Barrett, J., Mendis, A.H.W. & Butterworth, P.E. (1986*b*). Carbohydrate metabolism in *Brugia pahangi* (Nematoda:Filaroidea). *International Journal for Parasitology* **16**, 465–9.

Campbell, T., Rubin, N. & Komuniecki, R. (1989). Succinate-dependent energy generation in *Ascaris suum* mitochondria. *Molecular and Biochemical Parasitology* **33**, 1–12.

Daum, G., Thalhofer, H.P., Harris, B.G. & Hofer, H.W. (1986). Reversible activation and inactivation of phosphofructokinase from *Ascaris suum* by the action of tissue-homologous protein phosphorylating and dephosphorylating enzymes. *Biochemical and Biophysical Research Communications* **139**, 215–21.

Davies, K.P. & Köhler, P. (1990). The role of amino acids in the energy generating pathways of *Litomosoides carinii*. *Molecular and Biochemical Parasitology* (in press).

Fioravanti, C.F. (1981). Coupling of mitochondrial NADPH:NAD transhydrogenase with electron transport in adult *Hymenolepis diminuta*. *Journal of Parasitology* **67**, 823–31.

Fioravanti, C.F. & Kim, Y. (1988). Rhodoquinone requirement of the *Hymenolepis diminuta* mitochondrial electron transport system. *Molecular and Biochemical Parasitology* **28**, 129–34.

Fry, M. & Jenkins, D.C. (1984). Nematoda: Aerobic respiratory pathways of adult parasitic species. *Experimental Parasitology* **57**, 86–92.

Grieshaber, M.K. (ed.) (1986). *Stoffwechsel unter Extrembedingungen. (Zoologische Beiträge* (ed. W. Dohle, G. Weigmann & I. Zerbst-Boroffka), vol. 30.) (422pp.) Duncker & Humblot, Berlin. [In German.]

Hata-Tanaka, A., Kita, K., Furushima, R., Oya, H. & Itoh, S. (1988). ESR studies on iron–sulfur clusters of complex II in *Ascaris suum* mitochondria, which exhibits strong fumarate reductase activity. *FEBS Letters* **242**, 183–6.

Hayashi, E. & Terada, M. (1973). The influence of oxygen pressure on the survival time of *Ascaris lumbricoides suum*. 4. The electron transport system containing cytochrome *c* peroxidase in *Ascaris* muscle. *Japanese Journal of Parasitology* **22**, 1–12.

Hofer, H.W., Allen, B.L., Kaeini, M.R., Pette, D. & Harris, B.G. (1982). Phosphofructokinase from *Ascaris suum*. Regulatory kinetic studies and activity near physiological conditions. *Journal of Biological Chemistry* **257**, 3801–6.

Kita, K., Takamiya, S., Furushima, R., Ma, Y.-C., Suzuki, H., Ozawa, T. & Oya, H. (1988*a*). Electron transfer complexes of *Ascaris suum* muscle mitochondria. III. Composition and fumarate reductase activity of Complex II. *Biochimica et Biophysica Acta* **935**, 130–40.

Kita, K., Takamiya, S., Furushima, R., Ma, Y.-C. & Oya, H. (1988*b*). Complex II is a major component of the respiratory chain in the muscle mitochondria of *Ascaris suum* with high fumarate reductase activity. *Comparative Biochemistry and Physiology* **89B**, 31–4.

Kita, K., Takamiya, S., Furishima, R., Suzuki, H., Ozawa, T. & Oya, H. (1990). Indispensability of rhodoquinone in the NADH-fumarate reductase system of *Ascaris* muscle mitochondria. In *Biochemistry, Bioenergetics and Clinical Applications of Ubiquinone* (ed. L. Giorgio). London: Taylor & Francis Ltd. (in press).

Kmetec, E. & Bueding, E. (1961). Succinic and reduced diphosphopyridine nucleotide oxidase systems in *Ascaris* muscle. *Journal of Biological Chemistry* **236**, 584–91.

Köhler, P. (1980). The function of mitochondrial enzymes in parasitic helminths. In *Industrial and Clinical Enzymology* (ed. Lj. Vitale & V. Simeon), pp. 243–56. Oxford: Pergamon Press.

Köhler, P. (1982). Bioenergetics in parasitic protozoa and helminths. In *Parasites – Their World and Ours* (ed. D.F. Mettrick & S.S. Desser), pp. 101–12. Amsterdam: Elsevier Biomedical Press.

Köhler, P. (1985). The strategies of energy conservation in helminths. *Molecular and Biochemical Parasitology* **17**, 1–18.

Köhler, P. (1988) Nutrition and metabolism. In *Parasitology in Focus* (ed. H. Mehlhorn), pp. 412–53. Heidelberg: Springer-Verlag.

Köhler, P. & Bachmann, R. (1980). Mechanisms of respiration and phosphorylation in *Ascaris* muscle mitochondria. *Molecular and Biochemical Parasitology* **1**, 75–90.

Köhler, P., Bryant, C. & Behm, C.A. (1978). ATP synthesis in a succinate decarboxylase system from *Fasciola hepatica* mitochondria. *International Journal of Parasitology* **8**, 399–404.

Komuniecki, R., Campbell, T. & Rubin, N. (1987). Anaerobic metabolism in *Ascaris suum*: acyl CoA intermediates in isolated mitochondria synthesizing 2-methyl branched-chain fatty acids. *Molecular and Biochemical Parasitology* **24**, 147–54.

Komuniecki, R., Fekete, S. & Thissen, J. (1984*a*). 2-Methylbutyryl CoA dehydrogenase from the mitochondria of *Ascaris suum* and its relationship to NADH-dependent 2-methylcrotonyl CoA reduction. *Biochemical and Biophysical Research Communications* **118**, 783–8.

Komuniecki, R., Komuniecki, P.R. & Saz, H.J. (1979). Purification and properties of the *Ascaris* pyruvate dehydrogenase complex. *Biochimica et Biophysica Acta* **571**, 1–11.

Komuniecki, R., Komuniecki, P.R. & Saz, H.J. (1981). Pathway of formation of branched-chain volatile fatty acids in *Ascaris* mitochondria. *Journal of Parasitology* **67**, 841–6.

Komuniecki, R., Rioux, A. & Thissen, J. (1984*b*). NADH-dependent tiglyl-CoA reduction in disrupted mitochondria of *Ascaris suum*. *Molecular and Biochemical Parasitology* **10**, 25–32.

Komuniecki, P.R. & Vanover, L. (1987). Biochemical changes during the aerobic-anaerobic transition in *Ascaris suum* larvae. *Molecular and Biochemical Parasitology* **22**, 241–8.

Komuniecki, R., Wack, M. & Coulson, M. (1983). Regulation of *Ascaris suum* pyruvate dehydrogenase complex by phosphorylation and dephosphorylation. *Molecular and Biochemical Parasitology* **8**, 165–76.

Kulkarni, G., Rao, G.S.J., Srinivasan, N.G., Hofer, H.W., Yuan, P.M. & Harris, B.G. (1987). *Ascaris suum* phosphofructokinase. Phosphorylation by protein kinase and sequence of the phosphopeptide. *Journal of Biological Chemistry* **262**, 32–4.

McLaughlin, G.L., Saz, H.J. & deBruyn, B.S. (1986). Purification and properties of an acyl CoA transferase from *Ascaris suum* muscle mitochondria. *Comparative Biochemistry and Physiology* **83B**, 523–7.

Paget, T.A., Fry, M. & Lloyd, D. (1987). Effects of inhibitors on the oxygen kinetics of *Nippostrongylus brasiliensis*. *Molecular and Biochemical Parasitology* **22**, 125–33.

Pietrzak, S.M. & Saz, H.J. (1981) Succinate decarboxylation to propionate and the associated phosphorylation in *Fasciola hepatica* and *Spirometra mansonoides*. *Molecular and Biochemical Parasitology* **3**, 61–70.

Podesta, R.B., Mustafa, T., Moon, T.W., Hulbert, W.C. & Mettrick, D.F. (1976). Anaerobes in an aerobic environment: Role of CO_2 in energy metabolism of *Hymenolepis diminuta*. In *Biochemistry of Parasites and Host-Parasite Relationships* (ed. H. van den Bossche), pp. 81–8. Amsterdam: North-Holland.

Ramp, Th., Bachmann, R. & Köhler, P. (1985). Respiration and energy conservation in the filarial worm *Litomosoides carinii*. *Molecular and Biochemical Parasitology* **15**, 11–20.

Ramp, Th. & Köhler, P. (1984). Glucose and pyruvate catabolism in *Litomosoides carinii*. *Parasitology* **89**, 229–44.

Rees, M.J. & Comley, J.C.W. (1988) Radiorespirometry, a sensitive parameter for quantifying in vitro antifilarial drug responses. *Tropical Medicine and Parasitology* **39**, 80–1.

Ruff, V., Desai, S., Dubrul, E.F. & Komuniecki, R. (1988). In vitro synthesis and processing of components of the *Ascaris suum* pyruvate dehydrogenase complex. *Molecular and Biochemical Parasitology* **29**, 1–8.

Sato, M., Yamada, K. & Ozawa, H. (1972). Rhodoquinone specificity in the reactivation of succinoxidase activity of acetone-extracted *Ascaris* mitochondria. *Biochemical and Biophysical Research Communications* **46**, 578–82.

Saz, H.J. (1971). Anaerobic phosphorylation in *Ascaris* mitochondria and the effects of anthelminthics. *Comparative Biochemistry and Physiology* **39B**, 627–37.

Saz, H.J. (1981). Energy metabolism of parasitic helminths: Adaptations to parasitism. *Annual Review of Physiology* **43**, 323–41.

Saz, H.J. & deBruyn, B.S. (1987). Separation and functions of two acyl CoA transferases from *Ascaris lumbricoides* mitochondria. *Journal of Experimental Zoology* **242**, 241–5.

Saz, H.J. & Hubbard, J.A. (1957). The oxidative decarboxylation of malate by *Ascaris lumbricoides*. *Journal of Biological Chemistry* **225**, 921–33.

Saz, H.J. & Pietrzak, S.M. (1980). Phosphorylation associated with succinate decarboxylation to propionate in *Ascaris* mitochondria. *Archives in Biochemistry and Biophysics* **202**, 399–404.

Starling, J.A., Allen, B.L., Kaeini, M.R., Payne, D.M., Blytt, H.J., Hofer, H.W. & Harris, B.G. (1982). Phosphofructokinase from *Ascaris suum*. Purification and properties. *Journal of Biological Chemistry* **257**, 3795–800.

Takamiya, S., Furushima, R. & Oya, H. (1984). Electron transfer complexes of *Ascaris suum* muscle mitochondria. I. Characterization of NADH-cytochrome *c* reductase (complex I–III), with special reference to cytochrome localization. *Molecular and Biochemical Parasitology* **13**, 121–34.

Takamiya, S., Furushima, R. & Oya, H. (1986). Electron transfer complexes of *Ascaris suum* muscle mitochondria. II. Succinate-coenzyme Q reductase (complex II) associated with substrate-reducible cytochrome *b*-558. *Biochimica et Biophysica Acta* **848**, 99–107.

Thalhofer, H.P., Daum, G., Harris, B.G. & Hofer, H.W. (1988). Identification of two different phosphofructokinase-phosphorylating protein kinases from *Ascaris suum* muscle. *Journal of Biological Chemistry* **263**, 952–7.

Thissen, J., Surbhi, D., McCartney, P. & Komuniecki, R. (1986). Improved purification of the pyruvate dehydrogenase complex from *Ascaris suum* body wall muscle and characterization of PDH kinase activity. *Molecular and Biochemical Parasitology* **21**, 129–38.

Thompson, D.P., Morrison, D.D., Pax, R.A. & Bennett, J.L. (1984). Changes in glucose metabolism and cyanide sensitivity in *Schistosoma mansoni* during development. *Molecular and Biochemical Parasitology* **13**, 39–51.

Tielens, A.G.M., van den Heuvel, J.M. & van den Bergh, S.G. (1984). The energy metabolism of *Fasciola hepatica* during its development in the final host. *Molecular and Biochemical Parasitology* **13**, 301–7.

Tielens, A.G.M., van den Heuvel, J.M. & van den Bergh, S.G. (1987). Differences in intermediary metabolism between juvenile and adult *Fasciola hepatica*. *Molecular and Biochemical Parasitology* **24**, 273–81.

Tielens, A.G.M. & van den Bergh, S.G. (1987). Glycogen metabolism in *Schistosoma mansoni* worms after their isolation from the host. *Molecular and Biochemical Parasitology* **24**, 247–54.

VandeWaa, E.A., MacKenzie, N.E., Geary, T.G., Bennett, J.L. & Williams, J.F. (1988). The role of glutamine in the intermediary metabolism of filarial parasites and effects of glutamine antagonists. *Tropical Medicine and Parasitology* **39**, 85.

Van Oordt, B.E.P., van den Heuvel, J.M., Tielens, A.G.M. & van den Bergh, S.G. (1985). The energy production of the adult *Schistosoma mansoni* is for a large part aerobic. *Molecular and Biochemical Parasitology* **16**, 117–26.

Ward, P.F.V. (1982). Aspects of helminth metabolism. *Parasitology* **84**, 177–94.

GEROLF GROS

The role of carbonic anhydrase within the tissues, with special reference to mammalian striated muscle

In this chapter, some of the roles of carbonic anhydrase (CA) in various tissue functions are reviewed. The scope of the article is limited to functions of CA that have been established or proposed for mammalian striated muscle. These functional roles in muscle are then compared to similar functions in other tissues.

The carbonic anhydrase present in muscle and other mammalian tissues

Which CA isozymes are present in the main tissues considered here? Briefly, many red blood cells, those of the human and rabbit for example, have two cytosolic isozymes, CA I, which is largely inhibited by intraerythrocytic Cl^- and whose function is not known, and CA II, which is the isoenzyme with the higher turnover number ($ca.$ 10^6 s^{-1}) and the higher sensitivity towards sulphonamides (inhibition constant $K_I = 10^{-8}$ M for acetazolamide) and which is responsible for most of the red cell CA activity. Kidney tissue also possesses cytosolic CA II and in addition a membrane-bound form called CA IV (Maren, 1980). The same holds for lung tissue, where a cytosolic CA and a membrane-bound enzyme have been demonstrated (Whitney & Briggle, 1982; Henry $et\ al.$, 1986). In striated muscles, the situation is more complicated, as may be seen in Table 1. In rabbit muscles that contain almost solely slow oxidative (SO) fibres, as soleus does, one finds high concentrations of the cytosolic CA III, an isoenzyme of very low sensitivity towards sulphonamides ($K_I = 10^{-4}$ M for acetazolamide) and of very low catalytic efficiency (turnover number 10^3 s^{-1}) that occurs almost exclusively in mammalian skeletal muscle. There is no cytosolic CA II in soleus muscle but there is Triton-extractable membrane-bound CA activity.[1] In heart muscle, we find no cytosolic CA at

---

1 Activity units given (U) indicate the factor (minus 1) by which the CO_2 hydration velocity is accelerated owing to CA. Triton-extracted activity is given in units per concentration of extracted proteins in milligrams per millilitre (see Gros & Dodgson, 1988).

Table 1. *Carbonic anhydrase activities in homogenate supernatant and particulate fraction of various striated rabbit muscles*[a]

Muscle	Sulphonamide-sensitive CA activity in supernatant (CA II) (U)	Sulphonamide-sensitive CA activity in supernatant (CA III) (U)	CA activity in Triton extract of homogenate pellet (U ml mg^{-1} protein)
Heart	0	0	17.5
Soleus	0	400	2.1
Vastus intermedius and semimembranosus proprius	50	360	2.5
Tibialis anterior	390	8	2.8
Extensor digitorum longus	250	16	—
Gastrocnemius (white)	290	—	4.1
Gastrocnemius intermedius	170	—	3.1
Masseter	—	0	6.7

[a]Sulphonamide-resistant activity was determined in the presence of 2×10^{-6} to 10^{-5} M acetazolamide; sulphonamide-sensitive activity was obtained as the difference between total and sulphonamide-resistant activity. One activity unit (1 U) indicates the enzyme concentration necessary to cut the reaction time of CO_2 hydration under the conditions of Maren's (1960) assay down to one half.
Data from C. Geers *et al.*, quoted from Gros & Dodgson (1988). With permission.

all, but a rather high specific activity of a membrane-bound enzyme. In extensor digitorum longus muscle (EDL, composed of fast glycolytic (FG) and fast oxidative-glycolytic (FOG) fibres) and in masseter muscle there is practically no CA III, but EDL has a substantial activity of cytosolic CA II. Both muscles have the membrane-bound form. Simplifying somewhat, it appears that, although all striated muscles possess a membrane-bound CA, fast skeletal muscle fibres also have cytosolic CA II whereas slow fibres have cytosolic CA III.

Where is the membrane-bound CA localized within the muscle tissue? There is functional and histochemical evidence arguing against its presence in the endothelial cell membranes (Geers *et al.*, 1985; Jeffery *et al.*, 1986) and we have previously shown that rabbit muscle mitochondria contain little or no CA (Bruns *et al.*, 1986). As shown in Table 2, membrane-bound

Table 2. *Carbonic anhydrase activity and marker enzymes in skeletal muscle membrane preparations*

	Yield (mg protein per 50 g muscle)	Ouabain-sensitive Na$^+$–K$^+$-ATPase (μmol P$_i$ mg^{-1} h^{-1})	Mg^{2+}-ATPase (μmol P$_i$ mg^{-1} min^{-1})	Ca^{2+}-ATPase (μmol P$_i$ mg^{-1} min^{-1})	Cholesterol (μmol mg^{-1})	CA (U ml mg^{-1})
White muscle[a]						
Microsomes	101	1.5	0.09	2.24	0.08	2.6
Sarcoplasmic reticulum	32.1	0.1 (n.s.)	0.01 (n.s.)	2.27	0.05	0.7
Sarcolemma	0.4	13.1	1.18	0	1.33	38.0
Red muscle[a]						
Microsomes	48	4.1	0.36	0.59	0.20	3.3
Sarcoplasmic reticulum	3.5	2.4	0.17	0. 41	0.09	2.4
Sarcolemma	0.5	9.6	0.90	0	0.59	15.6

[a]Membrane fractions from hindlimb muscles of the rabbit; n.s. indicates statistically not significantly different from zero. Data from Wetzel & Gros (1990).

CA of white and red (the latter represented by pooled solei, vasti inter-medii and semimembranosi proprii) rabbit muscles is found in the purified fractions of the sarcoplasmic reticulum (SR) and the sarcolemma (SL). The marker substances ouabain-sensitive Na^+–K^+-ATPase, Mg^{2+} (basal)-ATPase, Ca^{2+}-ATPase and cholesterol indicate, especially for white mus-cle, a high degree of purity of both membrane fractions. SR has a specific CA activity of 1–2 U ml (mg protein)$^{-1}$, as previously shown by Bruns *et al.* (1986); SL exhibits a much higher activity, 20–40 U ml (mg protein)$^{-1}$. The localization of membrane-bound CA to SR and SL has been confirmed by histochemical studies using dansylsulphonamide (Dermietzel *et al.*, 1985; Geers *et al.*, 1985; Bruns *et al.*, 1986) and by immunocytochemical tech-niques (Jeffery *et al.*, 1986). Figure 1 shows fluorescence micrographs of dansylsulphonamide-stained cryosections from rabbit masseter muscle showing the stained plasma membranes and SR; the latter gives the appearance of longitudinally oriented lines in longitudinal sections and of intracellular spots in cross-sectioned fibres.

In the following four sections I shall discuss four different possible func-tions of the carbonic anhydrases of skeletal muscle and compare them to similar functions of CA in other tissues: facilitated CO_2 diffusion, extracel-lular catalysis of CO_2 transport, intra- and extracellular buffering, and the role of CA in providing a rapid source of H^+ or HCO_3^-.

Facilitated diffusion of CO_2

Facilitated diffusion of CO_2 results from a simultaneous diffusion of HCO_3^- and H^+, the latter being mediated by mobile buffers (Gros & Moll, 1974; Gros *et al.*, 1976). When the layer in which this process is to occur is less than 1 mm thick, carbonic anhydrase must be present; it has amply been demonstrated that a CA homogeneously distributed in such a layer can fulfil the task of facilitating CO_2 transport. With cells less than 100 μm in diameter, therefore, an intracellular facilitation of CO_2 diffusion can only occur when CA is present; it is usually thought that this must be a cytosolic CA.

In view of the variety of CAs that occur in striated muscle, we were interested in the question of which ones would produce facilitated CO_2 diffusion and which types of muscle (or fibre) would exhibit facilitation. We therefore chose four muscles with quite different CA isozyme patterns (Table 1): soleus, heart, EDL and masseter (G. Gros, F. Romanowski & W. Schierenbeck, unpublished). We cut 1 mm thick sections from the central portions of the muscles (the septum in the heart), as parallel to the longitudinal fibre axes as possible. These sections were mounted between

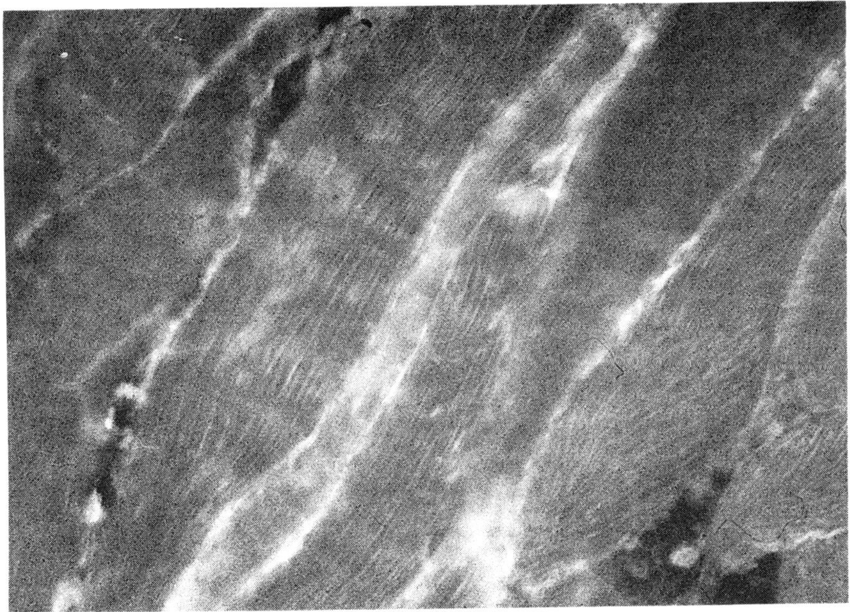

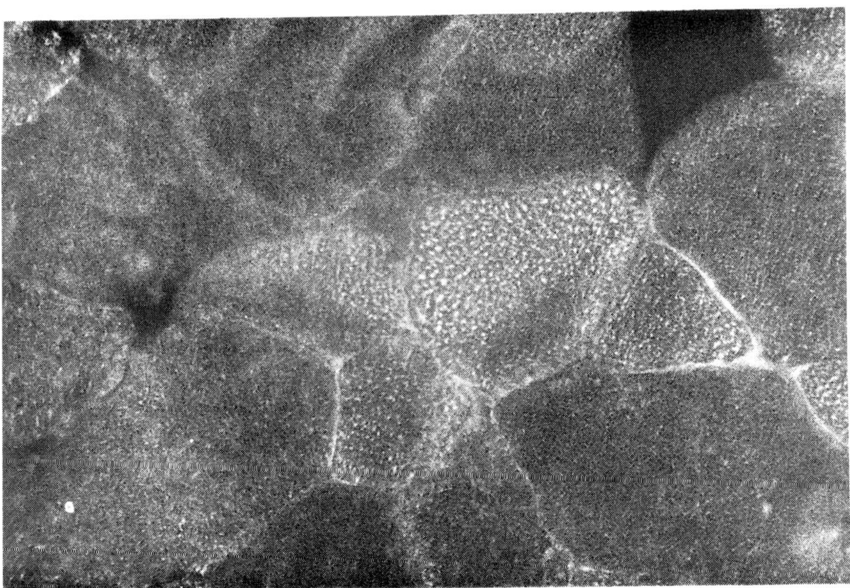

Figure 1. Fluorescence micrographs of cryosections from rabbit masscter incubated with 1×10^{-4} M dansylsulfonamide as described by Dermietzel *et al.* (1985). Above, longitudinally sectioned fibres; below, cross-sectioned fibres.

two chambers (a,b) that were both filled with a mixture of 95% O_2 and 5% CO_2. Into one chamber (a) a small amount of $^{14}CO_2$ and of acetylene was introduced. By taking gas samples from both chambers in regular time intervals it could be established that within the 4 h studied the concentrations of $^{14}CO_2$ and acetylene in chamber (a) remained practically constant while there was a linear increase with time in both gas concentrations in chamber (b), although these never exceeded 1% of the levels present in (a). Under these quasi-steady state conditions the amount of gas diffusing across the muscle layer per unit time, dm/dt, was obtained from the slope of the concentrations in (b) over time, and the partial pressure difference driving the diffusion process, ΔP, was approximated by the gas concentration in (a). These data allowed us to calculate the ratio of the diffusion constant for CO_2, K_{CO_2}, to the diffusion constant for acetylene, $K_{\text{acetylene}}$:

$$\frac{K_{CO_2}}{K_{\text{acetylene}}} = \frac{dm_{CO_2}/dt}{dm_{\text{acetylene}}/dt} \times \frac{\Delta P_{\text{acetylene}}}{\Delta P_{CO_2}}.$$

This ratio has a value of about 0.9 when no facilitation of CO_2 diffusion occurs; any increase above this value thus indicates that facilitated CO_2 diffusion has occurred. The advantage of using this ratio of diffusion constants is that the geometry of the muscle section (which has a rather inhomogeneous thickness) need not be known to establish facilitated diffusion.

Table 3 shows $K_{CO_2}/K_{\text{acetylene}}$ values for the four muscles studied. All are significantly greater than 0.9, indicating that facilitated CO_2 diffusion occurred. After incubation of the muscle layers in 0.15 M sodium lactate, pH 4, facilitation is completely suppressed ($K_{CO_2}/K_{\text{acetylene}}$ is approximately 0.9). After preincubation in $1-5\times10^{-3}$ M Diamox (acetazolamide), all ratios are drastically diminished, illustrating the dependence of facilitated CO_2 diffusion on the presence of active carbonic anhydrase. The 'facilitation factors' indicate that, in the muscles studied, intracellular CO_2 transport is accelerated by a factor of 2–4.

Physiologically, it appears to make sense that the two muscles with the highest rates of oxygen consumption and CO_2 production, soleus and heart, exhibit the greatest facilitation factors, 3–4, whereas EDL, which is known to have a lower $\dot{V}_{O_2}$, shows a facilitation factor of only 2. The high degree of facilitation in the heart, which has no cytosolic CA, shows clearly that membrane-bound CAs can also produce facilitated CO_2 diffusion. In the soleus, on the other hand, which has a high activity of CA III and a relatively low activity of the membrane-bound enzyme, it appears likely that the cytosolic CA III is mainly responsible for facilitation. This should

Table 3. *Facilitated CO_2 diffusion in 1 mm sections from striated muscles*[a]

	$K_{CO_2}/K_{acetylene}$	$K_{CO_2}/K_{acetylene}$ (pH 4)	$K_{CO_2}/K_{acetylene}$ ($1-5\times10^{-3}$ M Diamox)	Facilitation factor
Soleus	3.5±0.5	0.9±0.2	1.5±0.6	4.0
Heart	2.5±0.4	0.9±0.2	1.4±0.2	2.8
EDL	1.8±0.2	0.9±0.1	1.0±0.1	2.1
Masseter	1.9±0.3	1.0	1.3	1.9

[a]All muscles are from the rabbit. K_{CO_2} is the CO_2 diffusion constant (cm^2 min^{-1} atm^{-1}); $K_{acetylene}$, the acetylene diffusion constant (cm^2 min^{-1} atm^{-1}). The facilitation factor is the ratio of $(K_{CO_2}/K_{acetylene}):(K_{CO_2}/K_{acetylene})$ after preincubation of the muscle layer at pH 4. Values are means±s.d. Temperature 22 °C. Data from G. Gros, F. Romanowski & J. Schierenbeck (unpublished).

be in agreement with previous results for rat abdominal muscle, where titration of facilitated CO_2 diffusion with acetazolamide indicated that CA III was the isozyme responsible (Gros & Dodgson, 1988).

In two other mammalian tissues, red cells and the lung, facilitated CO_2 diffusion has been shown to occur (Gros & Moll, 1972; Enns & Hill, 1983; Heming *et al.*, 1986). In red cells, the diffusivity of CO_2 was increased by 60%. Enns & Hill (1983) observed a 39% decrease in lung membrane diffusing capacity upon CA inhibition; Heming *et al.* (1986) observed a 25% decrease in CO_2 excretion of rat lungs, which they could attribute to inhibition of the intracellular (cytosolic) lung CA.

Extracellular catalysis of CO_2 transport

Since plasma does not contain CA activity, and the half-time of the uncatalysed CO_2 hydration is about 10s, the system $CO_2-HCO_3^--H^+$ is not expected to be in equilibrium when the plasma leaves the capillary bed after a fraction of a second or a few seconds at most. This 'pH disequilibrium' was theoretically predicted and experimentally demonstrated by Forster & Crandall (1975), Crandall *et al.* (1977), Hill *et al.* (1977), Crandall & O'Brasky (1978) and Bidani *et al.* (1983). Because proton exchange across the red cell membrane is slow, this phenomenon will prevent the buffer capacity of the plasma from being utilized for CO_2 uptake or release during capillary transit and thus reduce the amount of CO_2 taken up or released per arteriovenous P_{CO_2} difference.

It appears that in some organs, in skeletal muscle and in lung, this disadvantage is avoided by means of an extracellular CA that maintains

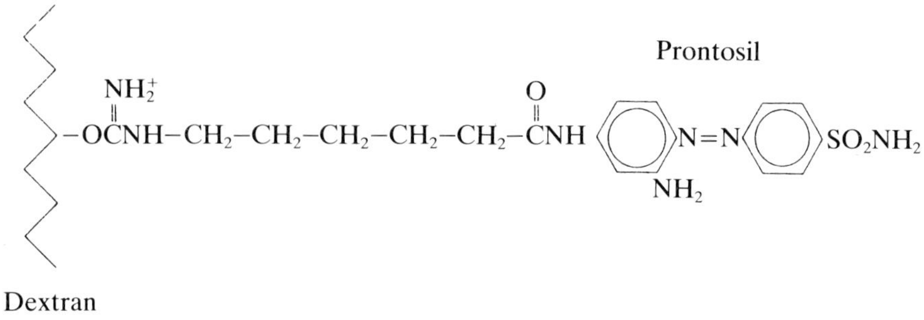

Prontosil

NH_2^+

$-O\overset{\text{O}}{\overset{\|}{C}}NH-CH_2-CH_2-CH_2-CH_2-CH_2-\overset{\text{O}}{\overset{\|}{C}}NH$ —⟨benzene⟩— $N=N$ —⟨benzene⟩— SO_2NH_2

NH_2

Dextran

Figure 2. Structure of Prontosil–dextran (PD).

CO_2, HCO_3^- and H^+ at equilibrium in the intracapillary or extracellular space. Indirect evidence for such an enzyme in skeletal muscle was first presented by Zborowska-Sluis *et al.* (1974). In order to study the localization of this CA, we observed the washout behaviour of $H^{14}CO_3^-$ in the isolated perfused rabbit hindlimb and the effect of CA inhibitors of various molecular sizes on it (Geers & Gros, 1984; Geers *et al.*, 1985). By coupling Prontosil to dextrans (Figure 2) we synthesized highly stable Prontosil–dextran (PD) macromolecular CA inhibitors of relative molecular mass (M_r) 5000 (PD 5000, expected to have access to the entire extracellular space) and of M_r 100 000 (PD 100 000, expected to be confined to the intravascular space). Their effects on $H^{14}CO_3^-$-washout were compared to that of the inhibitor acetazolamide, which has a low molecular mass.

Figure 3*a* (control) shows that the washout of ^{14}C after a bolus injection of $H^{14}CO_3^-$ and [^3H]dextran (M_r 80 000) is much slower than that of ^3H. If extracellular CA is present, this result is expected because then $H^{14}CO_3^-$ is rapidly converted to $^{14}CO_2$, which can readily enter the entire intracellular space and thus delay the washout of ^{14}C. The right-hand side of Figure 3*a* shows that in the presence of PD 5000 the washout of ^{14}C is markedly accelerated while that of ^3H remains unaltered. This effect of PD 5000 is quite similar to that of the low-M_r inhibitor acetazolamide. However, as shown in Figure 3.*b*, PD 100 000 clearly has no effect on ^{14}C-washout. We conclude that the CA that is involved here is accessible to PD 5000 but not to PD 100 000; in other words, it is located in the interstitial space. From the histochemical and biochemical evidence given above for a high CA activity in the sarcolemma, it appears likely that this interstitial CA is associated with the sarcolemma and with its active centre oriented towards the cell exterior.

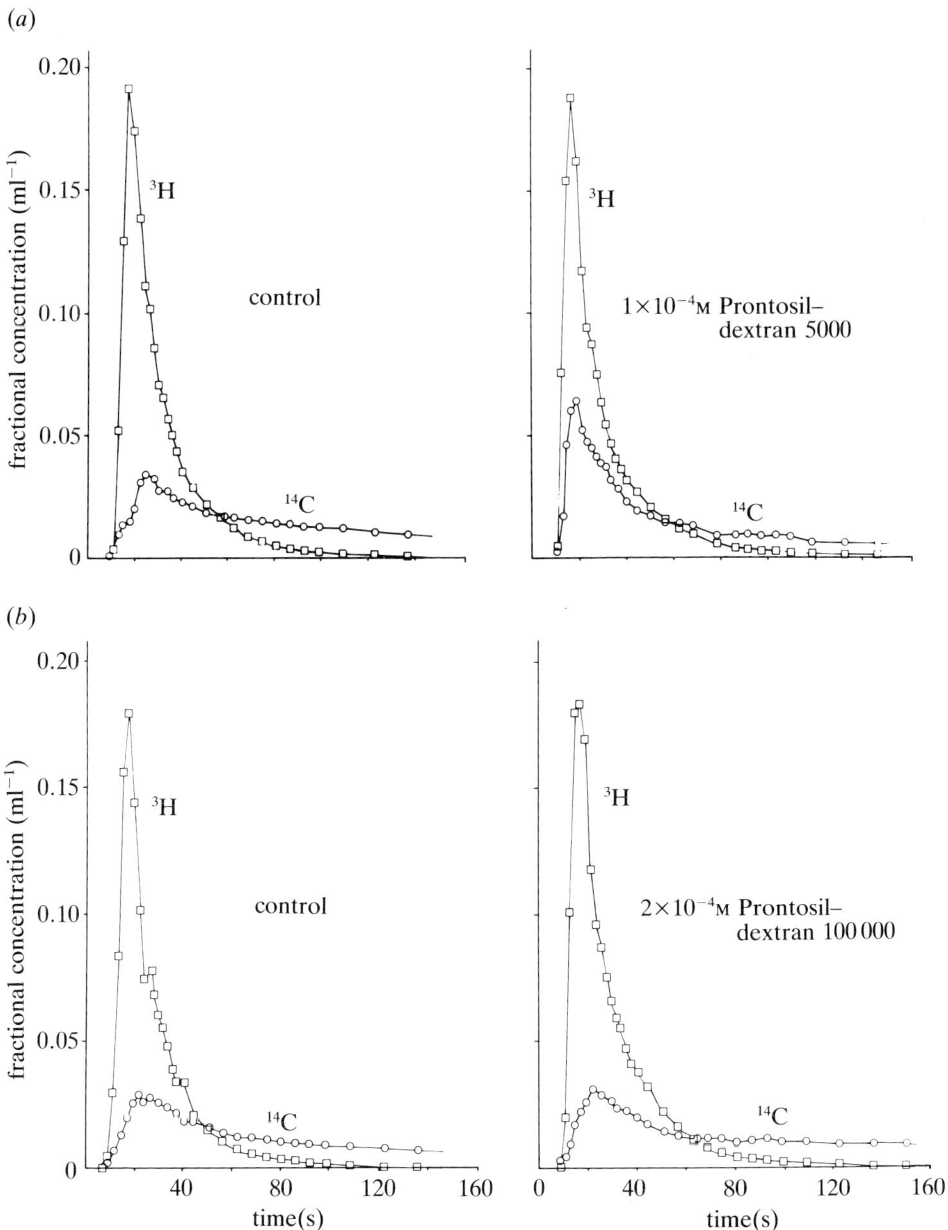

Figure 3. Washout curves of ^{14}C and ^{3}H from the isolated perfused rabbit hindlimb after bolus injection of $H^{14}CO_3^-$ and [^{3}H]dextran (M_r 80 000) into the blood-free perfusate entering the femoral artery. (a) Effect of PD 5000 on washout; (b) effect of PD 100 000 on washout. From Geers *et al.* (1985), with permission.

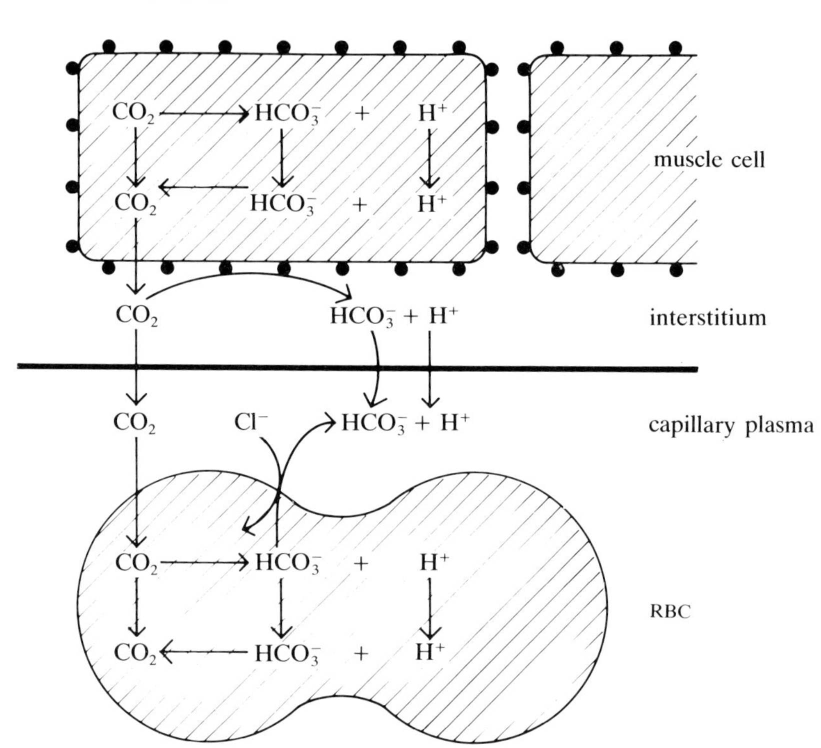

Figure 4. (*a*) Scheme illustrating the contributions of CAs to CO_2 exchange between muscle and blood; (*b*) analogous scheme of CO_2

Figure 4*a* shows a model of the CO_2 exchange between skeletal muscle and blood that incorporates the known contributions of CAs to this process. CO_2 exit from the cell occurs by free and facilitated CO_2 diffusion in the sarcoplasm; membrane permeation is only in the form of CO_2, as the permeability of the sarcolemma to HCO_3^- is very low. Just after CO_2 has left the cell it is partly converted to HCO_3^- and H^+, which apparently can both cross the capillary wall (this follows from the observation of O'Brasky & Crandall (1980) that the extracellular CA can establish at least partial pH equilibrium in the capillary) and be taken up in the plasma. A major part of the CO_2 leaving the muscle cells will diffuse to the red cell without undergoing chemical reaction. In the red cell the diffusion of CO_2 is again facilitated, owing to the presence of CA and to the haemoglobin acting as a mobile buffer. This diffusion step should not, however, be limiting for the speed of the overall process because K_{CO_2} is high and the intraerythrocytic

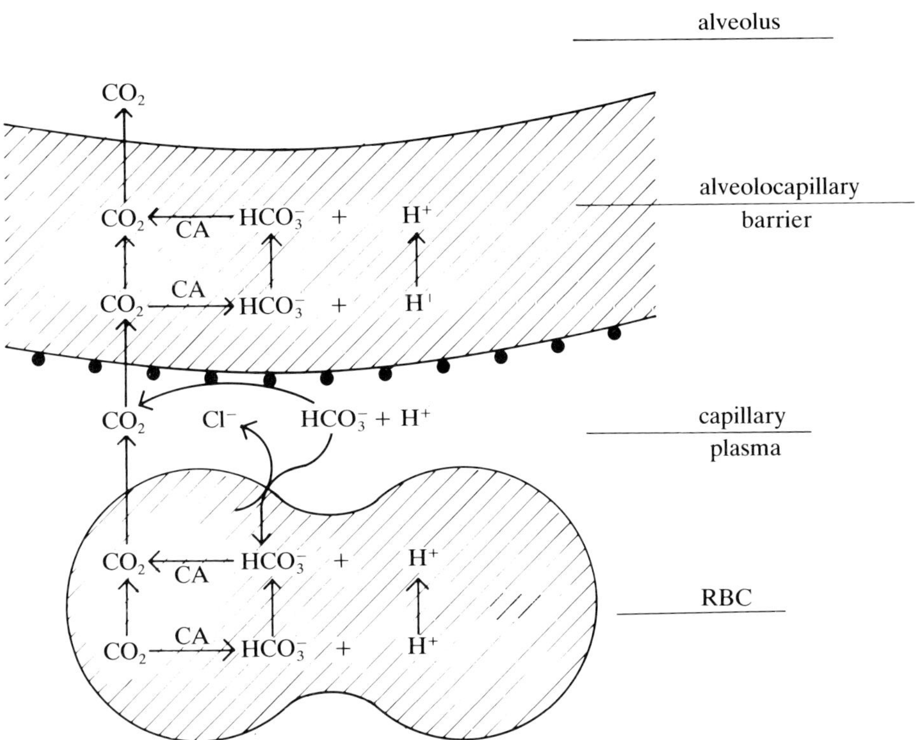

exchange between blood and lung. Filled circles on the outside of plasma membranes indicate extracellular membrane-bound CA.

diffusion distance is small. Similarly, the speed of the conversion of CO_2 to HCO_3^- and H^+ should not be critical as this reaction has a half-time of *ca.* 1 ms within the red cell owing to the high CA activity there ($A = 10\,000$ U). The limiting factors for CO_2 uptake by the red cell appear to be the Cl^-–HCO_3^- exchange across the membrane and the formation of haemoglobin carbamate (Klocke, 1988; Gros *et al.*, 1981).

Figure 4*b* shows the analogous model of CO_2 exchange between blood and alveoli. The processes within the red cell, of course, are all reversed. The major portion of plasma HCO_3^- enters the red cell, where it associates with haemoglobin-bound H^+, generating CO_2. A minor part, however, does not enter the red cell but combines with H^+ buffered in the plasma to give CO_2. This reaction is possible because there is also extracellular CA in the lung. However, this CA seems to be intravascular and localized at the capillary endothelial membranes. This is suggested by histochemical evidence (Ryan *et al.*, 1982) and by the results of functional studies using the macromolecular inhibitors PD 5000 and PD 100 000 (Heming *et al.*, 1986). As discussed above, transit of CO_2 through the alveolocapillary barrier is augmented by facilitated diffusion, although again this is probably not a critical step in view of the high 'membrane diffusing capacity' of the lung (Schuster, 1987).

Intra- and extracellular H^+ buffering

In neurons and skeletal muscle, respectively, there is experimental evidence, albeit sparse, for a role of cytoplasmic as well as extracellular carbonic anhydrase in the kinetics of H^+ buffering. It is well known that, at constant P_{CO_2}, the CO_2–HCO_3^- system makes a major contribution to the intracellular buffer capacity. However, without CA this system may be too slow when protons appear in the cytoplasm very rapidly. For example, it is likely that intracellular lactic acid rises to 20–30 mM within considerably less than 1 min (Hirche *et al.*, 1975; Sahlin *et al.*, 1978). The CO_2 hydration–dehydration reaction, however, requires minutes for near completion.

The following consideration illustrates the loss in intracellular buffering power when the CO_2–HCO_3^-, for such a reason, does not participate. Let us assume an intracellular 'non-bicarbonate' buffer factor of 45 mM/ΔpH, $P_{CO_2}=45$ Torr[1], $pH_i=7.1$, $[HCO_3^-]_i=14.3\times10^{-3}$M (Aickin & Thomas, 1977). When intracellular lactic acid concentration rises to 30 mM and the CO_2–HCO_3^- reacts while P_{CO_2} remains constant, pH_i will fall to 6.65; if the CO_2–HCO_3^- system does not participate in buffering, pH_i will fall to 6.43. Thus, one third of the intracellular buffering power under these conditions is due to HCO_3^-.

The results of an experiment by Thomas (1984), demonstrating the importance of CA for intracellular buffering by CO_2–HCO_3^- in a snail neuron are shown in Figure 5. The lower trace is a recording of pH_i from a microelectrode; the middle trace shows when defined amounts of HCl were injected into the same neuron. It may be seen that each injection is initially followed by a considerable fall in pH_i when the preparation is superfused with Ringer solution equilibrated with 0.5% CO_2. When that solution at constant pH_o is equilibrated with 4% CO_2, the fall in pH_i after HCl injection is much smaller because intracellular $[HCO_3^-]$ and buffer power have increased. When acetazolamide is added to the superfusate, the fall in pH_i upon HCl injection becomes much larger again, indicating that in the

---

1 1 Torr=1 mmHg=133.3 N m^{-2}.

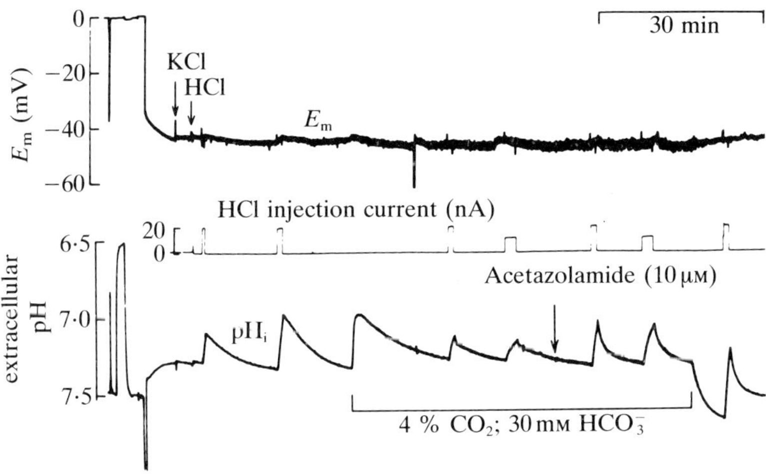

Figure 5. A pH microelectrode recording from a snail neuron with repeated pulses of HCl injection; E_M, membrane potential. The extracellular pH of the superfusate was 7.5. Superfusate was equilibrated with 0.5% CO_2 except in the period indicated, in which CO_2 was raised to 4%. The arrow indicates the addition of acetazolamide to the superfusate. Reproduced from Thomas (1984), with permission.

absence of active CA the buffering power provided by HCO_3^- cannot be fully utilized. It is of interest to note that this effect even becomes apparent on the relatively slow time scale of Figure 5.

Figure 6 shows results from an experiment by de Hemptinne *et al.* (1987) demonstrating the role of sarcolemmal CA in extracellular buffering. By using pH microelectrodes, pH_i (upper trace) and surface pH, pH_s (lower trace), were recorded from a rat soleus fibre. The left-hand part of this recording shows the intracellular acidification and surface alkalinization (due to unstirred layer effects despite constant superfusate pH) that accompany exposure to propionate and entry of propionic acid into the fibre. The same experiment was repeated on the right-hand part of the figure after a short pre-exposure of the muscle to 10^{-4} M acetazolamide. It is apparent that there are much larger changes in surface pH, presumably because inhibition of the extracellular, sarcolemmal CA prevents rapid utilization of CO_2–HCO_3^- as an extracellular buffer. It may in addition be noted that under these conditions the recovery of pH_i is also slowed down.

It may be noted that, in fast muscles, with their high capacity of lactic

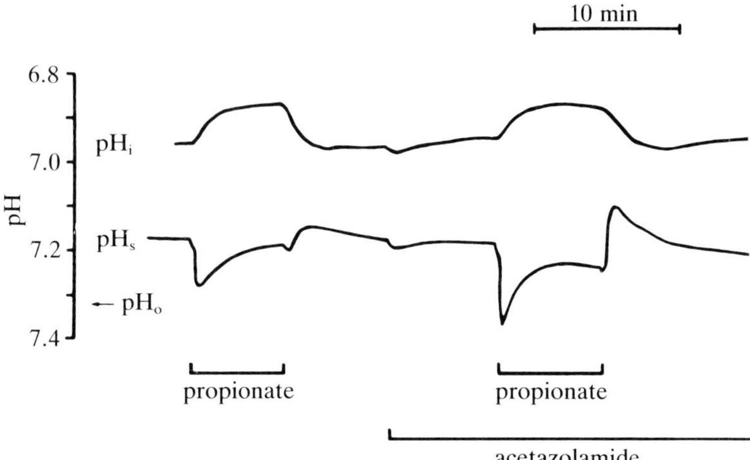

Figure 6. pH microelectrode recordings of intracellular pH (pH$_i$) and surface pH (pH$_s$) from rat soleus. Cl$^-$ (20 mM) was replaced by 20 mM propionate before and after addition of 0.1 mM acetazolamide to the superfusate. The superfusate had a pH (pH$_o$) of 7.3 throughout and a HCO$_3^-$ concentration of 21 mM, and was equilibrated with 5% CO$_2$. Reproduced from de Hemptinne *et al.* (1987), with permission.

acid production, there is (i) cytosolic CA II which does not occur in slow and oxidative muscles (soleus, heart), and may play a role in intracellular H$^+$ buffering, and (ii) a markedly higher level of sarcolemmal CA than in slow skeletal muscles such as soleus (Table 2). Since CO$_2$–HCO$_3^-$ is almost the only buffer system in the interstitial space, an important role of the sarcolemmal CA may be to prevent drastic acidification of the interstitium when lactic acid moves out of fast muscle fibres.

Rapid H$^+$ or HCO$_3^-$ source

The role of CA in providing H$^+$ or HCO$_3^-$ for secretory processes has long been recognized. It appears that there is CA wherever significant amounts of HCO$_3^-$ or H$^+$ move across cell membranes, a few examples being stomach, kidney, osteoclasts (H$^+$ secretion), muscle, platelets, red cells (Na$^+$–H$^+$ exchange and/or HCO$_3^-$–Cl$^-$ exchange), and pancreas, aqueous humour and cerebrospinal fluid (HCO$_3^-$ secretion).

Figure 7 shows the role of cytosolic CA in H$^+$ secretion in the stomach (top) and in HCO$_3^-$ secretion in pancreas, aqueous and CSF (bottom). The scheme in the middle shows the role of CAs in HCO$_3^-$ reabsorption in the proximal tubule of the kidney, probably the best understood of the three mechanisms. As in the stomach, there is a cytosolic CA providing the

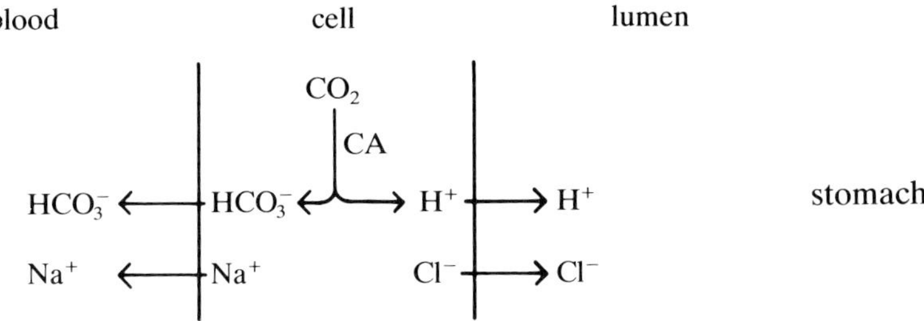

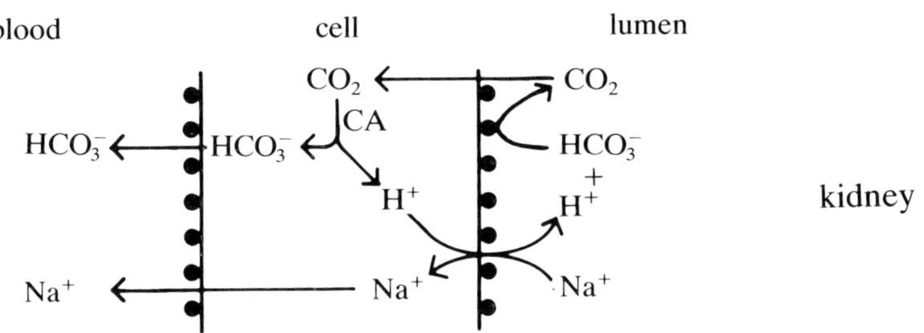

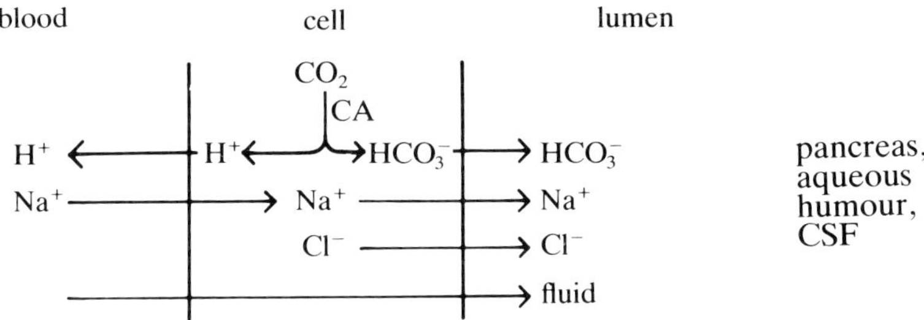

Figure 7. Schematic representation of the role of CA in H^+ and HCO_3^- secretion in various organs. Modified from Maren (1987).

protons secreted into the lumen. There is a second CA involved, the luminal membrane-bound CA that catalyses the formation of CO_2 from filtered HCO_3^- and secreted H^+. On the basolateral side there is HCO_3^- secretion from the cell into the interstitium in which the basolateral membrane-bound CA plays a central but not completely understood role; this process can be inhibited by exposing the basolateral side to the macromolecular inhibitor PD 5000, as well as by SITS (Burckhardt *et al.*, 1985).

Is CA necessary for the rates of H^+ or HCO_3^- secretion observed in these organs? The following consideration for the proximal kidney tubule illustrates this. Assuming the mass of the proximal epithelia involved in H^+ secretion to be 30% of total kidney mass (Brod, 1973), we estimate the rate of H^+ secretion to be 3×10^{-2} mmol H^+ min^{-1} g^{-1} cells. Using $k_{CO_2}=0.1$ s^{-1} and $P_{CO_2}=65$ Torr ($[CO_2]=2.1\times10^{-3}$ M) we calculate the maximal rate of H^+ production in the absence of CA by neglecting the back reaction from

$$(d[H^+]/dt)_{max}=k_{CO_2}\,[CO_2]$$

$$=1.3\times10^{-2} \text{ mmol } H^+ \text{ min}^{-1} \text{ ml}^{-1}.$$

Thus, even with an extreme disequilibrium of the CO_2–HCO_3^-–H^+ system, the necessary rate of H^+ production cannot be achieved in these cells. If, on the other hand, we have an intracellular CA activity of 100, we can produce the required amount of H^+ with only minor disequilibrium:

$$(d[H^+]/dt)=101\times k_{CO_2}\,([CO_2]-[H^+][HCO_3^-]/K_1')$$

where K_1' is the antilogarithm of $-pK_1'$ of the Henderson–Hasselbalch relation. The right-hand side of this equation assumes the required value of 3×10^{-2} mmol H^+ min^{-1}/ml^{-1} with pH$=7.21$ (equilibrium value: pH 7.20) and $[HCO_3^-]=26.3$ mM (equilibrium value: 26.6 mM).

I shall close this chapter by proposing the hypothesis that CA may play a similar role in muscle by providing a rapid H^+ source and sink within the sarcoplasmic reticulum. In the first part of my paper I have presented histochemical and biochemical evidence showing that CA is associated with the SR. The possible functional role of this CA is illustrated in Figure 8; there is considerable evidence that H^+ moves in exchange for Ca^{2+} across the SR membrane during Ca^{2+} release and re-uptake by the SR (Meissner & Young, 1980; Somlyo *et al.*, 1981; Meissner, 1981). Because of the unlimited availability of CO_2, CO_2–HCO_3^- would be an ideal H^+ donor and acceptor within the SR. But, since Ca^{2+} transients ($t_\frac{1}{2}\approx50$ ms) are much faster than the half-time of the uncatalysed hydration–dehydration reaction ($t_\frac{1}{2}\approx10$ s), this requires CA. We have estimated the intra-SR CA activity to be *ca.* 1000 U, which would cut down the half-time of the reaction to *ca.* 10 ms (Bruns *et al.*, 1986), sufficiently fast to match Ca^{2+} kinetics.

From this hypothesis one would predict that the Ca^{2+} transients and consequently the rise time and relaxation kinetics of isometric force are slowed down when SR–CA is inhibited. Dr C. Geers and I have studied the contractile parameters of isolated rat solei under isometric conditions (Geers & Gros, 1988) and looked at the effect of CA inhibition by chlorzolamide (5×10^{-4} M) and NaCNO (10^{-2} M). It may be seen from

Table 4. *Contractile parameters of isolated, directly stimulated rat solei[a]*

	Incubation time (h)	Tetanic force (N)	Single twitch		
			Time-to-peak (ms)	Half-relax. time (ms)	n
Control	6	0.59±0.06	137±13	207±25	16
CLZ		0.48±0.07*	168±11*	387±26*	
Control	4	1.29±0.09	133± 5	183±16	6
Na CNO		0.66±0.02*	155±10	240±14*	

[a]The isolated solei were kept in Ringer solution equilibrated with 95% O_2 : 5% CO_2 at 20 °C. Control muscles are compared with muscles exposed to 5×10^{-4} M chlorzolamide (CLZ) or 10^{-2} M CNO^-, respectively. Asterisks indicate a statistically significant difference from the control value.
Data from C. Geers & G. Gros (unpublished).

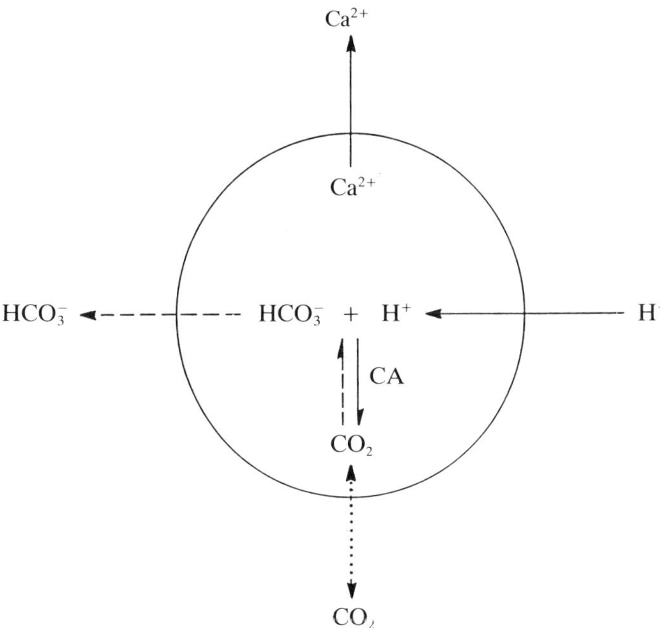

Figure 8. Schematic illustration of the hypothetical role of the CA of sarcoplasmic reticulum for H^+ uptake (and release) during Ca^{2+} release (and reuptake) by the SR. As an alternative to H^+ moving as a counter-ion against Ca^{2+}, HCO_3^- could move as a co-ion.

Table 4 that CA inhibition by chlorzolamide, besides a reduction of tetanic force, causes significant increases in time-to-peak and half-relaxation time. We were able to show that these changes could be explained neither by an effect of CA inhibition on pH_i nor by the effects of intracellular phosphates. Thus, these data seem compatible with the hypothesis of Figure 8, but more direct evidence will be required to ascertain such a role of SR–CA in intracellular Ca^{2+} mobilization.

References

Aickin, C. & Thomas, R.C. (1977). Micro-electrode measurement of the intracellular pH and buffering power of mouse soleus muscle fibres. *Journal of Physiology* **267**, 791–810.

Bidani, A., Mathew, S.J. & Crandall, E.D. (1983). Pulmonary vascular carbonic anhydrase activity. *Journal of Applied Physiology* **55**, 75–83.

Brod, J. (1973). *The Kidney*. Butterworth, London.

Bruns, W., Dermietzel, R. & Gros, G. (1986). Carbonic anhydrase in the sacroplasmic reticulum of rabbit skeletal muscle. *Journal of Physiology* **371**, 351–64.

Burckhardt, B.-Ch., Geers, C. & Frömter, E. (1985). Role of membrane-bound carbonic anhydrases in HCO_3-transport across rat renal proximal tubular cell membranes. *Pflügers Archiv* **405**, R31.

Crandall, E.D., Bidani, A. & Forster, R.E. (1977). Postcapillary changes in blood pH *in vivo* during carbonic anhydrase inhibition. *Journal of Applied Physiology* **43**, 582–90.

Crandall, E.D. & O'Brasky, J.E. (1978). Direct evidence for participation of rat lung carbonic anhydrase in CO_2 reactions. *Journal of Clinical Investigation* **62**, 618–22.

de Hemptinne, A., Marannes, R. & Vanheel, B. (1987). Surface pH and the control of intracellular pH in cardiac and skeletal muscle. *Canadian Journal of Physiology and Pharmacology* **65**, 970–7.

Dermietzel, R., Leibstein, A., Siffert, W., Zamboglou, N. & Gros, G. (1985). A fast screening method for histochemical localization of carbonic anhydrase. *Journal of Histochemistry and Cytochemistry* **33**, 93–8.

Enns, T. & Hill, E.P. (1983) CO_2 diffusing capacitry in isolated dog lung lobes and the role of carbonic anhydrase. *Journal of Applied Physiology* **54**, 483–90.

Forster, R.E. & Crandall, E.D. (1975). Time course of exchanges between red cells and extracellular fluid during CO_2 uptake. *Journal of Applied Physiology* **38**, 710–18.

Geers, C. & Gros, G. (1984). Inhibition properties and inhibition kinetics of an extracellular carbonic anhydrase in perfused skeletal muscle. *Respiration Physiology* **56**, 269–87.

Geers, C. & Gros, G. (1988). Carbonic anhydrase inhibition affects contraction of directly stimulated rat soleus. *Life Sciences* **42**, 37–45.

Geers, C., Gros, G. & Gärtner, A. (1985). Extracellular carbonic anhydrase of skeletal muscle associated with the sarcolemma. *Journal of Applied Physiology* **56**, 548–58.

Gros, G. & Dodgson, S.J. (1988). Velocity of CO_2 exchange in muscle and liver. *Annual Review of Physiology* **50**, 669–94.

Gros, G. & Moll, W. (1972). The facilitated diffusion of CO_2 in hemoglobin solutions and phosphate solutions. In *Oxygen Affinity of Hemoglobin and Red Cell Acid Base Status* (ed. M. Rorth & P. Astrup), pp. 484–92. Academic Press, New York.

Gros, G. & Moll, W. (1974). Facilitated diffusion of CO_2 across albumin solutions. *Journal of General Physiology* **64**, 356–71.

Gros, G., Moll, W., Hoppe, H. & Gros, H. (1976). Proton transport by phosphate diffusion – a mechanism of facilitated CO_2 transfer. *Journal of General Physiology* **67**, 773–90.

Gros, G., Wittmann, B., Guggenburger, L. (1981). Carbamate kinetics of blood proteins. *Progress in Respiration Research* **16**, 205–10.

Heming, T.A., Geers, C., Gros, G., Bidani, A. & Crandall, E.D. (1986). Effects of dextran-bound inhibitors on carbonic anhydrase activity in isolated rat lungs. *Journal of Applied Physiology* **61**, 1849–56.

Henry, R.P., Dodgson, S.J., Forster, R.E. & Storey, B.T. (1986). Rat lung carbonic anhydrase: activity, localization, and isozymes. *Journal of Applied Physiology* **60**, 638–45.

Hill, E.P., Power, G.G. & Gilbert, R.D. (1977). Rate of pH changes in blood plasma *in vitro* and *in vivo*. *Journal of Applied Physiology* **42**, 928–34.

Hirche, H., Hombach, V., Langohr, H.D., Wacker, R. & Busse, J. (1975). Lactic acid permeation rate in working gastrocnemii of dogs during metabolic alkalosis and acidosis. *Pflügers Archiv* **356**, 209–22.

Jeffery, S., Carter, N.D. & Smith, A. (1986). Immunocytochemical localization of carbonic anhydrase isozymes I, II, and III in rat skeletal muscle. *Journal of Histochemistry and Cytochemistry* **34**, 513–16.

Klocke, R.A. (1988). Velocity of CO_2 exchange in blood. *Annual Review of Physiology* **50**, 625–37.

Maren, T.H. (1960). A simplified micromethod for the determination of carbonic anhydrase and its inhibitors. *Journal of Pharmacology and Experimental Therapeutics* **130**, 26–9.

Maren, T.H. (1980). Current status of membrane-bound carbonic anhydrase. *Annals of the New York Academy of Sciences* **341**, 246–58.

Maren, T.H. (1987). Carbonic anhydrase: General perspectives and advances in glaucoma research. *Drug Development Research* **10**, 255–76.

Meissner, G. (1981). Calcium transport and monovalent cation and proton fluxes in sarcoplasmic reticulum vesicles. *Journal of Biological Chemistry* **256**, 636–43.

Meissner, G. & Young, R.C. (1980). Proton permeability of sarcoplasmic reticulum vesicles. *Journal of Biological Chemistry* **255**, 6814–19.

O'Brasky, J. E. & Crandall, E.D. (1980). Organ and species differences in tissue vascular carbonic anhydrase activity. *Journal of Applied Physiology* **49**, 211–17.

Ryan, U.S., Whitney, P.L. & Ryan, J.W. (1982). Localization of carbonic anhydrase in pulmonary artery endothelial cells in culture. *Journal of Applied Physiology* **53**, 914–19.

Sahlin, K.A., Alvestrand, A., Brandt, R. & Hultman, E. (1978). Intracellular pH and bicarbonate concentration in human muscle during recovery from exercise. *Journal of Applied Physiology* **45**, 474–80.

Schuster, K.-D. (1987). Diffusion limitation and limitation by chemical reactions during alveolar-capillary transfer of oxygen-labeled CO_2. *Respiration Physiology* **67**, 13–22.

Somlyo, A.V., Gonzalez-Serratos, H., Shuman, H., McClellan, G. & Somlyo, A.P. (1981). Calcium release and ionic changes in the sarcoplasmic reticulum of tetanized muscle: an electron-probe study. *Journal of Cell Biology* **90**, 577–94.

Thomas, R.C. (1984). Experimental displacement of intracellular pH and the mechanism of its subsequent recovery. *Journal of Physiology* **354**, 3P–22P.

Wetzel, P. & Gros, G. (1990). Sarcolemmal carbonic anhydrase in red and white rabbit skeletal muscle. *Archives of Biochemistry and Biophysics* (submitted).

Whitney, P.L. & Briggle, T.V. (1982). Membrane-associated carbonic anhydrase purified from bovine lung. *Journal of Biological Chemistry* **257**, 12056–9.

Zborowska-Sluis, D.T., L'Abbate, A. & Klassen, G.A. (1974). Evidence of carbonic anhydrase activity in skeletal muscle: a role for facilitated carbon dioxide transport. *Respiration Physiology* **21**, 341–50.

FRANK B. JENSEN

Multiple strategies in oxygen and carbon dioxide transport by haemoglobin

Introduction

In vertebrates, haemoglobin (Hb) plays a central role both in the transport of oxygen from the respiratory surfaces to the tissues and in the transport of carbon dioxide in the opposite direction. This dual function is accomplished through reversible and cooperative binding of O_2 to the four haem groups in tetrameric Hb, and by binding of H^+ and CO_2 to specific amino acid residues in the globin (protein) part of the molecule. The reversible binding of these ligands are interrelated by allosteric interactions between their binding sites, and are affected by further heterotropic interactions with binding sites for other erythrocytic cofactors, mainly organic phosphates. The Bohr–Haldane effect is the classical example of these interrelations. Lowering pH decreases HbO_2 affinity (the Bohr effect), whereas lowering HbO_2 saturation (Y) increases H^+-ion affinity (the Haldane effect). This interdependence of ligand binding makes Hb elegantly constructed to fulfil the transport needs of the blood.

The demands on the ligand-transporting properties of haemoglobin vary both between species and within species, owing to differences in metabolic rates and variations in living conditions imposed by the environment. As a consequence, major interspecific differences in the functional properties of haemoglobins have evolved, which adapt to long-term exploitation of particular ecological niches. Intraspecific adaptations, permitting short-term adjustments of Hb function, include a cascade of organismic and cellular mechanisms that ultimately affect function by changing the chemical microenvironment of Hb within the red cell.

The purpose of this review is to unravel the main strategies that are exploited by vertebrates to optimize O_2, CO_2 and H^+ transport in the blood in resting steady states, and as responses to changing internal and external milieu factors (with emphasis on exercise and hypoxia). The molecular basis for inter- and intraspecific adaptations is integrated with the organismic and cellular responses that adjust red cell Hb function to varying physiological needs. Equal attention is paid to O_2 and CO_2 transport and to effects that originate in the basic molecular linkage phenomena. As environmental variability applies to aquatic environments in particular, the respiratory properties of fish blood and Hb will be treated in detail. However, digressions to other vertebrate groups will be made whenever appropriate.

Molecular aspects of oxygen transport

Fish haemoglobins are $\alpha_2\beta_2$ tetramers with a tertiary and quaternary structure similar to that of other vertebrate haemoglobins. The amino acid sequences, however, differ widely within teleosts and between different systematic groups (Kleinschmidt & Sgouros, 1987). Many such differences are functionally neutral, and only a few substitutions in key positions appear to be responsible for differences in functional properties (Perutz, 1983).

The tetrameric Hb molecule can assume two quaternary structures. The T-structure, which characterizes the deoxy-state, is stabilized by inter- and intrasubunit salt bridges and has a low O_2 affinity. In the sequential oxygenation process, salt bonds are broken and the tetramer acquires the R conformation (i.e. undergoes the T→R allosteric transition), which has a high affinity for O_2. Heterotropic effectors (ATP, GTP, H^+, Cl^-, CO_2) bind preferentially to the T structure, which they stabilize by introducing additional bonds, thereby lowering HbO_2 affinity (Perutz, 1983). Figure 1 shows representative O_2 equilibria in tench Hb and illustrates the O_2 affinity modulation by the most potent modifiers, hydrogen ions (pH) and organic phosphates. The cooperativity of O_2 binding (reflected by a slope larger than unity in the middle range of Hill plots: cf. Figure 1) is, in general, lower in fish than in mammalian Hbs (n_{max} close to 2 and 3, respectively) (Weber *et al.*, 1987; Imai, 1982). Organic phosphates slightly elevate cooperativity of fish Hb at physiological red cell pH values, but decrease it at low pH (see, for example, Jensen & Weber, 1985).

The nucleoside triphosphates (NTP) in fish red cells (mainly ATP and GTP) bind to the Hb at the entrance to the central cavity between the two β-chains in the T structure. Two key amino acid substitutions, as compared with the DPG site in mammalian Hbs, make this site stereochemically complementary to ATP and GTP (Perutz & Brunori, 1982; Perutz, 1983). Bohr protons bind to several specific amino acid residues, with His HC3β being a major group in most vertebrate Hbs (Riggs, 1988). In tench Hb at *in vivo* NTP/Hb ratios and typical red cell pH values, ATP and GTP as well as protons lower HbO_2 affinity by decreasing the association constant

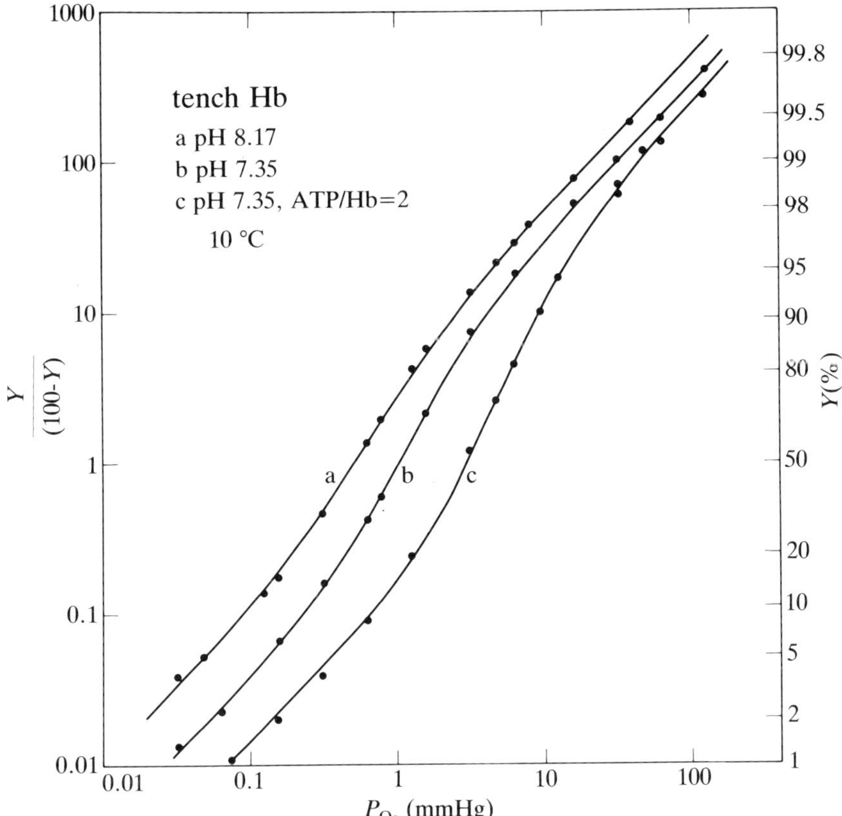

Figure 1. Hill plots of HbO_2 equilibria in tench haemoglobin (taken from Jensen & Weber, 1987).

for the T state (K_T) (move the bottom asymptote in the Hill plot to the right; see Figure 1) without significantly changing that for the R state (K_R), and by increasing the allosteric constant L (i.e. shifting the allosteric equilibrium towards the T state) (Weber *et al.*, 1987; Jensen & Weber, 1987). This resembles the effect of DPG and pH on human Hb (Imai, 1982). In the successive oxygenation process, NTPs lower the affinity (Adair constants, k_i ($i=1$–4)) of the first three O_2 molecules bound, whereupon NTP release is completed, and the affinity for the fourth O_2 is unchanged. At high NTP/Hb ratios, however, K_R and k_4 are also lowered (Weber *et al.*, 1987). The effect of GTP on P_{50} (O_2 tension at half-saturation), K_T, K_R and k_i is larger than that of ATP (Weber, 1982; Weber *et al.*, 1987), making GTP a more potent modifier of O_2 affinity. This depends on the ability of GTP to form an additional hydrogen bond to the globin, compared with ATP (see, for example, Perutz, 1983).

The alkaline Bohr effect (reduction in O_2 affinity upon reduction in pH) reflects an oxygenation-linked H^+ binding. Its magnitude depends on the number of Bohr groups and their pK shifts, and is further modulated (increased) by anions, notably organic phosphates (see, for example, Weber & Lykkeboe, 1978). In some amphibians a pH-dependent association of tetramers upon deoxygenation represents an alternative mechanism to elevate Bohr effects (Riggs, 1988). At very low pH (below 6.5–7, depending on cofactor conditions) many fish Hbs show a major decrease in cooperativity in addition to the affinity decrease, with Hills n_{50}-values approaching 1 (see, for example Binotti *et al.*, 1971; Jensen & Weber, 1985; Brittain, 1987). This unique feature of fish Hbs underlies the Root effect, and can be interpreted as an extension of the T state to very high HbO_2 saturations. The replacement of Cys F9β (mammalian Hbs) with serine (teleosts) may be important for this property, although the character of neighbouring groups influences its actual expression (Perutz, 1983). Thus lungfish and *Xenopus* have Ser at position F9, but do not exhibit Root effects (Rodewald *et al.*, 1984; Bridges *et al.*, 1985). In whole blood the Root effect causes a decrease in O_2 capacity as pH is reduced, and functions in O_2 secretion into the swimbladder and the retina of the eye (see Brittain, 1987), but it may hamper arterial O_2 loading in stress.

Many fish, including eel, trout and some catfish, possess multiple Hb systems, in which cathodal components exhibit no Bohr effect, or a slight reverse one. This may safeguard arterial O_2 loading in acidotic stress situations where anodal Hbs (Bohr–Root effect components) have poor loading capacity (Powers, 1972; Weber *et al.*, 1976). Structurally, the absence of a Bohr effect in cathodal components correlates with the replacement of the major Bohr group His HC3β with phenylalanine and the acetylation of α-amino groups in α-chains (Powers, 1972; Gillen & Riggs, 1973). This strategy is, however, not universal. Thus carp possess multiple Hbs that all have large Bohr effects (Weber & Lykkeboe, 1978) and in tench a major Bohr-Root effect component completely dominates the haemolysate properties (Jensen & Weber, 1982).

The specific effect of CO_2 on O_2 affinity (due to oxygenation-linked carbamate formation) is small or absent in most fish (Weber & Lykkeboe, 1978; Farmer, 1979; Jensen & Weber, 1982; Dobson & Baldwin, 1982) as α-amino groups in the α-chains are acetylated, and since organic phosphates compete effectively for binding to those of the β-chains. Lack of a specific CO_2 effect is beneficial in fish acclimated to combined hypoxia and hypercapnia, as its presence would counteract the phosphate-mediated increase in O_2 affinity (Jensen & Weber, 1982). It is therefore of interest that the same functional property applies to the mole, which experiences hypoxia–hypercapnia in its subterranean habitat (Jelkmann *et al.*, 1981), and to divers such as the grey seal and sea turtles (Lapennas, 1983; Lutz & Lapennas, 1982).

Apart from heterotropic ligands, temperature also influences HbO_2 affinity. Oxygenation is exothermic, and an increase in temperature decreases O_2 affinity. In ectotherms this facilitates O_2 offloading in parallel with the increased O_2 demand at higher temperatures. The actual magnitude of the affinity shift is, however, controlled by heterotropic ligands (cf. Weber & Jensen, 1988). Release of Bohr protons and phosphates during oxygenation is endothermic and accordingly reduces the temperature sensitivity of O_2 equilibria (Ikeda-Saito *et al.*, 1983; Jensen & Weber, 1987).

Cellular and organismic aspects of O_2 transport

The amount of O_2 delivered to the tissues by blood per unit time $(\dot{M}_{O_2})$, is a product of cardiac output $(\dot{Q}_h)$, the O_2 tension difference between arterial and mixed venous blood, and the blood O_2 capacitance coefficient (β_{O_2}):

$$\dot{M}_{O_2} = \dot{Q}_h \, \beta_{O_2} \, (P_{aO_2} - P_{vO_2}). \tag{1}$$

The capacitance coefficient is the slope of the concentration versus tension O_2 equilibrium curve (the mean value given by the line connecting the arterial and venous points), and it thus reflects the haemoglobin's O_2-transporting properties in whole blood. These properties depend on the concentration of O_2-binding sites as well as the affinity of these for the ligand, and thus can be adjusted both by quantitative (via [Hb]) and qualitative means (modulation of O_2 affinity).

Adjustments of Hb concentration

A high haemoglobin concentration in the blood secures a high β_{O_2} and decreases the pumping requirement of the heart. Since in most vertebrates red cell [Hb] is about 5 mM, which is close to the solubility limit (see Riggs, 1976), elevated blood [Hb] must proceed via elevated haematocrit values (i.e. greater ratios of red cells to plasma). This provides an upper limit to the quantitative adaptation strategy, owing to high blood viscosity at high haematocrits. In most ectothermic vertebrates the blood [Hb] is lower than in homeotherms (about 1 and 2 mM tetramers, respect-

60   *F.B. Jensen*

ively). This, however, correlates with a lower resting metabolic rate in ectotherms, whereby their circulatory requirement ($\dot{Q}_h/\dot{M}_{O_2}$) is normally only slightly higher than that in mammals. When blood [Hb] is reduced further, the circulatory requirement strongly increases, reaching its extreme in antarctic icefish that lack Hb (Wood & Lenfant, 1979). Very low blood β_{O_2} values also can be caused by oxidation of haem groups. Thus, environmental nitrite contamination can lead to detrimental methaemoglobinaemia and O_2 shortage in fish (Jensen *et al.*, 1987).

Blood [Hb] is ubiquitously elevated in hypoxic mammals (Lenfant, 1973) but the response is slow (i.e. days), depending on an erythropoietin-stimulated differentiation of stem cells to red cells. Among fishes, hypoxia or exercise-stress (exhausting bursts of exercise) strongly elevates blood [Hb] in trout, yellowtail and eel, but only slightly in tench, carp, striped bass and plaice (see Weber & Jensen, 1988). Some active fish (e.g. tuna species) may have relatively high blood [Hb] (Wood & Lenfant, 1979; Cech *et al.*, 1984; Jones *et al.*, 1986), that adapt to a high $\dot{M}_{O_2}$. Salmonids, however, have a similar [Hb] to that of less active cyprinid species. In sustained exercise in trout, [Hb] is elevated only when the maximal swimming speed is approached (Kiceniuk & Jones, 1977), or if arterial P_{O_2} decreases (Thomas *et al.*, 1987). When they occur, rapid [Hb] elevations in exercised and hypoxic fish can result from a catecholamine-mediated release of stored erythrocytes via spleenic contractions, water shifts to lactate-loaded muscles, and increased diuresis (Yamamoto *et al.*, 1980; Yamamoto, 1987; Jensen, 1987; Milligan & Wood, 1986).

Inter- and intraspecific adjustments of O_2 affinity

The role of the blood O_2 equilibrium curve is dual and dynamic, encompassing the needs for sufficient O_2 loading at the respiratory surfaces and O_2 unloading in the tissues, at a reasonably high unloading P_{O_2} (which sets the pressure head for diffusion). An evaluation of the physiological role of the curve thus has to take into consideration its shape and position as well as the arterial and venous blood O_2 tensions. These blood P_{O_2} values in turn are determined by a complex interplay of HbO_2 affinity *per se*, the ventilation/perfusion ratio, the diffusion conductance at the respiratory surfaces and in the tissues, and by the tissue O_2 demand.

Active fish species living in well-aerated waters will benefit from a relatively low blood O_2 affinity and cooperative Hb–O_2 binding, as arterial P_{O_2} can be kept high to secure O_2 loading, and extensive O_2 unloading can occur on the steep portion of the O_2 equilibrium curve, at a high mean tissue capillary P_{O_2}. A similar argument applies to the higher P_{50}-values often found in blood and haemoglobin solutions from small mammals in comparison with large mammals, since the former have larger mass-specific $\dot{M}_{O_2}$ values (Riggs, 1976; Schmidt-Nielsen, 1984). Within single species, however, a similar correlation between mass and blood P_{50} can be absent (Burggren *et al.*, 1987), just as the interspecific correlation can be overridden by environmental factors such as hypoxia.

Hypoxic-tolerant fishes are preadapted to low environmental P_{O_2} by having a higher blood O_2 affinity than fish exploiting well-aerated waters (Krogh & Leitch, 1919). The same evolutionary trend is found in burrowing mammals and mammals native to high altitudes, when compared with the normoxic mammalian allometric relation between P_{50} and body mass (Lenfant, 1973; Boggs *et al.*, 1984). A high blood O_2 affinity safeguards arterial loading and a high blood β_{O_2} at low P_{O_2}, but on the other hand dictates offloading at comparatively low tissue capillary P_{O_2}. Fish with a high blood O_2 affinity (e.g. carp and tench) will, however, have a large mean P_{O_2} diffusion gradient between water and blood that reduces ventilatory requirement and produces a high O_2 extraction coefficient from the water, when compared with fish with low blood O_2 affinity (e.g. trout) under normoxic conditions (Malte & Weber, 1987). In hypoxic carp, ventilatory requirement is increased, but high water O_2 extractions are preserved (Lomholt & Johansen, 1979), presumably as a result of a further increase in O_2 affinity (see below), and an increase in the gill diffusive conductance via lamellar recruitment and decreased diffusion distance (Randall & Daxboeck, 1984).

Some amphibians, reptiles and birds have blood with relatively high P_{50} values, but with n values increasing to over 4 with rising O_2 saturations; this may be a result of association of tetramers at low saturations (Lykkeboe & Johansen, 1978; Lutz, 1980; Riggs, 1988). This unique property protects arterial O_2 loading, while at the same time unloading P_{O_2} can be kept high.

Interspecific differences in the oxygen-binding properties of whole blood are a complex function of the intrinsic O_2 affinity of the Hb(s), the type of, and sensitivity to, heterotropic ligands, the concentration of these within the erythrocytic microenvironment, and competition between effectors for common binding sites. Species-specific adaptations in O_2 affinities can accordingly be achieved via several routes. The intrinsic O_2 affinity of tench Hb is much higher than that of trout Hb ($P_{50}=1.5$ and 15 mmHg, respectively, at a typical red cell pH_i of 7.35 and at 15 °C). Together with fine-tuning via NTPs (ATP+GTP in tench and ATP in trout, at NTP/Hb ratios of 2 and 1.5, respectively), this explains the whole-blood P_{50} values of 6 and 22 mmHg, respectively (Jensen & Weber, 1982; Weber *et al.*,

1976). In the hypoxia-tolerant llama, in contrast, the intrinsic O_2 affinity of the Hb is low, and a comparatively high blood O_2 affinity is achieved via a low affinity for the erythrocytic phosphate 2,3-diphosphoglycerate (2,3-DPG). This originates in the replacement of positively charged His NA2β (normally involved in phosphate binding) by neutral Asn, which weakens the Hb–DPG interaction (Bauer *et al.*, 1980).

Whereas the respiratory properties that are evolutionarily 'coded' in the Hb molecule contribute to an overall adaptation to particular ecological niches, intraspecific mechanisms are needed to adjust Hb function to the fluctuating requirements associated with variations in the internal and external environment of the animal. For example, in hypoxic fish where ΔP_{O_2} decreases and $\dot{Q}_h$ often changes only slightly (Randall & Daxboeck, 1984), equation 1 shows the universal requirement for an increase in β_{O_2}, although a decrease in O_2 demand *per se* (see, for example, Lomholt & Johansen, 1979; Boutilier *et al.*, 1988) may limit its required magnitude.

Intraspecific 'qualitative' adaptations of Hb function in fish involve three major mechanisms: (i) adjustment of red cell pH; (ii) adjustment of the erythrocytic content of organic phosphates; and (iii) red cell volume changes. A fourth possibility, alterations in the composition and relative abundance of isoHbs, seems less frequently used (Weber & Jensen, 1988).

Red cell pH (pH_i) is normally governed by a passive Donnan-like distribution of H^+ across the red cell membrane (Hladky & Rink, 1977; Heming *et al.*, 1986; Jensen, 1988), whereby any change in extracellular pH (pH_e) tends to be transferred to the red cell and change O_2 affinity via the Bohr effect. The immediate increase in ventilation in hypoxic fish (Dejours, 1973) accordingly benefits arterial loading both by limiting the arterial P_{O_2} decrease and by decreasing arterial P_{CO_2}, as the resulting respiratory alkalosis increases blood O_2 affinity. The magnitude of the affinity change depends on the magnitudes of Bohr factors (ϕ) and ΔpH values. In carp, arterial blood pH_e can increase from 8.06 to 8.36 when environmental P_{O_2} decreases from 140 to 25 mmHg within 20 minutes (F.B. Jensen, N.A. Andersen & N. Heisler, unpublished results), which halves P_{50} from 7 to 3.5 mmHg with a ϕ value of -1 (Weber & Lykkeboe, 1978). The response may, however, be counteracted by a lactacidosis, depending on species (i.e. hypoxia tolerance) and severity of O_2 paucity. Also, the benefits of the increase in pH caused by hyperventilation can be fully exploited only in normocapnic hypoxia. In combined hypoxia and hypercapnia, which prevails in many O_2-depleted habitats (Jensen & Weber, 1982; Bridges *et al.*, 1984), elevated environmental P_{CO_2} values will counteract the blood alkalosis and may even lead to a respiratory acidosis.

This will raise the requirement for acid–base regulation in parallel with an increased need for optimizing blood O_2 transport. In tench the elevated arterial P_{CO_2} is slowly compensated by HCO_3^- accumulation, which in the fully acclimated state allows for an increase in blood O_2 affinity (compared with normoxic controls) via reduced red cell GTP concentrations (Jensen & Weber, 1982).

Reduced erythrocytic NTP content is an important adaptation in fish to increase O_2 affinity in all conditions with a low blood P_{O_2}, whereas the response is absent when O_2 shortage results from increased O_2 demand (exercise) or methaemoglobinaemia (Weber, 1982; Weber & Jensen, 1988; Jensen *et al.*, 1987). Whereas an 'oxyphilic' fish such as trout uses ATP to modulate O_2 affinity (Tetens & Lykkeboe, 1981), many fish that often encounter hypoxia during their life selectively reduce their GTP content (Johansen *et al.*, 1976; Weber & Lykkeboe, 1978; Jensen & Weber, 1982). This is an elegant strategy in view of the larger allosteric effect of GTP than of ATP. In addition, the reduction in NTP elevates red cell pH by increasing the Donnan distribution ratio of H^+ ($[H^+]_e/[H^+]_i$) (Wood & Lenfant, 1979; Jensen, 1988), which acts synergistically to the reduced direct allosteric effect of NTP, and raises O_2 affinity. Full manifestation of the integrated O_2 affinity response depends on the time course of NTP reduction (hours to days, depending on species, degree of hypoxia and acuteness of exposure) and on the recovery kinetics of a simultaneous extracellular acidosis where present. When manifested, the affinity increase may allow a reduction in the ventilatory requirement (Lomholt & Johansen, 1979). In contrast to the response of nucleated fish and amphibian red cells, anucleated mammalian red cells elevate their phosphate (DPG) content in hypoxia, which eventually decreases O_2 affinity (see, for example, Wood & Lenfant, 1979). This affinity change is beneficial only in moderate hypoxia where actual arteriovenous $C_{O_2}-P_{O_2}$ values reflect an increased β_{O_2}. In severe hypoxia (as typically encountered by fish) mammals, like fish, are better suited with a left-shifted O_2 equilibrium curve (see, for example, Turek *et al.*, 1973). In view of this, natural selection may have favoured the reduced DPG sensitivity seen in some mammals that live permanently in hypoxic conditions (Bauer *et al.*, 1980; Jelkmann *et al.*, 1981).

Fish red cells swell to a remarkable extent in hypoxia and/or hypercapnia. The resulting dilution of Hb and NTP at constant NTP/Hb ratios reduces NTP–Hb complexing and increases O_2 affinity, as shown for carp Hb by Lykkeboe & Weber (1978). Appreciable red cell swelling results passively from the increase in Donnan distribution ratios of Cl^- ($[Cl^-]_i/[Cl^-]_e$) and HCO_3^- ($[HCO_3^-]_i/[HCO_3^-]_e$) associated with reductions

tions in pH, HbO_2 saturation and NTP (Jensen, 1988; Salama & Nikinmaa, 1988). Additionally, red cell volume and pH can be under β-adrenergic control (Nikinmaa, 1982).

In fish erythrocytes, catecholamines liberated during stress bind to the membrane β-receptors and stimulate a cAMP-dependent Na^+–H^+ antiporter, whereby Na^+ enters the red cell (down its chemical gradient) in exchange for H^+. This alkalinizes the red cell, which subsequently stimulates HCO_3^- efflux and Cl^- influx (Nikinmaa, 1986; Cossins & Richardson, 1985; Borgese *et al.*, 1986). The net result is an elevation of red cell $[Na^+]$ and $[Cl^-]$ that causes swelling, and an elevation of red cell pH above the normal steady-state value. Thus catecholamines are indirect modulators of O_2 affinity (Nikinmaa, 1983). In trout, this mechanism is important to protect pH_i and thus arterial O_2 content in acidotic exercise-stress (Primmett *et al.*, 1986), hypoxia (Tetens & Christensen, 1987; Boutilier *et al.*, 1988), hypercapnia (Perry, 1986), and during acid injections (Boutilier *et al.*, 1986). Tench, on the other hand, lack the β-adrenergic response in normoxic exercise-stress, and its red cell pH decreases along with the extracellular pH (Jensen, 1987). The alternative strategy in this species is a pronounced increase in arterial P_{O_2} that protects arterial loading and allows an increased arteriovenous difference in O_2 content via the Bohr–Root shift (Jensen *et al.*, 1983). This mechanism is not available to exercised trout, as its high resting arterial P_{O_2} offers no reserve for an increase. Carp red cells, like tench red cells, do not exhibit β-adrenergic swelling under normoxic conditions, but do so in hypoxia (Salama & Nikinmaa, 1988). Interestingly, the β-adrenergic responses of deoxygenated trout cells also exceed those of oxygenated cells, suggesting a saturation-dependent interaction of haemoglobin with membrane proteins and Na^+–H^+ transport (Motais *et al.*, 1987).

In contrast to the case of exercise-stress, where a large anaerobic component contributes to a severe extracellular acidosis, arterial pH can remain unchanged in sustained aerobic exercise, and catecholamines are not released (Butler *et al.*, 1986). In this situation trout optimize arteriovenous O_2 transport via an elevated $\dot{Q}_h$ and decreased venous P_{O_2} (Kiceniuk & Jones, 1977). Increased capillarization of tissues then represents a mechanism to compensate reduced off-loading pressures, whereas redistribution of blood flow can direct O_2 to where it is most needed.

Blood CO_2 transport

The basic principles of CO_2 transport in fish are similar to those in other vertebrates (Randall & Daxboeck, 1984; Perry, 1986). The bulk of

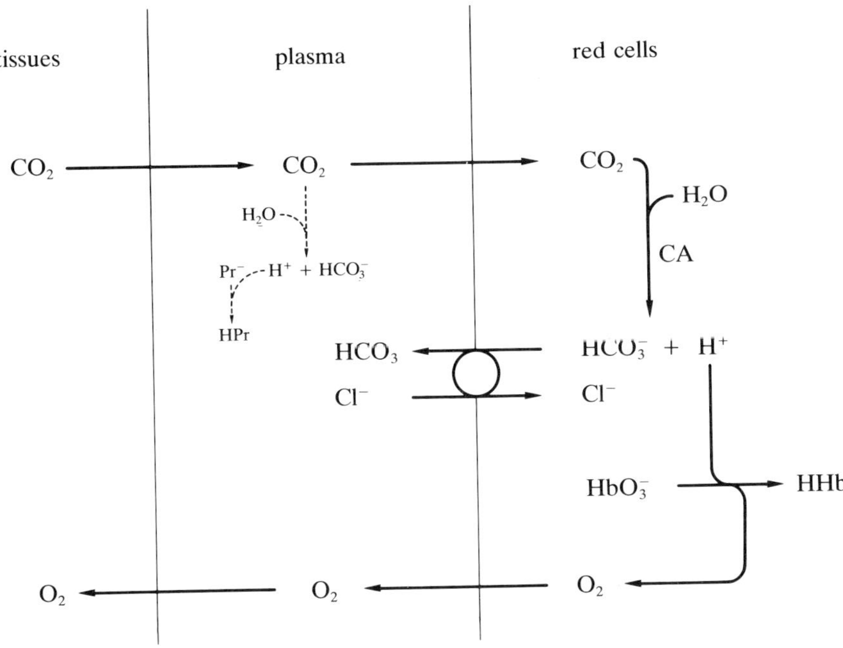

Figure 2. Overview of the main processes and chemical reactions when CO_2 is added to the blood in tissues.

CO_2 produced in tissues diffuses into the blood plasma and further into the red cells. Only a minor amount of CO_2 is hydrated at the uncatalysed rate in plasma, where in addition the buffer value is low. In the red cell, however, carbonic anhydrase (CA) rapidly catalyses the hydration of CO_2 to H^+ and HCO_3^-. This equilibrium reaction is displaced far in the direction of HCO_3^- formation by extensive binding of produced H^+ to the Hb. This binding of H^+ is linked to off-loading of O_2 via the Bohr–Haldane effect. Red cell HCO_3^- is subsequently shifted into the plasma in exchange for Cl^- via the membrane-bound 'Band 3' anion exchanger in order to re-establish Donnan distribution ratios for the diffusive ions. These ratios increase as a result of a reduced non-permeable negative charge on Hb (Figure 2). In the gills the reactions proceed in the opposite direction and CO_2 is eliminated mainly as molecular CO_2 by diffusion, with a small amount of CO_2 being hydrated by gill CA to produce counterions for the branchial ion exchangers. The blood total CO_2 content (mainly HCO_3^-) decreases by about 20% during blood passage through the gills (Randall & Daxboeck, 1984). The extent of diffusion limitation for gas exchange varies, but it may be significant, and actually higher for CO_2 than for O_2, as reflected by a low ratio between CO_2 diffusion and perfusion conductances in dogfish (Piiper & Scheid, 1982). Oxylabile carbamate, which accounts for about 13% of total CO_2 exchange in humans (Klocke, 1988), probably has little significance in fish. A small overall carbamate content in fish blood results from low P_{CO_2} values, acetylation of α-amino groups, and the fact that ε-amino groups of lysine have pK values of about 10 and thus will be protonated and not available for carbamino formation.

Haemoglobin–hydrogen ion equilibria

The molecular exchanges of protons between the haemoglobin and the solvent are determined by several factors: (i) the basic H^+-binding properties of Hb as given by amino acid composition, and the pK values of dissociable amino acid residues and their accessibility to the solvent; (ii) the conformation shifts associated with altered oxygenation degree, which changes the molecular surroundings of certain amino acid residues and thereby their pK (i.e. the Haldane effect); and (iii) allosteric regulation of H^+ binding, notably via the NTPs.

The Hb–H^+ equilibria in teleosts differ significantly from those in other vertebrate groups (Jensen, 1989). At a fixed oxygenation degree the buffer values of teleost Hbs (Figure 3) (Breepoel *et al.*, 1980; Jensen & Weber, 1985) are much lower than those of elasmobranch or mammalian Hbs (Figure 3). This originates in a strongly reduced number of titratable groups in teleost Hbs compared with other vertebrate Hbs. At physiological pH the main buffer groups are the imidazole group of histidine residues and the terminal α-amino groups (i.e. groups with pK values close to physiological pH). Teleost Hbs have only 14–24 His residues per tetramer, whereas in other phylogenetic groups of fish (elasmobranchs and dipnoans), and in amphibians, reptiles, birds and mammals, the His content is 32–48 per tetramer (see Jensen (1989) for details). Additionally, in teleost Hbs the N-terminal amino acids of the α-chains are acetylated, which contributes to lower buffer values as acetylation excludes proton exchange at physiological pH. Not all histidine residues within a Hb molecule are titratable: some will be buried in the interior of the molecule. In carp Hb it appears that 7 histidine residues, out of a total of 18 histidines per tetramer, are titratable (Jensen, 1989) compared with 20 titratable

---

Figure 3. H^+ titration curves of stripped oxygenated and deoxygenated Hbs from two teleosts (carp and rainbow trout), one elasmobranch (spiny dogfish) and one mammal (pig), at 15 °C and in 0.1 M KCl. Buffer values (mol H^+ per mol tetramer per pH unit) are given by the slope of individual curves; the fixed-acid Haldane effect (mol H^+ taken up per mol tetramer upon deoxygenation) by the vertical distance between oxy and deoxy curves (from Jensen, 1989).

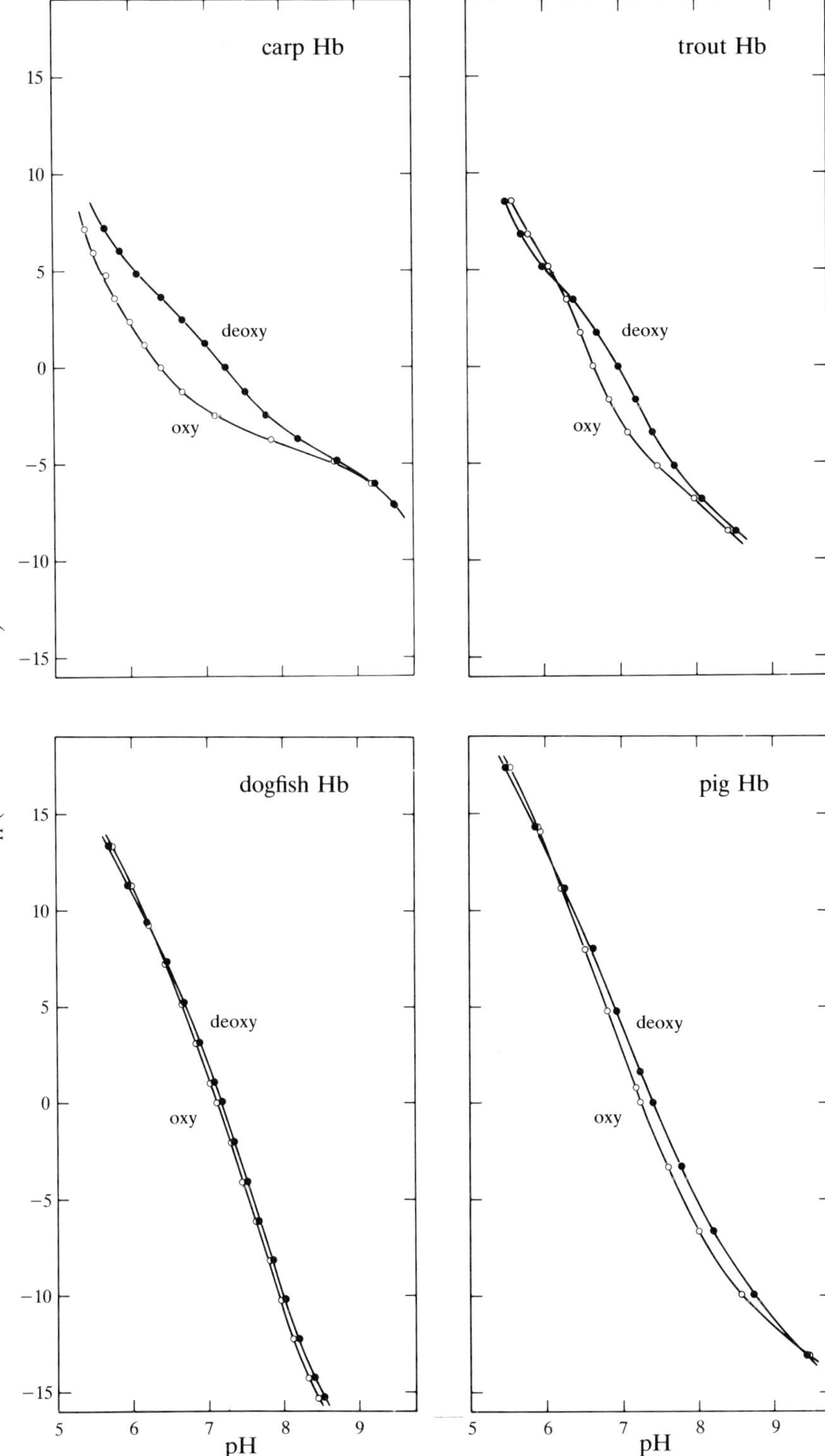

histidines out of a total of 38 histidines in human Hb (Janssen *et al.*, 1970).

Low intrinsic buffer values in teleost Hbs are maladaptive with regard to blood H^+ buffering and CO_2 transport, but can be compensated for by the possession of large Haldane effects. Indeed, there is an inverse relationship between the magnitudes of buffer values and fixed acid Haldane effects, with the largest Haldane effects in teleost Hbs (Figure 3) (Jensen, 1989). Two contrasting piscine strategies to safeguard the H^+ buffering essential to CO_2 transport accordingly emerge: (i) very high intrinsic Hb buffer values and an insignificant Haldane effect in the elasmobranch dogfish; and (ii) low buffer values but high Haldane effects in teleosts, notably carp.

The Haldane effect results from an increase in pK of specific amino acid residues upon deoxygenation. The magnitude of this additional H^+ uptake at an unchanged pH, and the pH value of its maximum, increases with addition of organic phosphates (see Jensen & Weber, 1985), as does the Bohr effect. This favours a high H^+ and thus CO_2 uptake in the tissue capillaries under *in vivo* conditions. Furthermore, the low-oxygenation-linked carbamino formation helps to keep the Haldane effect high, as the uptake of Bohr–Haldane protons upon deoxygenation is not countered by H^+ release from carbamino formation.

Whole blood CO_2 transporting properties

The blood convective CO_2 transport is described by a Fick equation equivalent to that for O_2 transport:

$$\dot{M}_{CO_2} = \dot{Q}_h\, \beta_{CO_2}\, (Pv_{CO_2} - Pa_{CO_2}), \tag{2}$$

the modulating role of Hb being to optimize the capacitance coefficient β_{CO_2}.

The coefficient β_{CO_2} increases with increasing non-bicarbonate buffer values. An increase in blood [Hb] accordingly represents a useful strategy for limiting acid–base disturbances and safeguarding CO_2 transport during stress. This strategy is, however, not ubiquitously exploited by fish, as is evident from the variance in the blood [Hb] response during hypoxia and exercise (see above). The low [Hb] and intrinsic buffer values in teleost blood tend to reduce β_{CO_2} compared with values in mammals. However, since the slope of CO_2 equilibrium curves (i.e. total CO_2 as function of P_{CO_2}) is larger at low than at high P_{CO_2}, this effect is counteracted by the low P_{CO_2} in fish blood. Also, the Haldane effect (which increases β_{CO_2}) is of greater importance for veno-arterial CO_2 transport in teleosts than in elasmobranchs and mammals: the physiological β_{CO_2} in teleosts may actually exceed that in mammals. The high P_{CO_2} in hypercapnic fish tends to reduce β_{CO_2}. This may be compensated for by the larger total CO_2 difference between deoxygenated and oxygenated blood at high P_{CO_2} (see, for example, Bridges *et al.*, 1984), and by the pH regulatory elevation of $[HCO_3^-]$ that tends to elevate β_{CO_2} (Piiper & Scheid, 1982). A rise in temperature, however, reduces both β_{CO_2} at constant oxygenation and the Bohr–Haldane effect (Bridges *et al.*, 1984). This predicts elevated Pv_{CO_2} and ΔP_{CO_2} values, and an increased $\dot{Q}_h$, to mediate the larger CO_2 flux needed at high temperatures.

A large Haldane effect allows a high H^+ uptake without a pH_i decrease, when CO_2 is added in the tissue capillaries. According to the mass law equation for CO_2 hydration, this gives greater HCO_3^- formation than when the Haldane effect is lower but the buffer value higher (since here the same CO_2 addition and H^+ uptake would only be achieved with a pH_i fall). In situations where HCO_3^-/Cl^- exchange is rate-limiting, this could compensate for a slow HCO_3^- movement to plasma (which, when rapid, also increases red cell CO_2 hydration and HCO_3^- formation). HCO_3^-/Cl^- exchange may indeed show differences between species (preliminary experiments suggest that it is slower in tench, which has a large Haldane effect, than in trout) and become rate-limiting when cardiac output goes up and transit time in capillaries goes down, as in exercise. The chemical reactions would then continue after the blood has passed through the capillaries, which may contribute to the elevated arterial P_{CO_2} seen in exercised fish (Wood & Perry, 1985).

The characteristic low buffer values and high Haldane effects in teleosts causes an increase in red cell pH on deoxygenation; this increase can be tenfold larger than that known from mammals (Jensen, 1988). In tench, this pH_i change even predominates between 40 and 100% HbO_2 saturation, suggesting an almost full exploitation of the Haldane effect within normal physiological ranges of blood O_2 saturations (Jensen, 1986). This correlates with molecular data on tench Hb, showing that most Bohr–Haldane protons are released late in the oxygenation event (Weber *et al.*, 1987). Similar non-linear relations between proton release and oxygen saturation have been reported for other stripped fish Hbs (Saffran & Gibson, 1981; Ikeda-Saito *et al.*, 1983).

Linkage effects in blood O_2 and CO_2–H^+ transport

The multiple and interconnected equilibria between Hb and its ligands in general nicely match the physiological needs, but they also give rise to some compromises and secondary effects, briefly discussed in this section.

Just as the Haldane effect benefits CO_2 transport, the Bohr effect may augment O_2 unloading to tissues and O_2 loading in the gills, via the arterio-venous pH shifts resulting from CO_2 addition and removal. From the classical linkage equation between the Bohr coefficient and the Haldane coefficient of Hb (Wyman, 1964)

$$(\partial \log P_{O_2}/\partial \mathrm{pH})_y = (\partial \bar{H}^+/\partial Y)_{\mathrm{pH}} \tag{3}$$

it is evident that it is not possible to exploit both an optimal Bohr shift and an optimal oxygenation linked H^+ binding. A large Haldane effect tends to reduce the arteriovenous ΔpH, and thus the actual Bohr shift ($\Delta \log P_{50} = \phi \, \Delta$pH, where ϕ is the Bohr–Haldane coefficient). This raises the question whether actual magnitudes of Bohr–Haldane coefficients reflect primary concern for O_2 or for CO_2 transport in the arteriovenous cycle. In mammalian blood lacking a specific CO_2 effect, the right shift of O_2 equilibrium curves in the tissues is maximal if the numerical value of ϕ_e (Bohr factor referred to extracellular pH) is about one half of the respiratory quotient (RQ, 0.7–1.0), whereas pH compensation and CO_2 transport are optimized if $-\phi_e = $ RQ (Lapennas, 1983). The analogy with fish blood thus suggests that the high Bohr coefficients ($\Delta \log P_{50}/\Delta \mathrm{pH}_e$) in tench ($-0.72$) (Jensen & Weber, 1982) and in carp (-0.98) (Weber & Lykkeboe, 1978) mainly serve steady-state pH and CO_2 homeostasis. In trout, where $\phi_e = -0.52$ (Weber *et al.*, 1976), more emphasis appears to be placed on arteriovenous O_2 affinity shifts than in the cyprinids. It must be remembered, however, that the Bohr effect remains significant for all fish during acid–base disturbances, such as hyperventilatory alkaloses and lactacidoses.

Due to differences in red cell and plasma acid–base status, the magnitude of key blood respiratory parameters depends strongly on the compartment to which they are referred. Thus, the $\Delta \mathrm{pH}_i/\Delta \mathrm{pH}_e$ value of about 0.6 in blood from normoxic tench (Jensen & Weber, 1982) reflects that Bohr factors related to extracellular pH are lower than those related to red cell pH (since $\phi_e/\phi_i = \Delta \mathrm{pH}_i/\Delta \mathrm{pH}_e$). When red cell NTP content decreases, $\Delta \mathrm{pH}_i/\Delta \mathrm{pH}_e$ increases (Jensen, 1988), ϕ_i decreases (see above) and ϕ_e comes closer to ϕ_i. The actual magnitude of NTP reductions in hypoxic tench (about 18%) (see, for example, Jensen & Weber, 1982) accordingly may reflect a balance, which allows both an NTP–pH_i-mediated O_2 affinity increase and a ϕ_i that remains high to secure CO_2 homeostasis.

An interesting consequence of the large change in pH_i with HbO_2 saturation in fish is that whole blood O_2 equilibria related to constant pH_e (or constant P_{CO_2}) will proceed with a large change in pH_i (low pH_i at high

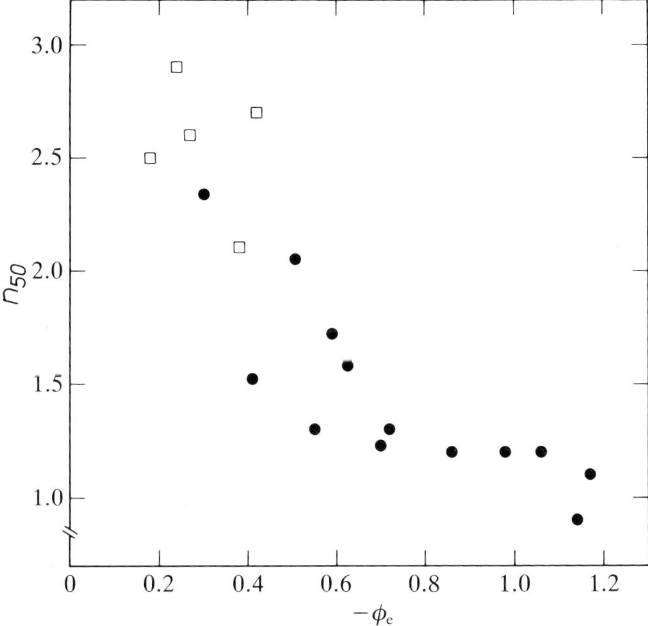

Figure 4. Whole blood n_{50} values as a function of the corresponding whole blood Bohr coefficient in different teleosts (filled circles) (at 10–15 °C), with inclusion of a few amphibian species (open squares). Sources of data: Albers *et al.* (1981); Boutilier & Toews (1981); Cech *et al.* (1979, 1984); Dobson & Baldwin, (1982); Hayden *et al.* (1975) (recalculated); Innes & Wells (1985); Jensen & Weber (1982); Johansen *et al.* (1978, 1980); Jokumsen & Weber (1980); Jones *et al.* (1986); Laursen *et al.* (1985); Tazawa *et al.* (1979); Tetens & Lykkeboe (1981); Weber & Lykkeboe (1978).

HbO_2 saturation and vice versa). This will produce curves with a lower apparent Hill n-value than curves related to constant pH_i. This probably explains why whole blood n_{50} values are about 1.3 in tench at physiological pH_e, whereas in Hb solutions at corresponding cofactor conditions and pH_i (held almost constant during oxygenation via buffer addition) n_{50} values are about 2 (Jensen & Weber, 1982). In human blood, where ΔpH_i (deoxy–oxy) is small, blood and Hb solutions on the other hand will have similar n-values under comparable conditions (Imai, 1982). As the large ΔpH_i (deoxy–oxy) in fish depends on high Haldane effects, and since Haldane and Bohr effects are linked, one might consequently predict an inverse relation between whole blood Bohr factors (determined as $\Delta \log P_{50}/\Delta pH_{e,50}$) and n_{50}-values in blood from different species. Indeed, from the data shown in Figure 4 there seems to be a clear tendency for low whole blood n_{50} values when ϕ_e is high and for high n_{50} values when ϕ_e is low, in spite of cooperativity differences not related to the Bohr effect and of different experimental techniques used.

Concluding remarks

Vertebrates exploit multiple strategies to optimize their blood O_2 and CO_2 transport. The functional properties of Hbs vary immensely between species with respect to affinity and cooperativity of O_2 binding, extent of carbamate formation, magnitude of Bohr–Haldane effects, magnitude of buffer values, and sensitivity to organic phosphates. Such differences, resulting from key amino acid substitutions, reflect an ultra-slow 'evolutionary' adaptation to milieux factors, and may contribute to acute limits of tolerance to environmental perturbations of homeostasis. Intraspecific adaptations via rapid organismic and cellular responses provide compensation of acute homeostatic disturbances and establish a fine-tuning of gas transporting properties to altered physiological needs. These responses have variable kinetics, and different 'routes of adaptation' can be used, depending on species and environment. Thus exploitation of qualitative versus quantitative means to adjust blood O_2 and CO_2 capacitance coefficients may vary, as may cardiovascular versus capacitance preferences for optimizing gas transport. The essential feature of many responses is, however, the precise control of the chemical microenvironment of Hb, made possible by the encapsulation of Hb in red cells.

References

Albers, C., Götz, K.H. & Welbers, P. (1981). Oxygen transport and acid-base balance in the blood of the sheatfish, *Silurus glanis. Respiration Physiology* **46**, 223–36.

Bauer, C., Rollema, H.S., Till, H.W. & Braunitzer, G. (1980). Phosphate binding by llama and camel hemoglobin. *Journal of Comparative Physiology* **136**, 67–70.

Binotti, I., Giovenco, S., Giardina, B., Antonini, E., Brunori, M. & Wyman, J. (1971). Studies on the functional properties of fish hemoglobins. II. The oxygen equilibrium of the isolated hemoglobin components from trout blood. *Archives of Biochemistry and Biophysics* **142**, 274–80.

Boggs, D.F., Kilgore, D.L. & Birchard, G.F. (1984). Respiratory physiology of burrowing mammals and birds. *Comparative Biochemistry and Physiology* **77A**, 1–7.

Borgese, F., Garcia-Romeu, F. & Motais, R. (1986). Catecholamine-induced transport systems in trout erythrocyte. Na^+/H^+ counter-transport or NaCl cotransport? *Journal of General Physiology* **87**, 551 66.

Boutilier, R.G., Dobson, G., Hoeger, U. & Randall, D.J. (1988). Acute exposure to graded levels of hypoxia in rainbow trout (*Salmo gair-*

dneri): Metabolic and respiratory adaptations. *Respiration Physiology* **71**, 69–82.

Boutilier, R.G., Iwama, G.K. & Randall, D.J. (1986). The promotion of catecholamine release in rainbow trout, *Salmo gairdneri*, by acute acidosis: Interactions between red cell pH and haemoglobin oxygen-carrying capacity. *Journal of Experimental Biology* **123**, 145–57.

Boutilier, R.G. & Toews, D.P. (1981). Respiratory properties of blood in a strictly aquatic and predominantly skin-breathing urodele, *Cryptobranchus alleganiensis*. *Respiration Physiology* **46**, 161–76.

Breepoel, P.M., Kreuzer, F. & Hazevoet, M. (1980). Studies of the hemoglobins of the eel (*Anguilla anguilla* L.). I. Proton binding of stripped hemolysate; separation and properties of two major components. *Comparative Biochemistry and Physiology* **65A**, 69–75.

Bridges, C.R., Pelster, B. & Scheid, P. (1985). Oxygen binding in blood of *Xenopus laevis* (amphibia) and evidence against Root effect. *Respiration Physiology* **61**, 125–36.

Bridges, C.R., Taylor, A.C., Morris, S.J. & Grieshaber, M.K. (1984). Ecophysiological adaptations in *Blennius pholis* (L.) blood to intertidal rockpool environments. *Journal of Experimental Marine Biology and Ecology* **77**, 151–67.

Brittain, T. (1987). The Root effect. *Comparative Biochemistry and Physiology* **86B**, 473–81.

Burggren, W.W., Dupré, R.K. & Wood, S.C. (1987). Allometry of red cell oxygen binding and hematology in larvae of the salamander, *Ambystoma tigrinum*. *Respiration Physiology* **70**, 73–84.

Butler, P.J., Metcalfe, J.D. & Ginley, S.A. (1986). Plasma catecholamines in the lesser spotted dogfish and rainbow trout at rest and during different levels of exercise. *Journal of Experimental Biology* **123**, 409–21.

Cech, J.J. Jr, Laurs, R.M. & Graham, J.B. (1984). Temperature-induced changes in blood gas equilibria in the albacore, *Thunnus alalunga*, a warm-bodied tuna. *Journal of Experimental Biology* **109**, 21–34.

Cech, J.J. Jr, Mitchell, S.J. & Massingill, M.J. (1979). Respiratory adaptations of sacramento blackfish, *Orthodon microlepidotus* (Ayres), for hypoxia. *Comparative Biochemistry and Physiology* **63A**, 411–15.

Cossins, A.R. & Richardson, P.A. (1985). Adrenalin-induced Na^+/H^+ exchange in trout erythrocytes and its effects upon oxygen-carrying capacity. *Journal of Experimental Biology* **118**, 229–46.

Dejours, P. (1973). Problems of control of breathing in fishes. In *Comparative Physiology* (ed. L. Bolis, K. Schmidt-Nielsen & S.H.P. Maddrel), pp. 117–33. North-Holland/American Elsevier, Amsterdam and New York.

Dobson, G.P. & Baldwin, J. (1982). Regulation of blood oxygen affinity in the Australian blackfish *Gadopsis marmoratus*. I. Correlations between oxygen-binding properties, habitat and swimming behaviour. *Journal of Experimental Biology* **99**, 223–43.

Farmer, M. (1979). The transition from water to air breathing. Effects of

CO_2 on hemoglobin function. *Comparative Biochemistry and Physiology* **62A**, 109–14.

Gillen, R.G. & Riggs, A. (1973). Structure and function of the isolated hemoglobins of the American eel, *Anguilla rostrata*. *Journal of Biological Chemistry* **248**, 1961–9.

Hayden, J.B., Cech, J.J. Jr & Bridges, D.W. (1975). Blood oxygen dissociation characteristics of the winter flounder, *Pseudopleuronectes americanus*. *Journal of the Fisheries Research Board of Canada* **32**, 1539–44.

Heming, T.A., Randall, D.J., Boutilier, R.G., Iwama, G.K. & Primmett, D. (1986). Ionic equilibria in red blood cells of rainbow trout (*Salmo gairdneri*): Cl^-, HCO_3^- and H^+. *Respiration Physiology* **65**, 223–34.

Hladky, S.B. & Rink, T.J. (1977). pH equilibrium across the red cell membrane. In *Membrane Transport in Red Cells* (ed. J.C. Ellory & V.L. Lew), pp. 115–35. Academic Press, London.

Ikeda-Saito, M., Yonetani, T. & Gibson, Q.H. (1983). Oxygen equilibrium studies on hemoglobin from the bluefin tuna (*Thunnus thynnus*). *Journal of Molecular Biology* **168**, 673–86.

Imai, K. (1982). *Allosteric Effects in Haemoglobin*. Cambridge University Press.

Innes, A.J. & Wells, R.M.G. (1985). Respiration and oxygen transport functions of the blood from an intertidal fish, *Helcogramma medium* (Tripterygiidae). *Environmental Biology of Fishes* **14**, 213–26.

Janssen, L.H.M., De Bruin, S.H. & van Os, G.A.J. (1970). H^+ titration studies of human hemoglobin. *Biochimica et Biophysica Acta* **221**, 214–27.

Jelkmann, W., Oberthur, W., Kleinschmidt, T. & Braunitzer, G. (1981). Adaptation of hemoglobin function to subterranean life in the mole, *Talpa europaea*. *Respiration Physiology* **46**, 7–16.

Jensen, F.B. (1986). Pronounced influence of $Hb–O_2$ saturation on red cell pH in tench blood in vivo and in vitro. *Journal of Experimental Zoology* **238**, 119–24.

Jensen, F.B. (1987). Influences of exercise-stress and adrenaline upon intra- and extracellular acid-base status, electrolyte composition and respiratory properties of blood in tench (*Tinca tinca*) at different seasons. *Journal of Comparative Physiology* **B157**, 51–60.

Jensen, F.B. (1988). Red cell pH in tench. Interacting effects of cellular nucleoside triphosphates, Hb-oxygenation and extracellular pH. *Acta Physiologica Scandinavica* **132**, 431–7.

Jensen, F.B. (1989). Hydrogen–ion equilibria in fish haemoglobins. *Journal of Experimental Biology* **143**, 225–34.

Jensen, F.B., Andersen, N.A. & Heisler, N. (1987). Effects of nitrite exposure on blood respiratory properties, acid-base and electrolyte regulation in the carp (*Cyprinus carpio*). *Journal of Comparative Physiology* **B157**, 533–41.

Jensen, F.B., Nikinmaa, M. & Weber, R.E. (1983). Effects of exercise stress on acid-base balance and respiratory function in blood of the teleost *Tinca tinca*. *Respiration Physiology* **51**, 291–301.

Jensen, F.B. & Weber, R.E. (1982). Respiratory properties of tench blood and hemoglobin. Adaptation to hypoxic-hypercapnic water. *Molecular Physiology* **2**, 235–50.

Jensen, F.B. & Weber, R.E. (1985). Proton and oxygen equilibria, their anion sensitivities and interrelationships in tench hemoglobin. *Molecular Physiology* **7**, 41–50.

Jensen, F.B. & Weber, R.E. (1987). Thermodynamic analysis of precisely measured oxygen equilibria of tench (*Tinca tinca*) hemoglobin and their dependence on ATP and protons. *Journal of Comparative Physiology* **B157**, 137–43.

Johansen, K., Lykkeboe, G., Korncrup, S. & Maloiy, G.M.O. (1980). Temperature insensitivity O_2 binding in blood of the tree frog *Chiromantis petersi*. *Journal of Comparative Physiology* **136**, 71–6.

Johansen, K., Lykkeboe, G., Weber, R.E. & Maloiy, G.M.O. (1976). Respiratory properties of blood in awake and estivating lungfish, *Protopterus amphibius*. *Respiration Physiology* **27**, 335–45.

Johansen, K., Mangum, C.P. & Weber, R.E. (1978). Reduced blood O_2 affinity associated with air breathing in osteoglossid fishes. *Canadian Journal of Zoology* **56**, 891–7.

Jokumsen, A. & Weber, R.E. (1980). Haemoglobin-oxygen binding properties in the blood of *Xenopus laevis*, with special reference to the influences of aestivation and of temperature and salinity acclimation. *Journal of Experimental Biology* **86**, 19–37.

Jones, D.R., Brill, R.W. & Mense, D.C. (1986). The influence of blood gas properties on gas tension and pH of ventral and dorsal aortic blood in free-swimming tuna, *Euthynnus affinis*. *Journal of Experimental Biology* **120**, 201–13.

Kiceniuk, J.W. & Jones, D.R. (1977). The oxygen transport system in trout (*Salmo gairdneri*) during sustained exercise. *Journal of Experimental Biology* **69**, 247–60.

Kleinschmidt, T. & Sgouros, J.G. (1987). Hemoglobin sequences. *Biological Chemistry Hoppe-Seyler* **368**, 579–615.

Klocke, R.A. (1988). Velocity of CO_2 exchange in blood. *Annual Review of Physiology* **50**, 625–37.

Krogh, A. & Leitch, I. (1919). The respiratory function of the blood in fishes. *Journal of Physiology* **52**, 288–300.

Lapennas, G.N. (1983). The magnitude of the Bohr coefficient: Optimal for oxygen delivery. *Respiration Physiology* **54**, 161–72.

Laursen, J.S., Andersen, N.A. & Lykkeboe, G. (1985). Temperature acclimation and oxygen binding properties of blood of the European eel, *Anguilla anguilla*. *Comparative Biochemistry and Physiology* **81A**, 79–86.

Lenfant, C. (1973). High altitude adaptation in mammals. *American Zoologist* **13**, 447–56.

Lomholt, J.P. & Johansen, K. (1979). Hypoxia acclimation in carp. How it affects O_2 uptake, ventilation, and O_2 extraction from water. *Physiological Zoology* **52**, 38–49.

Lutz, P.L. (1980). On the oxygen affinity of bird blood. *American Zoologist* **20**, 187–98.

Lutz, P.L. & Lapennas, G.N. (1982). Effects of pH, CO_2 and organic phosphates on oxygen affinity of sea turtle hemoglobins. *Respiration Physiology* **48**, 75–87.

Lykkeboe, G. & Johansen, K. (1978). An O_2–Hb 'paradox' in frog blood? (n-values exceeding 4). *Respiration Physiology* **35**, 119–27.

Lykkeboe, G. & Weber, R.E. (1978). Changes in the respiratory properties of the blood in the carp, *Cyprinus carpio*, induced by diurnal variation in ambient oxygen tension. *Journal of Comparative Physiology* **128**, 117–25.

Malte, H. & Weber, R.E. (1987). The effect of shape and position of the oxygen equilibrium curve on extraction and ventilation requirement in fishes. *Respiration Physiology* **70**, 221–8.

Milligan, C.L. & Wood, C.M. (1986). Intracellular and extracellular acid-base status and H^+ exchange with the environment after exhaustive exercise in the rainbow trout. *Journal of Experimental Biology* **123**, 93–121.

Motais, R., Garcia-Romeu, F. & Borgese, F. (1987). The control of Na^+/H^+ exchange by molecular oxygen in trout erythrocytes. A possible role of hemoglobin as a transducer. *Journal of General Physiology* **90**, 197–207.

Nikinmaa, M. (1982). Effects of adrenaline on red cell volume and concentration gradient of protons across the red cell membrane in the rainbow trout, *Salmo gairdneri*. *Molecular Physiology* **2**, 287–97.

Nikinmaa, M. (1983). Adrenergic regulation of haemoglobin oxygen affinity in rainbow trout red cells. *Journal of Comparative Physiology* **152**, 67–72.

Nikinmaa, M. (1986). Control of red cell pH in teleost fishes. *Annali Zoologici Fennici* **23**, 223–35.

Perry, S.F. (1986). Carbon dioxide excretion in fishes. *Canadian Journal of Zoology* **64**, 565–72.

Perutz, M.F. (1983). Species adaptation in a protein molecule. *Molecular Biology and Evolution* **1**, 1–28.

Perutz, M.F. & Brunori, M. (1982). Stereochemistry of cooperative effects in fish and amphibian haemoglobins. *Nature* **299**, 421–6.

Piiper, J. & Scheid, P. (1982). Models for a comparative functional analysis of gas exchange organs in vertebrates. *Journal of Applied Physiology* **53**, 1321–9.

Powers, D.A. (1972). Hemoglobin adaptation for fast and slow water habitats in sympatric catostomid fishes. *Science* **177**, 360–2.

Primmett, D.R.N., Randall, D.J., Mazeaud, M. & Boutilier, R.G. (1986). The role of catecholamines in erythrocyte pH regulation and oxygen transport in rainbow trout (*Salmo gairdneri*) during exercise. *Journal of Experimental Biology* **122**, 139–48.

Randall, D. & Daxboeck, C. (1984). Oxygen and carbon dioxide transfer across fish gills. In *Fish Physiology*, vol. XA (ed. W.S. Hoar & D.J. Randall), pp. 263–314. Academic Press, New York.

Riggs, A. (1976). Factors in the evolution of hemoglobin function. *Federation Proceedings* **35**, 2115–18.

Riggs, A. (1988). The Bohr effect. *Annual Review of Physiology* **50**, 181–204.

Rodewald, K., Stangl, A. & Braunitzer, G. (1984). Primary structure, biochemical and physiological aspects of hemoglobin from South American lungfish (*Lepidosiren paradoxus*, Dipnoi). *Hoppe-Seyler's Zeitschrift für Physiologische Chemie* **365**, 639–49.

Saffran, W.A. & Gibson, Q.H. (1981). Asynchronous ligand binding and proton release in a Root effect hemoglobin. *Journal of Biological Chemistry* **256**, 4551–6.

Salama, A. & Nikinmaa, M. (1988). The adrenergic responses of carp (*Cyprinus carpio*) red cells: Effects of pO_2 and pH. *Journal of Experimental Biology* **136**, 405–16.

Schmidt-Nielsen, K. (1984). *Scaling. Why is animal size so important?* Cambridge University Press.

Tazawa, H., Mochizuki, M. & Piiper, J. (1979). Blood oxygen dissociation curve of the frogs *Rana catesbeiana* and *Rana brevipoda*. *Journal of Comparative Physiology* **129**, 111–14.

Tetens, V. & Christensen, N.J. (1987). Beta-adrenergic control of blood oxygen affinity in acutely hypoxia exposed rainbow trout. *Journal of Comparative Physiology* **B157**, 667–75.

Tetens, V. & Lykkeboe, G. (1981). Blood respiratory properties of rainbow trout, *Salmo gairdneri*: Responses to hypoxia acclimation and anoxic incubation of blood in vitro. *Journal of Comparative Physiology* **145**, 117–25.

Thomas, S., Poupin, J., Lykkeboe, G. & Johansen, K. (1987). Effects of graded exercise on blood gas tensions and acid-base characteristics of rainbow trout. *Respiration Physiology* **68**, 85–97.

Turek, Z., Kreuzer, F. & Hoofd, L.J.C. (1973). Advantage or disadvantage of a decrease of blood oxygen affinity for tissue oxygen supply at hypoxia. A theoretical study comparing man and rat. *Pflügers Archiv* **342**, 185–97.

Weber, R.E. (1982). Intraspecific adaptation of hemoglobin function in fish to oxygen availability. In *Exogenous and Endogenous Influences on Metabolic and Neural Controls* (ed. A.D.F. Addink & N. Spronk), pp. 87–101. Pergamon Press, Oxford.

Weber, R.E. & Jensen, F.B. (1988). Functional adaptations in hemoglobins from ectothermic vertebrates. *Annual Review of Physiology* **50**, 161–79.

Weber, R.E., Jensen, F.B. & Cox, R.P. (1987). Analysis of teleost hemoglobin by Adair and Monod–Wyman–Changeux models. Effects of nucleoside triphosphates and pH on oxygenation of tench hemoglobin. *Journal of Comparative Physiology* **B157**, 145–52.

Weber, R.E. & Lykkeboe, G. (1978). Respiratory adaptations in carp blood. Influences of hypoxia, red cell organic phosphates, divalent cations and CO_2 on hemoglobin-oxygen affinity. *Journal of Comparative Physiology* **128**, 127–37.

Weber, R.E., Wood, S.C. & Lomholt, J.P. (1976). Temperature acclimation and oxygen binding properties of blood and multiple haemoglobins of rainbow trout. *Journal of Experimental Biology* **65**, 333–45.

Wood, C.M. & Perry, S.F. (1985). Respiratory, circulatory, and metabolic adjustments to exercise in fish. In *Circulation, Respiration, and Metabolism* (ed. R. Gilles), pp. 2–22. Springer-Verlag, Berlin.

Wood, S.C. & Lenfant, C. (1979). Oxygen transport and oxygen delivery. In *Evolution of Respiratory Processes. A Comparative Approach* (ed. S.C. Wood & C. Lenfant), pp. 193–223. Marcel Dekker, New York.

Wyman, J. (1964). Linked functions and reciprocal effects in hemoglobin: A second look. *Advances in Protein Chemistry* **19**, 223–86.

Yamamoto, K., Itazawa, Y. & Kobayashi, H. (1980). Supply of erythrocytes into the circulating blood from the spleen of exercised fish. *Comparative Biochemistry and Physiology* **65A**, 5–11.

Yamamoto, K. (1987). Contraction of spleen in exercised cyprinid. *Comparative Biochemistry and Physiology* **87A**, 1083–7.

E.W. TAYLOR, N.M. WHITELEY &
M.G. WHEATLY

Respiratory gas exchange and the regulation of acid–base status in decapodan crustaceans

Introduction

The decapodan crustaceans are a large and diverse group of primarily marine animals. However, some species live in the brackish waters of estuaries, the seashore or tropical mangrove swamps; others are freshwater and a few have ventured onto land to become air-breathers. Their most obvious characteristic is their rigid exoskeleton, manufactured from chitin, which is strengthened by calcium carbonate, except where it extends over the respiratory surfaces and the joints on the body and its appendages. The exoskeleton confers the advantages of physical resistance to mechanical damage and a relative impermeability to water and ions, which partly explains the ability of crustaceans to survive in coastal and even terrestrial habitats. The general impermeability restricts uptake of oxygen in water-breathers to the respiratory surfaces of the gills, which are enclosed in the branchial chambers on either side of the animal. The air-breathing species have evolved well perfused, elaborately folded or invaginated surfaces on the walls of these branchial chambers, which can function as lungs and are the primary site of oxygen uptake from air. Carbon dioxide excretion is chiefly over the gills, even in some air-breathers, and includes an element of electroneutral ion exchange, so that consideration of the mechanisms of exchange of CO_2 must include reference to regulation of acid–base and ion balance.

This account begins with a consideration of the role of the exoskeleton in limiting respiratory gas exchange and of the changes induced by moulting. This introduces the associated subjects of postmoult calcification, which involves deposition of CO_2 in $CaCO_3$, and subsequent remobilization of carapace carbonate to yield bicarbonate, which has an important role in compensating for systemic acidosis, such as that encountered during aerial exposure. It describes the uptake of oxygen by diffusion and the excretion of CO_2 by diffusion and electroneutral ion exchange, over the gills of water-breathers. This leads to an account of acid–base balance in the extracellular haemocoel and the intracellular compartments of various tissues. The effects of changes in oxygen, CO_2 and pH levels in water are described. Consideration of facultative air-breathing by primarily aquatic species leads to a discussion of the physiological adaptations involved in the commitment to air-breathing in land crabs. More detailed accounts of various aspects of crustacean respiratory physiology are available in a number of earlier reviews, notably those by Taylor (1982), Cameron & Mangum (1983), McMahon & Wilkens (1983), Mangum (1983a), Truchot (1983), Cameron (1986), McMahon & Burggren (1988) and Taylor & Innes (1988).

The exoskeleton and moulting

Recent interest from respiratory physiologists in the exoskeleton of crustaceans is in four separate though related areas: its potential as a barrier to diffusion of oxygen over the respiratory gas exchange surfaces; the control of moulting and influence of the moult cycle on respiratory gas exchange; the role of the exoskeleton in acid–base buffering and the acid–base properties of the carapace fluid, which appears to be strikingly different from other fluid compartments.

Krogh (1919) measured the diffusion rate of oxygen through insect cuticle and concluded that it was an order of magnitude lower than the rate through tissue such as frog muscle. Crustacean gills are covered in a thin (0.5–5 μm) cuticular layer which was thought to form a substantial barrier to the diffusion of oxygen (see, for example, Redmond, 1955; Jones, 1972; Taylor & Butler, 1978). However, direct measurement of the permeability to oxygen of portions of the uncalcified exoskeleton from freshly moulted crabs revealed that it was similar to Krogh's value for tissue (Mangum *et al.*, 1985). Recent observations of the diffusive conductance to oxygen of the gills and lungs of the land crab *Pseudothelphusa garmani* indicated that they resembled normal hydrated tissue (Innes & Taylor, 1986*b*). The conclusion must be that the exoskeleton does not restrict oxygen transfer over the respiratory exchange surfaces of crustaceans. Its presence may indeed provide some physical support, resisting the collapse of gills or diffuse lungs during air-breathing (see, for example, Taylor & Butler, 1978; Taylor & Innes, 1988).

Decapodan crustaceans spend their lives in a continuous cycle of moult-related events, either in preparation for or recovery from ecdysis when the old exoskeleton is shed to expose the new, soft and permeable cuticle below. Studies of the metabolic and physiological changes associated with

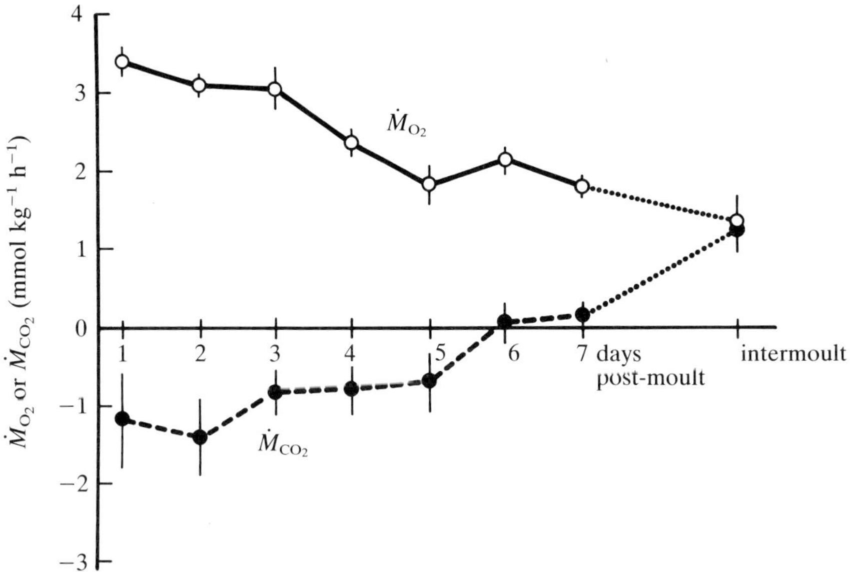

Figure 1. *Callinectes sapidus*, 15 °C. Rates of oxygen consumption ($\dot{M}_{O_2}$) and CO_2 excretion ($\dot{M}_{CO_2}$) in five crabs following the moult. After the old exoskeleton is shed, crabs show elevated $\dot{M}_{O_2}$ levels and a net negative CO_2 excretion rate, indicating net CO_2 uptake from the sea water. During the following five days $\dot{M}_{O_2}$ levels decline gradually and CO_2 uptake decreases to become net excretion seven days after the moult. Mean values are given ±s.e. (Taken from Cameron & Wood (1985).)

the moult cycle have concentrated on the premoult and immediate post-moult periods where the changes are most dramatic. In late premoult, respiratory gas exchange is impaired by the separation of old and new cuticle and oxygen transport is affected by a large reduction in circulating haemocyanin levels (Truchot, 1978). Despite these apparent disadvantages the marine blue crab (*Callinectes sapidus*) remains normoxic, possibly owing to active adjustments of internal and external convection systems and the maintenance of oxygen transport to the tissues by adjustments of haemocyanin oxygen affinity, via alterations in ionic conditions (Mangum *et al.*, 1985). An increase in the oxygen permeability of the general body surface after ecdysis leads to elevated Pa_{O_2} levels and an increase in oxygen uptake occurs during early postmoult (Figure 1).

A premoult alkalosis resulting from the accumulation of HCO_3^- compensates for a concomitant respiratory hypercapnia in *C. sapidus*, but the occurrence of anaerobic metabolism, due to restricted oxygen uptake during ecdysis, leads to a slight metabolic acidosis (Mangum *et al.*, 1985). The source of the base excess during premoult is thought to be uptake of HCO_3^- over the gills from seawater rather than mobilization of calcium carbonate from the exoskeleton, as changes in both free (Henry & Kormanik, 1985; Towle & Mangum, 1985) and total (Wheatly, 1985) calcium are negligible. Indeed, recent studies by Cameron (1989) have shown a sharp fall in haemolymph $[Ca^{2+}]$ immediately preceding ecdysis, which may result from the uptake of water occurring at this time.

The acid–base consequences of calcification of the exoskeleton following ecdysis have also been addressed in the blue crab. Calcium reabsorbed from the old exoskeleton prior to the moult is not retained but is excreted over the gills (Cameron & Wood, 1985). Deposition of $CaCO_3$ into the new exoskeleton is accompanied by apparent H^+ excretion and Ca^{2+} uptake from sea water at extremely high rates. Direct HCO_3^- uptake from sea water is implicated and CO_2 is deposited as $CaCO_3$ at rates in excess of its metabolic production; in fact, crabs show net CO_2 uptake for several days after edcysis (Figure 1). Cameron (1985) extended this study of post-moult calcification by manipulating external pH, $[HCO_3^-]$ and $[Ca^{2+}]$ in order to characterize the rapid net Ca^{2+} influx and H^+ efflux. He concluded that transfer occurs by indirectly linked electroneutral ion exchange, favouring the hypothesis that both external HCO_3^- and internal CO_2 react with incoming Ca^{2+} to form $CaCO_3$. His model predicts a $4H^+:2Ca^{2+}$ stoichiometry. Both fluxes were inhibited by a reduction in external Ca^{2+}, HCO_3^-, or pH. As the model predicts electrical and acid–base balance, any acid–base disturbance must be attributable to temporary inequalities in various fluxes. The most recent model indicates that Ca^{2+} and HCO_3^- ions are transported from water across the gills into the haemolymph and across the epithelium into the new cuticle, where pH values 0.3 to 0.4 units higher than the haemolymph favour the formation of $CaCO_3$ (Figure 2). Equivalent quantities of H^+ released from the reaction are transported in the opposite direction for excretion across the gills back to sea water (Cameron, 1989).

The mechanisms resulting in an apparent CO_2 uptake following the moult may be reflected in the unusually low respiratory quotient (RQ) reported from crustaceans in intermoult. Several authors have recorded RQ values of between 0.45 and 0.65, which imply that CO_2 excretion takes place at around half the rate of oxygen uptake. This could arise from a proportion of CO_2 being diverted to deposition of carbonates in the exoskeleton but further work is needed to clarify this matter (Wood & Boutilier, 1985).

It is still unclear how these physiological processes are controlled and

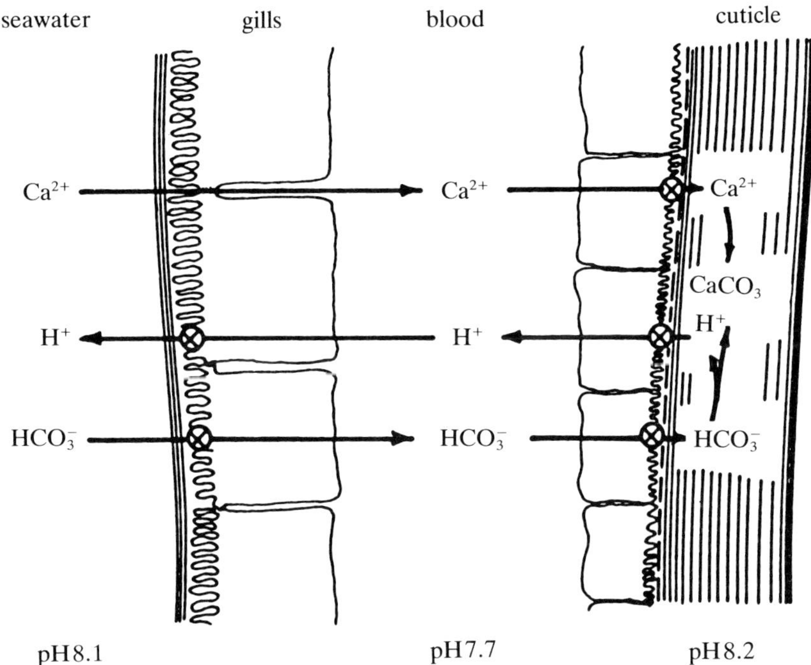

Figure 2. Schematic model taken from Cameron (1989) to show the transport processes involved in post-moult calcification of the exoskeleton in *Callinectes sapidus*. Both Ca^{2+} and HCO$_3^-$ are taken up from the sea water across the gills into the haemolymph and across the epithelium into the cuticle. H$^+$ produced from the reaction is transported in the opposite direction. Active transport processes (circles with crosses) predominate, with the exception of Ca^{2+} uptake, which is by diffusion across the gill from the sea water.

regulated during the moult cycle. Moulting hormones, such as the ecdysteroids 20-hydroxyecdysone and ponasterone A, have been implicated in stimulating the moult (Skinner, 1985) but their specific roles in the control and coordination of physiological events has yet to be demonstrated.

The dynamics of CaCO$_3$ deposition and reabsorption are of interest not only in relation to moulting *per se*. For many years indirect evidence has been collected, largely from terrestrial species, to suggest that the skeletal matrix can be eroded to buffer extracellular acidosis. A definitive experiment was performed by Cameron (1985) which illustrated that the carapace provided 8% of the HCO$_3^-$ accumulated extraceullularly in response to hyperoxia in blue crabs. This was observed as the erosion of radiolabelled Ca^{2+}, which had been loaded into the exoskeleton during a previous postmoult. Although the external sea water was the predominant source of HCO_3^- and the proton sink in this aquatic, marine crab, these experiments provided direct evidence for mobilization of carapace carbonate in response to acid–base disturbances. This mechanism undoubtedly plays an important role in the internal buffering of the potential respiratory and metabolic acidosis encountered by aquatic crustaceans venturing into air for prolonged periods, as described below.

Oxygen uptake and transport

In aquatic decapods, respiratory gas exchange during water-breathing takes place over the gills, which are enclosed on either side of the body in a branchial chamber ventilated by the rhythmic beating of the scaphognathite in the narrow exhalant channel (McMahon & Wilkens, 1983). The unidirectional flow of water through the branchial chambers is effective enough at high ventilation volumes in normoxia to maintain a P_{O_2} of 15.3–20 kPa (115–150 mmHg) at the respiratory surface (Taylor & Butler, 1978; Butler *et al.*, 1978; McMahon *et al.*, 1979; Wheatly & Taylor, 1981). This in turn can generate relatively high oxygen levels in the arterialized haemolymph: Pa_{O_2} of 10–13.3 kPa (75–100 mmHg), though settled levels are often lower (2.7–5.3 kPa, 20–40 mmHg) owing to reduced ventilation rates, which may include long periods of unilateral ventilation or ventilatory pauses (Butler *et al.*, 1978; McMahon & Wilkens, 1983). A relatively low oxygen level in the prebranchial (venous) haemolymph (Pv_{O_2}) of 0.7–2.7 kPa (5–20 mmHg) ensures a relatively high diffusion gradient for oxygen transfer (ΔPg_{O_2}) of about 12 kPa (90 mmHg) across the gills of actively ventilating crustaceans (Taylor & Wheatly, 1980). Despite this relatively low Pv_{O_2}, and a relatively low oxygen carrying capacity (0.5–1 mmol l^{-1}) conferred by their blood pigment haemocyanin (Mangum, 1979, 1983*a*), routinely active, water-breathing crustaceans maintain a substantial venous reserve of oxygen by virtue of the large volume of circulating haemolymph, which can constitute up to 40% of body volume (see Taylor *et al.*, 1987) and the relatively high affinity of haemocyanin for oxygen (P_{50}=0.7–2.7 kPa, 5–20 mmHg) (Mangum, 1983*a*). A range of values for these variables in both water and air-breathing species were listed by Innes & Taylor (1986*a*), Taylor & Innes (1988) and McMahon & Burggren (1988); others (many previously unpublished) are included in Table 1.

Oxygen transfer over the gills of fish and crustaceans is limited by diffusion; in crustaceans this limitation is particularly strong (Innes & Taylor, 1986*a*). Consequently, Pa_{O_2} rarely exceeds the oxygen levels in expired

Table 1. *Respiratory and acid–base variables in the arterialized haemolymph of a range of decapodan crustaceans*

The table includes aquatic decapods reliant on water as a respiratory medium (1, 2, 3); aquatic decapods with an ability to withstand aerial exposure with varying success (4,5); semi-terrestrial or terrestrial decapods which have developed a lung but still retain the gills for CO_2 excretion and ion regulation (6, 7); terrestrial crabs retaining gills but able to excrete CO_2 over the lungs in the absence of water (8, 9). W, in water; A, exposed in air; W/A, in air with access to standing water.

Species	t (°C)		Pa_{O_2} (kPa)	Ca_{O_2} (mmol l^{-1})	pHa	Pa_{CO_2} (kPa)	$CaCO_2$ (mmol l^{-1})	$[HCO_3^-]_a$ (mmol l^{-1})	Lactate (mmol l^{-1})	Reference
Maia squinado	15	W	8.3	0.20	7.72	0.3	6.5	6.4	1.4	E.W. Taylor
		A3 h	1.7	0.04	7.46	0.7	7.3	7.0	4.8	(unpublished)
Liocarcinus puber	20	W	8.6	0.70	7.98	0.2	7.2	7.0	0.9	E.W. Taylor, A.J. Watt &
		A3 h	4.7	0.20	7.18	0.6	3.3	3.1	6.8	N.M. Whiteley
										(unpublished)
Cancer pagurus	15	W	6.7	0.40	7.80	0.3	7.6	7.4	2.2	E.W. Taylor
		A3 h	2.0	0.30	7.71	0.8	14.5	14.3	4.4	(unpublished)
		A24 h	1.6	0.03	7.39	2.0	17.9	17.0	12.6	
Homarus gammarus	15	W	6.5	0.50	7.78	0.4	9.5	9.3	0.9	Taylor & Whiteley (1989)
		A3 h	1.6	0.20	7.64	0.7	11.0	10.7	1.8	
		A24 h	0.9	0.03	7.68	1.2	21.0	20.5	17.3	
Austropotamobius	15	W	4.4	0.30	7.89	0.4	7.1	6.8	0.5	Wheatly (1980)
pallipes		A3 h	1.3	0.14	7.45	1.2	8.0	7.2	2.9	Wheatly (1979)
		A24 h	1.3	0.28	7.79	1.0	14.8	13.7	0.6	Wheatly (1980)
Cardisoma carnifex	28	W	2.0	–	7.66	0.9	–	10.2	–	Cameron (1981)
	29	W/A	3.1	0.80	7.58	1.5	19.0	18.6	1.2*	Burggren & McMahon (1981)
										*Wood & Randall (1981)
	25	A36 h	8.7	–	7.6	1.5	–	21.0	0.7	Wood *et al.* (1986)
Ocypode saratan	25	W	2.7	0.30	7.74	1.0	16.8	16.5	9.2	Whiteley *et al.* (1990)
		W/A	9.0	0.90	7.76	1.7	20.9	20.4	0.8	
		A6 h	2.7	0.60	7.82	1.1	17.1	16.8	1.5	
Pseudothelphusa	25	W	3.5	0.30	7.65	0.7	–	9.0	–	A.J. Innes & E.W. Taylor
garmani		W/A	12.3	0.80	7.61	0.8	–	8.5	–	(unpublished)
		A35 h	17.6	0.60	7.70	1.3	–	18.6	–	
Birgus latro	25	A	5.8	1.10	7.73	1.0	13.6	–	0.2	Greenaway *et al.* (1988)

water, despite the proposed existence of a functional counter-current of haemolymph and water flow across the gill lamellae of crabs (Hughes *et al.*, 1969; Johansen *et al.*, 1970; Taylor, 1982).

The carrying capacity of haemocyanin in crustacean haemolymph is limited to less than 1 mmol O_2 l^{-1} in aquatic species, though it may rise to twice this value in some land crabs (McMahon & Burggren, 1988). This increases the capacity for oxygen above dissolved levels in the haemolymph of crustaceans by a factor of only 2 to 4 (Taylor, 1982; Mangum, 1983*a*; Cameron, 1986). This relatively low capacity for oxygen results from the presence of haemocyanin in colloidal suspension in the haemolymph, where it exerts a colloid osmotic pressure on the tissues which limits its concentration (Mangum & Johansen, 1975). This problem is avoided in vertebrates by their possession of erythrocytes, which confer much higher oxygen carrying capacities (up to 10 times higher) on their blood by packaging haemoglobin inside cells where it does not present osmotic problems. The binding of oxygen by the haemocyanin molecule is influenced by a number of factors. A predominant factor is $[H^+]$, as crustacean haemocyanin is generally characterized by a large positive Bohr effect (i.e. reduction in oxygen affinity with fall in pH) (Mangum, 1983*a*). Other modulators of oxygen affinity include inorganic ions (cf. Mangum, 1983*a*) and metabolites, particularly lactate ions (Truchot, 1980; Mangum, 1983*b*; Graham *et al.*, 1983; Bridges *et al.*, 1984; Morris & Bridges, 1985; Bridges & Morris, 1986) and certain neurohormones (Morris & McMahon, 1989). A recent study described in detail the interactive effects of Ca^{2+} and lactate on the oxygen affinity of crayfish haemocyanin (Morris *et al.*, 1986*a,b*). The importance of the lactate effect on haemocyanin oxygen affinity is that it counteracts the effects of protons so that accumulation of the acidic metabolic end products of crustacean anaerobiosis in the form of lactic acid may result in the potential Bohr shift being compensated by the lactate effect with no net change in oxygen affinity (A.J.S. Watt, unpublished observation). The importance of this effect is demonstrated when we consider facultative air-breathing.

CO₂ transport and excretion

The high ventilation rates of routinely active, aquatic gill-breathers relate to the low availability of oxygen in water (limited by its solubility, which decreases as temperature rises) and result in an effective hyperventilation with respect to CO_2 levels (Rahn, 1966). Consequently, circulating CO_2 levels are generally very low in water-breathers ($Pa_{CO_2} < 0.4$ kPa, 3 mmHg) (Dejours, 1981). Associated levels of HCO_3^- in the haemolymph range from 4 to 8 mmol l^{-1} depending on temperature (for reviews see Truchot, 1983; Cameron, 1986; Taylor & Innes, 1988). The non-bicarbonate buffering capacity varies from 5 to about 15 slykes between species and with acid load. Like oxygen capacity it is a function of haemolymph protein concentration (Wood & Randall, 1981), which is limited by its effects on colloid osmotic pressure.

The elimination of CO_2 over the gills of crustaceans is predominantly as molecular CO_2, as described for fish gills (see, for example, Randall, 1982). Metabolic CO_2 combines spontaneously with water to form carbonic acid whose dissociation yields H^+ and HCO_3^-, thereby influencing pH. The role of the enzyme carbonic anhydrase (CA) as a catalyst for this hydration–dehydration reaction in crustacean tissues has been studied in detail by a number of workers (compare reviews by Henry, 1984; Burnett *et al.*, 1985). Not only is the enzyme physiologically significant in facilitating CO_2 excretion but the provision of the counterions H^+ and HCO_3^- also has repercussions for acid–base balance, ion transport and calcification. Unlike fish, crustaceans do not have CA in their haemolymph because they lack erythrocytes, but its activity has been demonstrated on the gills (e.g. *Cardisoma carnifex*, Randall & Wood, 1981). It occurs at the gills in two functional forms: cytoplasmic CA in the gill epithelial cells, which is believed to be important in ion exchange (Henry & Cameron, 1983; Henry, 1984) and CA associated with the basolateral membranes of the cells (Burnett & McMahon, 1985), which is exposed to the haemolymph and may facilitate CO_2 excretion by catalysing the dehydration of the large haemolymph HCO_3^- pool to diffusable CO_2 (Figure 3). The role of membrane-associated CA has been elucidated by using membrane-impermeant CA inhibitors such as the dextran-bound inhibitor (DBI) in isolated perfused tissues, and quatenary ammonium sulfanilamide (QAS), which can be used *in vivo* (Henry, 1987*b*). DBI application to isolated gills reduced CO_2 efflux, emphasizing the importance of CA in maintaining the CO_2 gradient between haemolymph and water in situations where ambient CO_2 levels rise and where the branchial cells themselves are a source of metabolic CO_2 (Burnett & McMahon, 1985). QAS injection produces a respiratory acidosis *in vivo* in blue crabs (Henry, 1986, 1987*a*) which persists until the inhibitor is cleared, confirming the role of membrane-associated CA in CO_2 excretion.

The presence of CA activity is closely associated with the presence of salt absorbing cells and Na^+–K^+ ATPase activity. In marine species, these enzymes are concentrated in the posterior three or four gill pairs (Henry & Cameron, 1982). In the freshwater crayfish, however, both transport

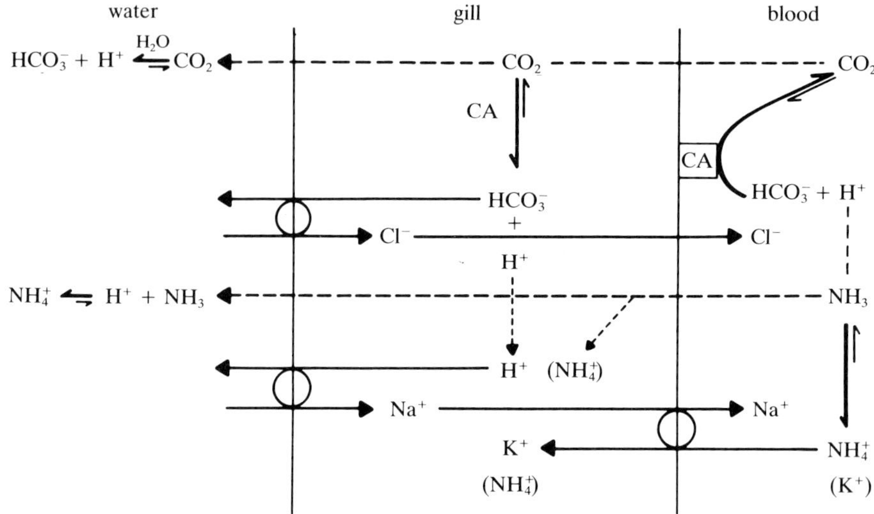

Figure 3. Schematic model of carbonic anhydrase (CA) function in the gill of the blue crab, *C. sapidus*, showing the physiological roles of both membrane-associated and cytoplasmic CA. Membrane-associated CA catalyses the conversion of HCO_3^- to CO_2 and is important in CO_2 excretion. Cytoplasmic CA catalyses the hydration–dehydration of CO_2 and is important in electroneutral branchial ion exchange. Broken lines represent diffusion; solid lines represent some form of coupled transport; and the large arrows represent the predominant direction of CA-catalysed reactions. (Taken from Henry (1987a).)

enzymes are homogeneously distributed throughout the gill chamber, implying that all gills are equally involved in ion and acid–base regulation (Wheatly & Henry, 1987).

Intracellular pH regulation

Intracellular pH regulation is believed to occur primarily by transmembrane electroneutral ion exchange (Thomas, 1989). Determinations on isolated invertebrate tissues had previously identified a mechanism that exchanged external Na^+ and HCO_3^- for internal Cl^- (Figure 4). In addition, Gaillard & Rodeau (1987) have recently identified a Na^+–H^+ exchanger sensitive to the specific inhibitor amiloride in crayfish neurons. This exchanger, which has similar kinetic properties to that found in vertebrates, depends on intracellular pH and extracellular $[Na^+]$, and may contribute 50% to overall cellular acid extrusion. The mechanism also depends on extracellular pH, as H^+ efflux is passive, and on intracellular $[Na^+]$, because of the pH dependency of Na^+–K^+ ATPase.

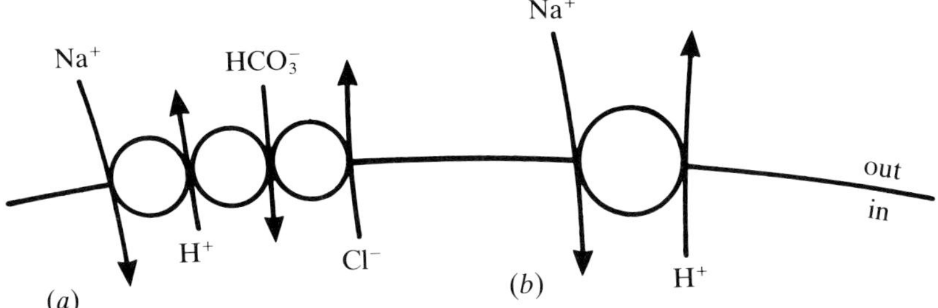

Figure 4. Diagram of the ion exchange mechanisms probably responsible for pH$_i$ regulation in crayfish neurons. The processes labelled (a) are blocked by SITS, those labelled (b) by amyloride. (Taken from Thomas (1989).)

The technique based on the distribution of the radiolabelled weak acid DMO has enabled pH$_i$ to be determined in various tissues in intact crustaceans under a variety of environmental conditions. Wood & Cameron (1985) found pH$_i$ values in soft body tissues that were generally 0.5 pH units below those in the haemolymph. However, pH$_i$ in the fluid compartment inside the carapace was about 0.5 pH units more alkaline than the haemolymph (Figure 5). This relatively alkaline pH is required to maintain the solid CaCO$_3$ phase in the calcified exoskeleton and the extent to which exoskeletal erosion is involved in H$^+$ buffering will depend on the degree of acidification of this fluid reservoir. In most tissues, change in pH$_i$ with acclimation temperature parallels the change in haemolymph pH and the pH at neutrality (pN), but the pH$_i$ in the alkaline reservoir of the carapace remained constant (Figure 5).

The bulk of total buffering capacity (about 90%) in animals is in the intracellular compartment and chiefly attributable to proteins (Heisler, 1984). In contrast the buffering capacity of the body fluids is relatively low; this is particularly the case in crustaceans, which have a low non-bicarbonate buffering capacity in the haemolymph because the concentration of circulating proteins is low. When exposed to an acidotic problem, animals will normally regulate intracellular pH (pH$_i$) in order to avoid deleterious disruption of cellular processes, such as metabolism. Following 30 min of strenuous exercise the blue crab, *Callinectes sapidus*, experienced a pronounced metabolic acidosis. Muscle pH$_i$ was restored to control levels within 1 h, whereas restoration of haemolymph pH was not complete until 4 h after exercise (Milligan *et al.*, 1989). In contrast, the initial acidosis on aerial exposure of the crayfish, *Austropotamobius pallipes*, was compensated in the haemolymph after 24 h in air, but pH$_i$ in abdominal muscle

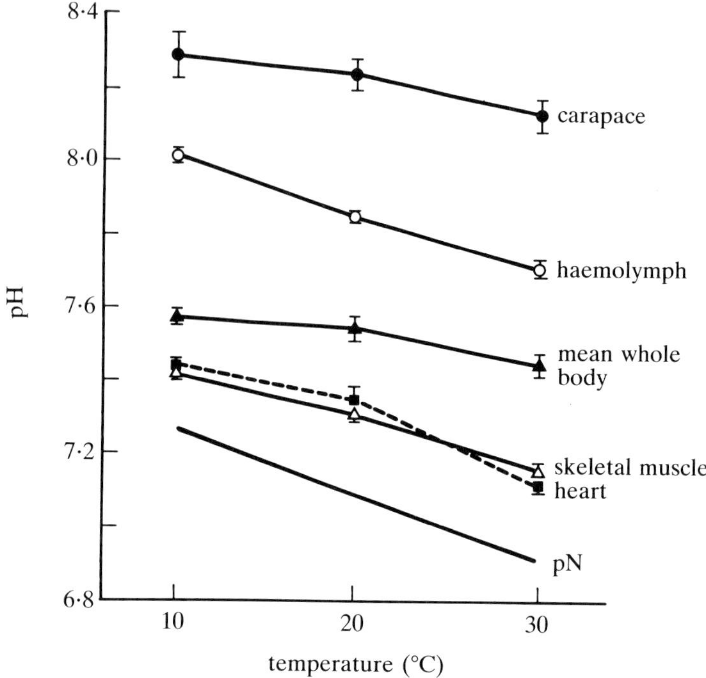

Figure 5. Intracellular pH variations in selected tissues of the blue crab, *C. sapidus*, acclimated to 10, 20 and 30 °C. Mean whole body pH_i, haemolymph pH and the pH of neutrality (pN) are also represented on this diagram. Mean values are given ±s.e. (Taken from Wood & Cameron (1985).)

remained uncompensated (Tyler-Jones & Taylor, 1988). This may indicate sequestration of protons in this highly buffered intracellular compartment. The predominantly glycolytic abdominal muscles may be able to tolerate retention of protons, enabling tissues requiring greater pH stability, such as the heart and CNS, to compensate pH_i by ionic exchange with the relatively poorly buffered haemolymph during aerial exposure.

Hyperoxia

Hyperoxia has been used as an experimental tool to study mechanisms of compensation for extracellular acidosis in crustaceans because the resultant hypoventilation causes CO_2 to accumulate in the haemolymph. Wheatly (1987) found that the initial hyperoxic acidosis in the rock crab, *Cancer irroratus*, was restored by HCO_3^- accumulation accompanied by excretion of acidic equivalents (primarily titratable) into the experimental water. The acid efflux and extracellular proton load were equivalent; P_{CO_2} partly recovered after 24 h. The base efflux during recovery greatly exceeded that required to restore haemolymph acid–base balance, suggesting that it originated intracellularly. Certain parallels were found during hyperoxia in freshwater crayfish (Wheatly, 1989). Again P_{CO_2} increased with time. The hyperoxic acidosis was similarly compensated by extracellular bicarbonate accumulation restulting from net H^+ output. This allegedly occurs via electroneutral ion exchange (Na^+–H^+ or Na^+–NH_4^+; Cl^-–HCO_3^-) at the gills. However, by working in fresh water it was possible to correlate transbranchial acidic equivalent exchange and unidirectional fluxes of the major electrolytes Na^+ and Cl^-. Net H^+ efflux was negatively correlated with net Na^+ flux and positively correlated with net Cl^- flux; however, net proton flux was not correlated with the difference between strong cation and anion fluxes. Unidirectional Cl^- fluxes were better correlated with net H^+ flux than were Na^+ fluxes. The primary response during hyperoxia is an increase in efflux and a reduction in influx. Even so, a large proportion of influx and efflux of both ions was via exchange diffusion (e.g. Na^+–Na^+) suggesting that only a small part occurs via electroneutral ion exchange. The majority of H^+ lost during hyperoxia originated in the tissue. Re-establishment of normoxia resulted in a very rapid reversal of these trends, implying that this contributed a more powerful stimulus of the branchial exchange mechanisms than the switch to hyperoxia. Unidirectional ion fluxes are far greater in marine species; this may explain why acid–base correction occurs more rapidly than in fresh water where ion exchanges may be limited by external concentrations.

The importance of exchange of acidic equivalents between the extracellular fluid (ECF) and experimental water in a variety of environmental conditions has already been stressed. Previously, no involvement of the antennal glands (kidney analogues) in acid–base regulation had been identified (Cameron, 1986), largely because experiments had only been performed on marine species, which produce isosmotic urine. Recently Wheatly & Toop (1989) reassessed this in the freshwater crayfish, which has the ability to produce a dilute urine owing to the high specific activity of transport enzymes such as Na^+–K^+ ATPase, and of CA in the antennal gland (Wheatly & Henry, 1987). The latter, which produces the counterions H^+ and HCO_3^- for ion reabsorption from metabolic CO_2, suggests a potential acid–base role of this organ. The crayfish antennal gland responded to extracellular hyperoxic acidosis by increasingly acidifying the urine (mainly by using non-titratable protons) although quantitatively the proton efflux was only one tenth of the value recorded at the gill. Net Na^+

and Cl^- were increasingly reabsorbed at the antennal gland, which is exactly counter to their reduced influx at the gills, and HCO_3^- was progressively reabsorbed increasing its extracellular accumulation.

Recently, Wheatly *et al.* (1990) studied the time-dependent changes in pH_i during compensation of hyperoxic extracellular acidosis in crayfish. Nerve pH_i remained unchanged; cardiac muscle underwent a small uncompensated respiratory acidosis, which was also seen in the carapace. Skeletal muscle, meanwhile, was able to partly compensate the intracellular acidosis by HCO_3^- accumulation.

Hypercapnia

Recent interest in hypercapnic effects has concentrated on progressive exposure of blue crabs to P_{CO_2} of 15, 30 and 45 mmHg (2–6 kPa) (Cameron & Iwama, 1987). Compensation of hypercapnic acidosis involved elevations in extracellular $[HCO_3^-]$ and strong ion difference (SID). $[HCO_3^-]$ levels of 50 meq l^{-1} were attained at the upper hypercapnic level, confirming low HCO_3^- permeability at the gills. Compensation only appears to be around 70% even though the animal is capable of further HCO_3^- accumulation. Presumably this new set-point represents a compromise between the opposing demands of pH and ion regulation at a mechanistic level.

Acid stress

A relatively new area of study is the acid–base response of invertebrates to acid or alkaline media. Wood & Rogano (1986) observed the effect of 5 days' exposure to pH 4 (H_2SO_4) in decarbonated soft water on the crayfish *Orconectes propinquus*. A marked net influx of protons occurred and a severe extracellular metabolic acidosis resulted, accompanied by reductions in Na^+ and Cl^-. The acidic equivalents taken up from the water, were predominantly sequestered outside the ECF, resulting in K^+ efflux from cells. When net Na^+ and Cl^- fluxes were resolved into their unidirectional components, it transpired that the loss of these ions was due to partial inhibition of influxes (effluxes were unaffected). Inhibition of Na^+ influx arose from competition between H^+ and Na^+ for a common carrier, whereas Cl^- inhibition resulted from conformational changes in the carrier or impairment of Cl^-–HCO_3^- exchange by low $[HCO_3^-]$. During continued exposure the Na^+ exchange diffusion component increased. At the same time circulating P_{CO_2} increased after 2 days exposure, owing to hypoventilation or a thickened diffusion barrier, and this contributed to the acidosis. A recent review by McMahon & Stuart

(1989) concluded that crayfish species varied in their tolerance of low pH, with some such as *Procambarus clarki* able to survive prolonged exposure to pH 4 and to compensate for initial disruption of haemolymph acid–base and ion balance.

Multiple environmental variables

Interest in the respiratory physiology of decapods in their natural environment has recently been revived; combined changes in experimental variables have been used to reproduce the complex changes often encountered in natural situations. Dejours *et al.* (1985) examined the effect of simultaneous changes in water temperature and oxygenation on the shore crab, *Carcinus maenas*. Temperature change at either isobaric normoxia or isoconcentration normoxia did not affect acid–base balance in *Carcinus*, unlike the crayfish, where an effect was seen (Dejours & Armand, 1983). This difference was attributed to the fact that acid–base balance is greatly influenced by [Cl$^-$], which is limited in fresh water. In another study, Truchot (1986) simulated the diurnal cyclical changes in temperature, P_{O_2} and P_{CO_2} that occur naturally in tide pools and discovered that the acid–base changes were less than those induced by single-factor laboratory experiments. At night, pools become hypoxic and hypercapnic with the reverse true in daytime conditions. The particular combination of changes in P_{O_2} and P_{CO_2} acted to minimize acid–base disturbances. For example, at night hypoxia served to increase extracellular pH via hyperventilation (whereas hypercapnia produces the inverse effect). Environmental factors were changing so rapidly that metabolic [HCO$_3^-$] accumulation did not occur. Because these individual variables have opposing effects on acid–base status, their combined effect was minimal. Interactions of this kind between the effects of environmental variables are probably instrumental in assuring survival in apparently hostile aquatic environments. An alternative strategy is to resort to air-breathing; it was the observation of shore-crabs bubbling air through their branchial chambers when exposed in rock pools that led to a study of the emersion response in *Carcinus maenas* (Taylor & Butler, 1973; Taylor *et al.*, 1973). Emersion involves essentially aquatic respiratory gas exchange as the gills remain submerged while water held in the branchial chambers is aerated. This improves oxygen delivery without leading to CO$_2$ accumulation so that there are no acid–base consequences (Wheatly & Taylor, 1979). However, in common with many other primarily aquatic species, *Carcinus* is able to leave water to become a facultative air-breather (Taylor & Wheatly, 1979).

Transition to air-breathing

Many primarily aquatic crustaceans are able to survive out of water, despite being chronically hypoxic and hypercapnic because their gill lamellae clump together and reduce their effectiveness as respiratory organs (Taylor, 1982; Taylor & Innes, 1988). The problems of clumping of the gills in air may be reduced in crustaceans because their gill lamellae do not bear secondary lamellae and are covered in an outer cuticular layer; in contrast, the diffuse and mucus-laden surfaces of fish gills adhere closely in air. Nevertheless, collapse of the gills in air limits both oxygen and CO_2 exchange. Fully aquatic crustaceans, which have relatively large gill areas, become hypoxic and accumulate both CO_2 and lactic acid in air. The combined respiratory and metabolic acidosis often remains uncompensated and results in death from asphyxia (see Table 1). However, many crustacean species are amphibious and able to maintain similar routine rates of oxygen uptake ($\dot{M}_{O_2}$) in water and in air (Taylor & Butler, 1978; Taylor & Wheatly, 1980) though this ability may vary with activity levels and other environmental variables (McMahon & Burggren, 1988). The relative rates of $\dot{M}_{O_2}$ of the primarily aquatic lobster *Homarus gammarus* in water and in air vary with temperature (Figure 6). At 10 °C the rates are similar but $\dot{M}_{O_2}$ in water increases with temperature up to 20 °C whereas it is limited in air, probably by diffusion over the collapsed gills. Consequently $\dot{M}_{O_2}$ in water is about 2.5 times $\dot{M}_{O_2}$ in air at 20 °C. The animals accumulate lactic acid in the haemolymph during aerial exposure, up to extremely high levels at 20 °C. The acid–base consequences of aerial exposure in *H. gammarus* at 10 and 20 °C have recently been described (Whiteley & Taylor, 1990).

An initial combined metabolic and respiratory acidosis on aerial exposure of the freshwater crayfish, *Austropotamobius pallipes*, is compensated by the sequestration of some H^+ in the muscles (Tyler–Jones & Taylor, 1988) and by progressive mobilization of buffer base ($[HCO_3^-]$ rose to about 15 meq l^{-1}) from the $CaCO_3$ in the exoskeleton, resulting in an associated increase in $[Ca^{2+}]$.

The predominant diffusion limitation of branchial exchange in air and consequent low Pa_{O_2} and Pv_{O_2} values (Figure 7) is matched by a low *in vivo* blood P_{50} (<10 mmHg), which maintains the degree of saturation of haemocyanin with oxygen at the gills relatively high (60–80%) during aerial gas exchange. The internal buffering by mobilization of HCO_3^- against a potential respiratory and metabolic acidosis results in restoration of a normal pH. This in turn reverses an initial Bohr shift on the haemocyanin

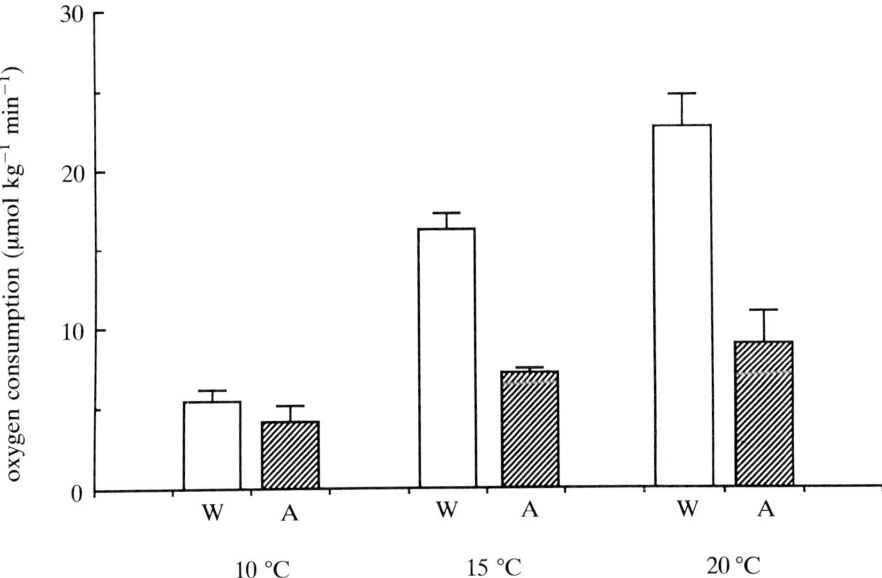

Figure 6. Rates of oxygen consumption of the lobster, *Homarus gammarus* in water (W) and after 6 h in air (A) at three temperatures (N. Whiteley & E. Taylor, unpublished data). Mean values are given ±s.e.

and its oxygen affinity is further increased, restoring oxygen transport, by the specific allosteric effects of calcium and lactate ions. These interactive responses to aerial exposure of the freshwater crayfish have been described in detail (Taylor & Wheatly, 1981; Morris *et al.*, 1986*b*; Taylor *et al.*, 1987; Tyler-Jones & Taylor, 1988) and are illustrated in Figure 7. They are common to many primarily aquatic crustaceans, exposed in air (e.g. the lobster) (Taylor & Whiteley, 1989) and may be considered as preadaptations for survival during aerial exposure which are utilized in submerged animals when active or exposed to acid conditions in water (see, for example, Wood & Randall, 1981; Booth *et al.*, 1984; Wood & Rogano, 1986).

In terms of the evolution of air-breathing, branchial gas exchange (i.e. across the gills) is not an effective means of aerial oxygen transfer, but as long as the animal can maintain a reservoir of water in contact with the gills it is an effective means of maintaining a low Pa_{CO_2} and consequent low $[HCO_3^-]_a$ (Taylor & Innes, 1988). Land crabs are characterized by having reduced gill areas with the gill lamellae widely spaced and fairly rigid, resisting collapse in air (McMahon & Burggren, 1988). This facilitates aerial gas exchange, though their effectiveness in air is restricted by a predominant diffusion limitation and their primary role seems to be ion

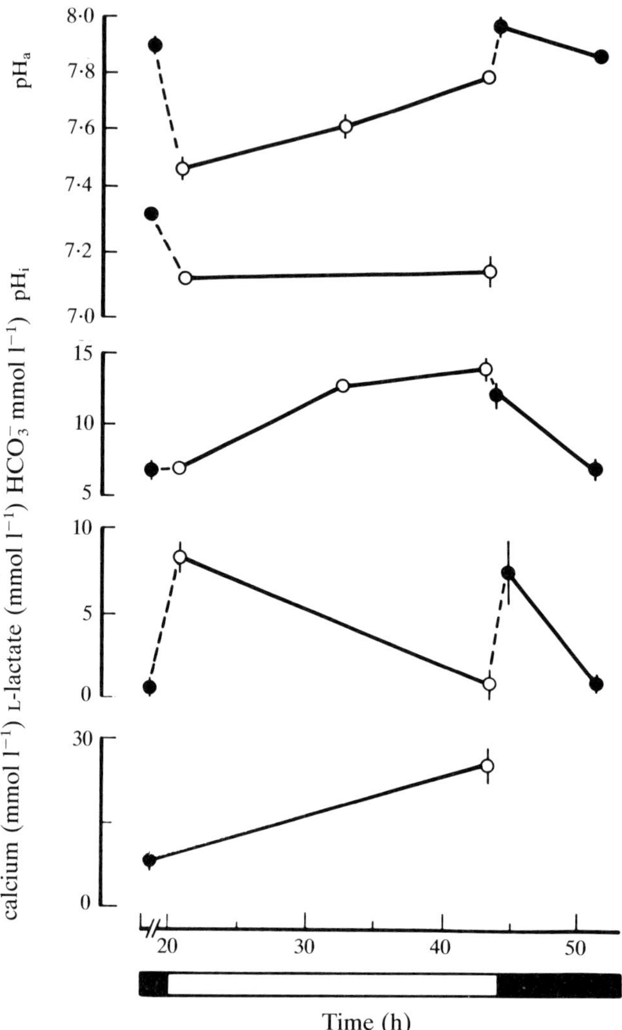

Figure 7. Changes in respiratory and acid–base variables during aerial exposure of the crayfish, *Austropotamobius pallipes*. On removal into air the animal is hypoxic (low Pa_{O_2}) and hypercapnic (high Pa_{CO_2}) and accumulates lactic acid rapidly. Haemolymph pH values (pH_a) are initially reduced but are restored on further aerial exposure owing to the mobilization of HCO_3^- from $CaCO_3$ in the exoskeleton, accompanied by a rise in haemolymph calcium ions and by a reduction of haemolymph lactate levels. Some of the potential H^+ load is sequestered in the skeletal muscles as evidenced by the sustained reduction in intracellular pH (pH_i).

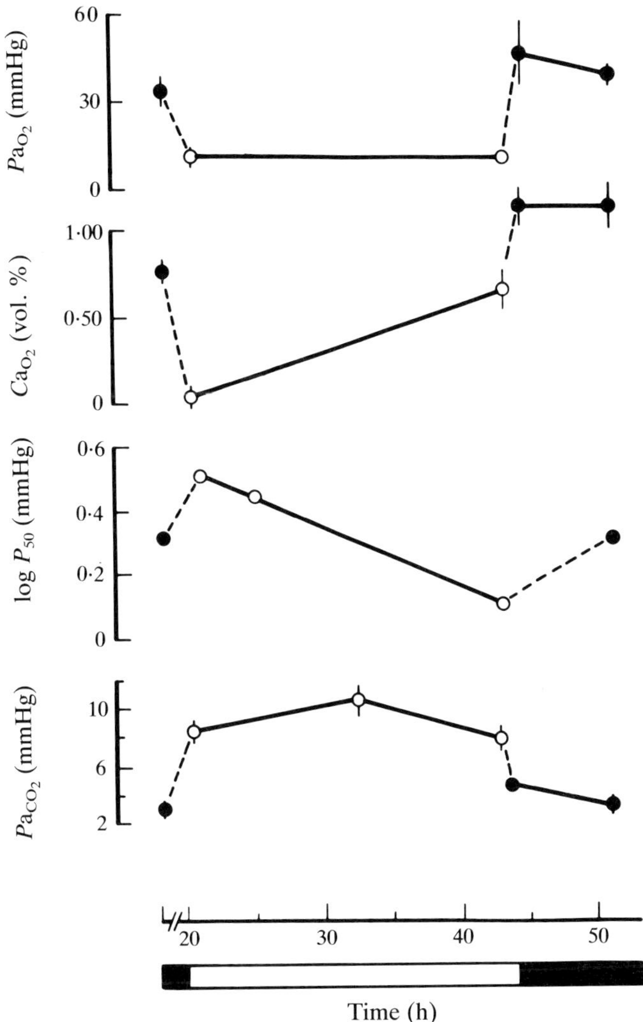

Restoration of pH$_a$ restores a Bohr shift on haemocyanin, which, together with the modulating effects of lactate and subsequently calcium, increases haemocyanin oxygen affinity as shown by changes in log P_{50}. Consequently, the initial hypoxaemia (low Ca_{O_2}) is compensated and oxygen delivery to the tissues is restored while the crayfish are in air. The initial filled symbol represents mean values for settled submerged animals, the open symbols represent values over a 24 h period of aerial exposure, and the final filled symbols represent subsequent recovery in water (data from Taylor & Wheatly, 1981; Taylor *et al.*, 1987; Tyler-Jones & Taylor, 1988).

regulation and acid–base balance. Recent research has considered the role of CO_2 excretion across the gills into water held in the branchial chambers, in the maintenance of a relatively low circulating P_{CO_2} in land crabs, compared with other air-breathers. Burnett & McMahon (1987) compared the role of the branchial water in acid-base balance in *Pachygrapsus crassipes*, a littoral crab capable of Cl^- regulation, with *Eurytium albidigitum*, an osmoconformer. The emersion-induced acidosis was compensated in *Pachygrapsus*, accompanied by an increase in the CO_2 content and titratable alkalinity of the branchial water. In *Eurytium* the acidosis remained uncompensated. The increase in $[HCO_3^-]$ in branchial water of *Pachygrapsus* would lower P_{CO_2}, thereby maximizing the outward diffusion gradient for CO_2 elimination. Acidic equivalent excretion into the branchial water appeared to be important in acid–base regulation and to occur via electroneutral ion exchange. Wood & Boutilier (1985) were able to correlate net H^+ flux in *Cardisoma carnifex* with the difference between strong cation and anion fluxes into fresh water held in the branchial chambers, confirming the operation of electroneutral ion exchange over the gills in amphibious species. A recent study by Greenaway *et al.* (1988) found that even though CO_2 elimination in *Birgus latro* was mainly across the gills (promoted by CA) in resting animals, half the output of CO_2 was pulmonary during exercise. This shows that CO_2 elimination can also take place across the lungs in obligate air-breathers, probably by diffusion when a P_{CO_2} gradient is present and the crab is hyperventilating.

Typical land crabs have evolved a lung consisting of the vascularized branchial chamber epithelium specialized for the uptake of oxygen from air. This is characterized by a folded or evaginated surface and a short diffusion path for respiratory gas exchange (blood–medium diffusion distance$=0.3$–$1.2\,\mu m$) over the thin, cuticularized processes of the epithelial cells (Greenaway & Taylor, 1976). Associated with the switch to lung breathing and results in elevated Pa_{O_2} levels (Table 1). These in turn are matched by a reduction in oxygen affinity of haemocyanin which maintains an adequate P_{O_2} gradient between haemolymph and tissues (Innes & Taylor, 1986a; Taylor & Innes, 1988).

sible for essentially all uptake of oxygen in the terrestrial robber crab *Birgus latro* (Greenaway *et al.*, 1988). The uptake of oxygen across the lung is highly effective when compared with gas exchange by aerial gill-breathing and results in elevated Pa_{O_2} levels (Table 1). These in turn are matched by a reduction in oxygen affinity of haemocyanin which maintains an adequate P_{O_2} gradient between haemolymph and tissues (Innes & Taylor, 1986a; Taylor & Innes, 1988).

The complex lung of the Trinidad mountain crab *Pseudothelphusa garmani* consists of diffuse, anastomosing respiratory airways, invaginated into a sinus containing circulating haemolymph (Innes *et al.*, 1987). When exposed in air with access to shallow water *Pseudothelphusa* uses its scaphognathites to bubble air through water held in contact with the gills while separately ventilating the lungs with air, with the result that Pa_{O_2} and Pv_{O_2} levels are relatively high and Pa_{CO_2} levels relatively low (Taylor & Innes, 1988). Following long-term exposure in air without access to water, which mimics the conditions during the dry season, slow ventilation of the invaginated lung raises pulmonary airway P_{O_2} in *Pseudothelphusa* to 19 kPa (141 mmHg), which generates a uniquely high Pa_{O_2} of 16–19 kPa (Table 1). Analysis of lung function and diffusion limitation reveals that the *Pseudothelphusa* lung is a perfusion-limited gas exchange organ and functionally resembles the complex, diffuse lung of higher vertebrates. The increase in oxygen partial pressure at the respiratory exchange surface and reduction in diffusion limitation are matched by a further reduction in oxygen affinity of haemocyanin (Innes & Taylor, 1986*a*). The elevation of Pa_{O_2} to around 18 kPa is accompanied by reduction of Pa_{CO_2} levels to below 2 kPa (Table 1). This is achieved by effective hyperventilation of the lung, which has a respiratory exchange ratio of about 1. This serves to 'blow off' carbon dioxide across the lung surface in the absence of an aquatic route across the gills. Despite this functional hyperventilation, actual rates of ventilation are low in quiescent *Pseudothelphusa* in order to reduce the rate of evaporative water loss (Innes & Taylor, 1986*a*).

Conclusion

Recent work has uncovered the crustacean answers to the problems of respiratory gas exchange and acid–base regulation, assumed to be associated with their reliance on growth by moulting and the presence of the exoskeleton over their respiratory gas exchange surfaces. Most decapodan crustaceans are fully marine and rely on diffusion-limited gills for oxygen uptake and CO_2 excretion. This latter function remains a primary role for the gills even among most land crabs, which have evolved lungs for oxygen uptake. Excretion of CO_2 over the gills is by diffusion and electroneutral ion exchange, implicating acid–base and ion regulation. Electroneutral ion exchange is employed to counter hyperoxic and hypercapnic acidosis and to regulate pH_i; it also enables ion regulation by the green gland in freshwater crustaceans to contribute to acid–base balance. Primarily aquatic crustaceans are hypoxic and hypercapnic in air, owing to the collapse of the gills and loss of their contact with water for ion exchange. Amphibious species can compensate a potential metabolic and respiratory acidosis by mobilization of an internal source of buffer base as HCO_3^-, from the calcium carbonate in the exoskeleton. Mobilization of the the exoskeleton may be seen as a predaptation for buffering against acidosis, conferred upon crustaceans by their moult cycle. Increased levels of lactate and calcium in the haemolymph cause a beneficial increase in the oxygen affinity of haemocyanin in some species, which can restore oxygen transport and aerobic metabolism. Land crabs are characterized by relatively high oxygen and low CO_2 partial pressures in the haemolymph, generated by exchange over the lungs and gills respectively, though in the absence of water CO_2 may be eliminated over the lungs.

References

Booth, C.E., McMahon, B.R., de Fur, P.L. & Wilkes, P.R.H. (1984). Acid-base regulation during exercise and recovery in the blue crab, *Callinectes sapidus*. *Respiration Physiology* **58**, 359–76.

Bridges, C.R. & Morris, S. (1986). Modulation of haemocyanin oxygen affinity by L-lactate. A role for other cofactors. In *Invertebrate Oxygen Carriers* (ed. B. Linzen), pp. 341–52. Springer–Verlag, Berlin.

Bridges, C.R., Morris, S. & Grieshaber, M.K. (1984). Modulation of haemocyanin oxygen affinity in the intertidal prawn *Palaemon elegans* (Rathke). *Respiration Physiology* **57**, 189–200.

Burggren, W.W. & McMahon, B.R. (1981). Hemolymph oxygen transport, acid-base status, and hydromineral regulation during dehydration in three terrestial crabs, *Cadisoma, Birgus* and *Coenibita*. *Journal of Experimental Zoology* **218**, 53–64.

Burnett, L.E. & McMahon, B.R. (1985). Facilitation of CO_2 excretion by carbonic anhydrase located on the surface of the basal membrane of crab gill epithelium. *Respiration Physiology* **62**, 341–8.

Burnett, L.E. & McMahon, B.R. (1987). Gas exchange, haemolymph acid-base status, and the role of branchial water stores during air exposure in three littoral crab species. *Physiological Zoology* **60**, 27–36.

Burnett, L.E., Dunn, T.N. & Infantino, R.L. (1985). The function of carbonic anhydrase in crustacean gills. In *Transport Processes, Iono- and Osmoregulation* (ed. R. Gilles & M. Gilles-Baillien), pp. 159–168. Springer-Verlag, Berlin.

Butler, P.J., Taylor, E.W. & McMahon, B.R. (1978). Respiratory and circulatory changes in the lobster (*Homarus vulgaris*) during long term exposure to moderate hypoxia. *Journal of Experimental Biology* **73**, 131–46.

Cameron, J.N. (1981). Acid-base responses to changes in CO_2 in two Pacific crabs: the coconut crab, *Birgus latro*, and a mangrove crab, *Cardisoma carnifex*. *Journal of Experimental Zoology* **218**, 65–73.

Cameron, J.N. (1985). Compensation of hypercapnic acidosis in the aquatic blue crab, *Callinectes sapidus*: the predominance of external sea water over carapace carbonate as the proton sink. *Journal of Experimental Biology* **114**, 197–206.

Cameron, J.N. (1986). Acid-base equlibria in invertebrates. In *Acid-Base Regulation in Animals* (ed. N. Heisler), pp. 357–394. Elsevier, New York.

Cameron, J.N. (1989). Post-moult calcification in the blue crab, *Callinectes sapidus*: timing and mechanism. *Journal of Experimental Biology* **143**, 285–304.

Cameron, J.N. & Iwama, G.K. (1987). Compensation of progressive hypercapnia in channel catfish and blue crabs. *Journal of Experimental Biology* **133**, 183–97.

Cameron, J.N. & Mangum, C.P. (1983). Environmental adaptations of the respiratory system: ventilation, circulation and oxygen transport. In *The Biology of Crustacea* (ed. F.J. Vernberg & W.B. Vernberg), vol. 8, pp. 43–65. Academic Press, New York.

Cameron, J.N. & Wood, C.M. (1985). Apparent H^+ excretion and CO_2 dynamics accompanying carapace mineralisation in the blue crab (*Callinectes sapidus*) following moulting. *Journal of Experimental Biology* **114**, 181–96.

Dejours, P. (1981). *Principles of Comparative Respiratory Physiology*. Elsevier, Amsterdam.

Dejours, P. & Armand, J. (1983). Acid-base balance of crayfish hemolymph: effects of simultaneous changes of ambient temperature and water oxygenation. *Journal of Comparative Physiology* **149**, 463–8.

Dejours, P., Toulmond, A. & Truchot, J.-P. (1985). Effects of simultaneous changes of water temperature and oxygenation on the acid-base balance of the shore crab, *Carcinus maenas*. *Comparative Biochemistry and Physiology* **81**A, 259–62.

Gaillard, S. & Rodeau, J.L. (1987). Na^+/H^+ exchange in crayfish neurones: dependence on extracellular sodium and pH. *Journal of Comparative Physiology* **157**, 435–44.

Graham, R.A., Mangum, C.P., Terwilliger, R.C. & Terwilliger, N.B. (1983). The effect of organic acids on oxygen binding of haemocyanins from the crab *Cancer magister*. *Comparative Biochemistry and Physiology* **74**A, 45–50.

Greenaway, P. & Taylor, H.H. (1976). Aerial gas exchange in the Australian arid-zone crab *Parathelphusa transversa* Von Martens. *Nature* **262**, 711–13.

Greenaway, P., Morris, S. & McMahon, B.R. (1988). Adaptations to a terrestrial existence by the robber crab *Birgus latro* II. *In vivo* respiratory gas exchange and transport. *Journal of Experimental Biology* **140**, 493–509.

Heisler, N. (1984). Acid-base regulation in fishes. In *Fish Physiology* (ed. W.S. Hoar & D.J. Randall), vol. 10A, pp. 315–340. Academic Press, New York.

Henry, R.P. (1984). The role of carbonic anhydrase in blood ion and acid-base regulation. *American Zoologist* **24**, 241–51.

Henry, R.P. (1986). Subcellular distribution of carbonic anhydrase in the gills of the blue crab, *Callinectes sapidus. American Zoologist* **26**, 216–23.

Henry, R.P. (1987*a*). Membrane associated carbonic anhydrase in the gills of the blue crab *Callinectes sapidus. American Journal of Physiology* **252**, 966–71.

Henry, R.P. (1987*b*). Quaternary ammonium sulfanilamide: a membrane-impermeant carbonic anhydrase inhibitor. *American Journal of Physiology* **252**, 959–65.

Henry, R.P. & Cameron, J.N. (1982). Acid-base balance in *Callinectes sapidus* during acclimation to low salinity. *Journal of Experimental Biology* **101**, 255–64.

Henry, R.P. & Cameron, J.N. (1983). The role of carbonic anhydrase in respiration, ion regulation and acid-base balance in the aquatic crab, *Callinectes sapidus*, and the terrestrial crab, *Gercarcinus lateralis. Journal of Experimental Biology* **101**, 205–23.

Henry, R.P. & Kormanik, G.A. (1985). Carbonic anhydrase activity and calcium deposition during the molt cycle of the blue crab *Callinectes sapidus. Journal of Crustacean Biology* **5**, 234–41.

Hughes, G.M., Knights, B. & Scammel, C.A. (1969). The distribution of PO_2 and hydrostatic pressure changes within branchial chambers of the shore crab *Carcinus maenas. Journal of Experimental Biology* **51**, 203–20.

Innes, A.J. & Taylor, E.W. (1986*a*). The evolution of air-breathing in crustaceans: a functional analysis of branchial, cutaneous and pulmonary gas exchange. *Comparative Biochemistry and Physiology* **85**A, 621–37.

Innes, A.J. & Taylor, E.W. (1986*b*). August Krogh and K. for crab chitin. *Journal of Physiology* **378**, 74P.

Johansen, K., Lenfant, C. & Mecklenberg, T.A. (1970). Respiration in the crab, *Cancer magister. Zeitschrift für vergleicher Physiologie* **70**, 1–19.

Jones, J.D. (1972). *Comparative Physiology of Respiration*. Arnold, London.

Krogh, A. (1919). The rate of diffusion of gases through animal tissues with some remarks on the coefficient of invasion. *Journal of Physiology* **52**, 391–408.

McMahon, B.R. & Burggren, W.W. (1988). Respiration. In *Biology of Land Crabs* (ed. W.W. Burggren & B.R. McMahon), pp. 249–297. Cambridge University Press.

McMahon, B.R. & Wilkens, J.L. (1983). Ventilation, perfusion and oxygen uptake. In *The Biology of Crustacea* (ed. L.H. Mantel), vol. 5, pp. 290–372. Academic Press, New York.

McMahon, B.R., McDonald, D.G. & Wood, C.M. (1979). Ventilation oxygen uptake and haemolymph oxygen transport following exhausting activity in the Dungeness crab *Cancer magister. Journal of Experimental Biology* **80**, 271–85.

McMahon, B.R. & Stuart, S. (1989). The physiological problems of crayfish in acid waters. In *Acid Toxicity and Aquatic Animals* (ed. R. Morris, E.W. Taylor, D.J.A. Brown & J.A. Brown), pp. 171–200. Cambridge University Press.

Mangum, C.P. (1979). The respiratory function of haemocyanins. *American Zoologist* **20**, 29–38.

Mangum, C.P. (1983*a*). Oxygen transport in the blood. In *The Biology of Crustacea* (ed. L.H. Mantel), vol. 5, pp. 373–429. Academic Press, New York.

Mangum, C.P. (1983*b*). On the distribution of lactate sensitivity among the hemocyanins. *Marine Biology Letters* **4**, 139–49.

Mangum, C.P. & Johansen, K. (1975). The colloid osmotic pressures of invertebrate body fluids. *Journal of Experimental Biology* **63**, 661–71.

Mangum, C.P., McMahon, B.R., de Fur, P.L. & Wheatly, M.G. (1985). Gas exchange, acid-base balance and the oxygen supply to the tissues during a moult of the blue crab, *Callinectes sapidus*. *Journal of Crustacean Biology* **5**, 188–206.

Milligan, C.L., Walsh, P.J., Booth, C.E. & McDonald, D.G. (1989). Intracellular acid–base regulation during recovery from locomotor activity in the blue crab (*Callinectes sapidus*). *Physiological Zoology* **62**, 621–38.

Morris, S. & Bridges, C.R. (1985). An investigation of haemocyanin oxygen affinity in the semiterrestrial crab *Ocypode saratan* Forsk. *Journal of Experimental Biology* **117**, 119–32.

Morris, S. & McMahon, B.R. (1989). Neurohumor effects on crustacean haemocyanin oxygen affinity. *Journal of Experimental Biology* **249**, 334–7.

Morris, S., Tyler-Jones, R. & Taylor, E.W. (1986*a*). The regulation of haemocyanin oxygen affinity during emersion of the crayfish *Austropotamobius pallipes*. I. An *in vitro* investigation of the interactive effects of calcium and L-lactate on oxygen affinity. *Journal of Experimental Biology* **121**, 315–26.

Morris, S., Tyler-Jones, R., Bridges, C.R. & Taylor, E.W. (1986*b*). The regulation of haemocyanin oxygen-affinity during emersion of the crayfish *Austropotamobius pallipes* II. An investigation of *in vivo* changes in oxygen affinity. *Journal of Experimental Biology* **121**, 327–37.

Rahn, H. (1966). Aquatic gas exchange: Theory. *Respiration Physiology* **1**, 1–12.

Randall, D.J. (1982). The control of respiration and circulation in fish during exercise and hypoxia. *Journal of Experimental Biology* **100**, 275–88.

Randall, D.J. & Wood, C.M. (1981). Carbon dioxide excretion in the land crab (*Cardisoma carnifex*). *Journal of Experimental Zoology* **218**, 37–44.

Redmond, J.R. (1985). The respiratory function of haemocyanin in Crustacea. *Journal of Cellular and Comparative Physiology* **46**, 209–47.

Skinner, D.M. (1985). Molting and regeneration. In *The Biology of*

Crustacea (ed. D.E. Bliss & L.H. Mantel) vol. 9, pp. 44–146. Academic Press, New York.

Taylor, E.W. (1982). Control and coordination of ventilation and circulation in crustaceans: responses to hypoxia and exercise. *Journal of Experimental Biology* **100**, 289–319.

Taylor, E.W. & Butler, P.J. (1973). The behavioural and physiological responses of the shore crab *Carcinus maenas* during changes in enironmental oxygen tension. *Netherlands Journal of Sea Research* **7**, 496–505.

Taylor, E.W. & Butler, P.J. (1978). Aquatic and aerial respiration in the shore crab, *Carcinus maenas* (L.) acclimated to 15 °C. *Journal of Comparative Physiology* **127**, 315–23.

Taylor, E.W., Butler, P.J. & Sherlock, P.J. (1973). The respiratory and cardiovascular changes associated with the emersion response of *Carcinus maenas* (L.) during environmental hypoxia, at three different temperatures. *Journal of Comparative Physiology* **86**, 95–115.

Taylor, E.W. & Innes, A.J. (1988). A functional analysis of the shift from gill- to lung-breathing during the evolution of land crabs (Crustacea. Decapoda). *Biological Journal of the Linnaean Society* **33**, 229–47.

Taylor, E.W., Tyler-Jones, R. & Wheatly, M.G. (1987). The effects of aerial exposure on the distribution of body water and ions in the freshwater crayfish *Austropotamobius pallipes* (Lereboullet). *Journal of Experimental Biology* **128**, 307–22.

Taylor, E.W. & Wheatly, M.G. (1979). The behaviour and respiratory physiology of the shore crab, *Carcinus maenas* (L.) at moderately high temperatures. *Journal of Comparative Physiology* **130**, 309–16.

Taylor, E.W. & Wheatly, M.G. (1980). Ventilation, heart rate and respiratory gas exchange in the crayfish *Austropotamobius pallipes* (Lereboullet) submerged in normoxic water and following 3 h exposure in air at 15 °C. *Journal of Comparative Physiology* **138**, 67–78.

Taylor, E.W. & Wheatly, M.G. (1981). The effect of long-term aerial exposure on heart rate, ventilation, respiratory gas exchange and acid-base status in the crayfish *Austropotamobius pallipes*. *Journal of Experimental Biology* **92**, 109–24.

Taylor, E.W. & Whiteley, N.M. (1989). Oxygen transport and acid-base balance in the haemolymph of the lobster, *Homarus gammarus*, during aerial exposure and resubmersion. *Journal of Experimental Biology* **144**, 417–36.

Taylor, H.H. & Greenway, P. (1984). The role of the gills and branchiostegites in gas exchange in a bimodally breathing crab, *Holthuisana transversa*: evidence for a facultative change in the distribution of the respiratory circulation. *Journal of Experimental Biology* **111**, 103–21.

Taylor, H.H. & Taylor, E.W. (1986). Observations of valve-like structures and evidence for rectification of flow within the gill lamellae of the crab *Carcinus maenas* (Crustacca, Decapoda). *Zoomorphology* **106**, 1–11.

Thomas, R.C. (1989). Intracellular pH regulation and the effects of external acidification. In *Acid Toxicity and Aquatic Animals* (ed. R. Morris, E.W. Taylor, D.J.A. Brown & J.A. Brown), pp. 113–24. SEB Seminar Series No. 34. Cambridge University Press.

Towle, D.W. & Mangum, C.P. (1985). Ionic regulation and transport ATPase activities during the molt cycle in the blue crab *Callinectes sapidus*. *Journal of Crustacean Biology* **5**, 216–22.

Truchot, J.-P. (1978). Variation de la concentration sanguine d'hémocyanine fonctionelle au cours du cycle d'intermue chez le crabe *Carcinus maenas* (L.). *Archives de Zoologie Experimentale et Generale* **119**, 265–82.

Truchot, J.-P. (1980). Lactate increases the oxygen affinity of crab haemocyanin. *Journal of Experimental Zoology* **214**, 205–8.

Truchot, J.-P. (1983). Regulation of acid-base balance. In *The Biology of Crustacea* (cd. L.H. Mantel), vol. 5, pp. 431–57. Academic Press, New York.

Truchot, J.-P. (1986). Changes in the hemolymph acid-base state of the shore crab *Carcinus maenas* exposed to simulated tide-pool conditions. *Biological Bulletin* **170**, 506–18.

Tyler-Jones, R. & Taylor, E.W. (1988). Analysis of haemolymph and muscle acid-base status during aerial exposure in the crayfish *Austropotamobius pallipes*. *Journal of Experimental Biology* **134**, 409–22.

Wheatly, M.G. (1979). Bimodal respiration in two species of decapod crustaceans. M.Sc. thesis, University of Birmingham.

Wheatly, M.G. (1980). The problems of respiratory gas exchange and acid-base balance in water and air in two amphibious decapodan crustaceans. Ph.D. thesis, University of Birmingham.

Wheatly, M.G. (1985). Free amino acid and inorganic ion regulation in the whole muscle and hemolymph of the blue crab *Callinectes sapidus* Rathbun in relation to the molting cycle. *Journal of Crustacean Biology* **5**, 223–33.

Wheatly, M.G. (1987). Physiological responses of the rock crab *Cancer irroratus* (Say) to environmental hyperoxia. I. Acid-base regulation. *Physiological Zoology* **60**, 398–405.

Wheatly, M.G. (1989). Physiological responses of the crayfish *Pacifastacus leniusculus* to environmental hyperoxia. I. Extracellular acid-base and electrolyte status and transbranchial exchange. *Journal of Experimental Biology* **143**, 33–51.

Wheatly, M.G. & Henry, R.P. (1987). Branchial and antennal gland Na^+/K^+ dependent ATPase and carbonic anhydrase activity during salinity acclimation of the euryhaline crayfish *Pacifastacus leniusculus*. *Journal of Experimental Biology* **133**, 73–86.

Wheatly, M.G., Morrison, R., Yow, L.C. and Toop, T. (1990). The mechanisms of acid-base and ionoregulation in the freshwater crayfish during environmental hyperoxia and recovery III. Intracellular acid-base balance. *Physiological zoology* (in press).

Wheatly, M.G. & Taylor, E.W. (1979). Oxygen levels, acid-base status and heart rate during emersion of the shore crab *Carcinus maenas* (L.) into air. *Journal of Comparative Physiology* **132**, 305–11.

Wheatly, M.G. & Taylor, E.W. (1981). The effect of progressive hypoxia on heart rates, ventilation, respiratory gas exchange and acid-base status in the crayfish *Austropotamobius pallipes*. *Journal of Experimental Biology* **92**, 125–41.

Wheatly, M.G. & Toop, T. (1989). Physiological responses of the crayfish *Pacifastacus leniusculus* to environmental hyperoxia. II. Role of the antennal gland in acid-base and ion regulation. *Journal of Experimental Biology* **143**, 53–70.

Whiteley, N.M. & Taylor, E.W. (1990). The acid-base consequences of aerial exposure in the lobster, *Homarus gammarus* (L.) at 10 and 20 °C. *Journal of Thermal Biology* **15**(1), 47–56.

Whiteley, N.M., Innes, A.J., Al-Wassia, A.H. & Taylor, E.W. (1990). Aerial and aquatic respiration in the ghost crab, *Ocypode saratan*. II. Respiratory gas exchange and transport in the haemolymph. *Marine Behaviour and Physiology* **16**, 261–74.

Wood, C.M. & Boutilier, R.G. (1985). Osmoregulation, ionic exchange, blood chemistry, and nitrogenous waste excretion in the land crab *Cardisoma carnifex*: a field and laboratory study. *Biological Bulletin* **169**, 267–90.

Wood, C.M., Boutilier, R.G. & Randall, D.J. (1986). The physiology of dehydration stress in the land crab, *Cardisoma carnifex*: respiration, ionoregulation. *Journal of Experimental Biology* **126**, 271–96.

Wood, C.M. & Cameron, J.N. (1985). Temperature and the physiology of intracellular and extracellular acid-base regulation in the blue crab *Callinectes sapidus*. *Journal of Experimental Biology* **114**, 151–79.

Wood, C.M. & Randall, D.J. (1981). Haemolymph gas transport, acid-base regulation and anaerobic metabolism during exercise in the land crab (*Cardisoma carnifex*). *Journal of Experimental Biology* **218**, 23–36.

Wood, C.M. & Rogano, M.S. (1986). Physiological responses to acid stress in crayfish (*Orconectes propinquus*): haemolymph ions, acid-base status, and exchanges with the environment. *Canadian Journal of Fisheries and Aquatic Science* **43**, 1017–26.

STEPHEN C. WOOD & MOGENS L. GLASS

Respiration and thermoregulation of amphibians and reptiles

Introduction

This paper highlights some physiological and behavioural factors controlling gas transport and temperature regulation of amphibians and reptiles. The topics covered in gas transport are the physiological mechanisms controlling the steps of the 'O_2 cascade', i.e.

(1) ventilation and its control;

(2) diffusion across the air-blood barrier;

(3) circulation of blood;

(4) oxygen delivery to tissues.

Topics covered in temperature regulation deal with behavioural mechanisms to conserve oxygen during hypoxic or hypercapnic stress.

Ventilation and its control

The control of breathing is aimed at minimizing work and optimizing stability of blood gases and pH. Chemical control requires ventilatory responses to altered O_2 and acid–base status of blood and cerebrospinal fluid. Most of the concepts of mammalian respiratory control systems apply as well to amphibians and reptiles. A major exception is sinusoidal breathing. This regular breathing pattern is characteristic of mammals. In contrast, most amphibians and reptiles exhibit periodic breathing with ventilatory periods interrupted by periods of apnoea. This breathing pattern provides a new variable for assessing the level of ventilation, i.e. the duration of the non-ventilatory period.

Ventilatory response to hypoxia

Amphibians and reptiles typically show a ventilatory response to hypoxia. Progressive hypoxia markedly influences breathing activity in the toad *Bufo marinus* (Boutilier & Toews, 1977). When this species becomes hypoxic at 22 °C, buccal cycles are less frequent but the frequency of pulmonary ventilation increases.

The effect of hypoxia on ventilation of amphibians and reptiles is temperature-dependent. When *B. paracnemis* is exposed to hypoxia at 25 and 32 °C there is a pronounced increase in the number of breathing episodes per unit time, augmenting inspired ventilation. In addition, inflation and deflation sequences become dominant components of the breathing patterns. In contrast, these responses are depressed by decreasing temperature and are absent at 15 °C (Kruhøffer, *et al.*, 1987).

A depression of ventilatory responses to hypoxia at low temperatures was established by Jackson (1973) for the turtle *Pseudemys scripta*. He suggested that because turtles have a large extraction of O_2 at high temperature they cannot further increase extraction during hypoxia. Consequently, O_2 uptake must be accomplished by means of increased ventilation. Conversely, he suggested that the low O_2 extraction at decreased temperature allows enhanced extraction as an alternative to increased ventilation. However, pulmonary O_2 extraction should probably be viewed as a dependent variable of ventilation: for example, for a given O_2 uptake, extraction will increase as ventilation decreases.

The high O_2 affinity of the blood at low temperature will also assist in maintaining O_2 transport in spite of a decreased alveolar and arterial P_{O_2}. For example, the turtle *Chrysemys picta bellii* does not increase ventilation relative to normoxic values at 10 °C until arterial P_{O_2} has fallen to only 5 Torr. At 30 °C, however, a ventilatory response occurs with a decrease to 30 Torr from a normoxic value of 60 Torr. This shift in the threshold for the ventilatory response correlated with an increase of P_{50} for the O_2 dissociation curve (ODC) from 5 Torr (at 10 °C) to 28 Torr (at 30 °C) (Glass *et al.*, 1983).

The correlation of ventilatory threshold with P_{50} is suggestive that arterial saturation rather than Pa_{O_2} is the regulated variable (see Wood & Hicks, 1985). This certainly applies from a functional point of view. However, it is not necessary that O_2 receptors be monitoring saturation. For example, the output of the carotid body in the cat is depressed by decrease of temperature with increasingly severe hypoxic conditions required to cause an elevated impulse frequency. Nevertheless, the carotid body is considered a P_{O_2} receptor (Lahiri & Gelfand, 1981). Recent studies have indicated that chemoreceptors are located in the aortic and pulmonary arches and the truncal region of turtles and some of these are O_2 receptors (Ishii *et al.*, 1985*a*). The specific stimulus (P_{O_2}, S_{O_2}, $[O_2]$) has not been determined. O_2 receptors are located in the aortic arch of the toad *Bufo vulgaris*. In addition, most amphibians possess an organ, the carotid labyrinth, that may also function in chemoreception (Ishii *et al.*, 1985*b*).

Acid–base regulation

The literature on acid–base regulation in amphibians and reptiles has been substantially influenced by the concepts of constant relative alkalinity (Rahn, 1967) and alphastat regulation (see Reeves, 1972). These concepts were intended to explain the significance of the inverse relationship between arterial pH and body temperature of ethothermic vertebrates. Reeves (1972) proposed that 'alphastat'-receptors are involved in adjustments of ventilation relative to pulmonary CO_2 output. The decrease of arterial pH required for alphastat regulation is approximately -0.019 units K^{-1} (range 5–20 °C). However, significantly smaller changes are common and the average temperature coefficient for 70 heterothermic species is $\Delta pH/\Delta T = -0.011$ units K^{-1} (Heisler, 1986).

Although the fundamental mechanisms underlying acid–base regulation in ectothermic vertebrates remain uncertain, substantial progress has been made in describing the outputs of the control system. It is well established that both amphibians and reptiles primarily defend their acid–base status by respiratory compensation in response to hypercapnia. In contrast, renal compensation of hypercapnia by means of actively increased plasma $[HCO_3^-]$ is small or absent in these groups (Boutilier *et al.*, 1979; Silver & Jackson, 1986; Glass & Heisler, 1986).

The receptor systems involved in responses to hypercapnia have been studied to some extent in reptiles. Hitzig & Jackson (1978) and Hitzig (1982) report that $[H^+]$ receptors are screening the cerebrospinal fluid in the red-eared turtle, *Pseudemys scripta*, and that information from these receptors overrides the peripheral receptor input to the central nervous system. The study by Hitzig (1982) required experimental conditions that may have elevated control values for ventilation, in particular at low temperature (see Glass *et al.*, 1985). However, an involvement of central receptors responsive to hypercapnia is supported by a study in which *Pseudemys* was exposed to a 30 minute period of anoxia (Davies & Sexton, 1987). Anoxia caused an acidosis from which the arterial blood recovered before the acid–base status of the extracellular cerebral fluid (ECF) returned to normal. An elevated ventilation was maintained until $[H^+]$ of the ECF decreased to control values.

The existence of peripheral receptors responsive to hypercapnia has also been established in reptiles. In fact, pulmonary or upper airway receptors are the only documented CO_2 receptors in lizards (see Ballam, 1985). The ventilatory responses to hypercapnia of some lizards and snakes are paradoxical. For example, inspiration of CO_2 causes a larger tidal volume but also increases the duration between breaths (see Ballam, 1985). This response seems inefficient since elimination of CO_2 is certainly not promoted by a decreased breathing frequency. As pointed out by Glass & Wood (1983), this could indicate a different effect of environmental hypercapnia on ventilation from that of metabolically produced acidosis. This seems confirmed for the tegu lizard, *Tupinambis nigropunctatus*, in which an elevated CO_2 concentration in the mouth and nasal cavities causes an increased breath-hold duration leading to 50% reduction in ventilation. Conversely, the loading of CO_2 into the blood via the gut increases breathing frequency and tidal volume (see Ballam, 1985). Cutaneous and nasal mucosa receptors in the bullfrog are also CO_2-sensitive and involved in depression of ventilation. These receptors are stimulated by a 0.5% increase in environmental CO_2 (Sakakibara, 1978).

The recent studies on control of breathing in amphibians and reptiles clearly indicate that the regulatory systems are highly complicated and advanced rather than primitive. Further work is needed to identify and quantify receptor systems and their roles.

Pulmonary and cutaneous gas exchange

Pulmonary diffusing capacity

The diffusing capacity of the lung (D_L) has been measured in a number of reptiles and amphibians (Glass *et al.*, 1981*a,b*; Gatz *et al.*, 1987; Lutcavage *et al.*, 1987). The values of D_L are similar for reptilian and amphibian lungs but considerably lower than the D_L measured in similar-sized birds or mammals. The low arterial P_{O_2} characteristic of amphibians and reptiles has been attributed to their relatively 'primitive' lungs and low diffusing capacity. However, recent studies have established that the D_L of reptilian and amphibian lungs is well matched to their relatively low O_2 uptake. The large O_2 gradients between alveolar gas and arterial blood that are common in amphibians and reptiles are due primarily to venous admixture (right-to-left shunt) and not to a high resistance to pulmonary gas transfer (see Glass & Wood, 1983).

With increasing body temperature, a matching between D_L and O_2 uptake is also evident, requiring a Q_{10} of about 2 for D_L. This is higher than the expected Q_{10} of 1.1 for diffusive gas transfer, indicating that capillary recruitment and/or reduction in functional inhomogeneities may account for part of the measured increases in D_L with rising temperature (see Glass, 1988).

Diffusion and perfusion in cutaneous gas exchange

Cutaneous gas exchange is important in amphibians and its regulation is an active area of research. Some of the most interesting studies have used lungless salamanders, an excellent example of the August Krogh Principle (Krogh, 1929), i.e. 'For a large number of problems there will be some animals of choice or a few such animals on which it can most conveniently be studied.' Piiper *et al.* (1976) showed that O_2 uptake and CO_2 output of lungless salamanders are both limited primarily by diffusion. Having no lungs means no ability to vary the rate of ventilation when O_2 demand increases. At first glance, this would seem to place a lungless salamander in the same category as a chicken egg in terms of exercise capacity. However, these salamanders are sometimes active and can increase their O_2 uptake several-fold during exercise (Feder, 1985; Full, 1985). This implies a several-fold increase in the gas conductance across the skin (see below).

Control of cutaneous gas exchange

One factor that controls cutaneous gas exchange is distribution of blood flow. In the bullfrog, *Rana catesbeiana*, 20% of total O_2 uptake and 80% of CO_2 output may occur cutaneously (Burggren & Moalli, 1984). As in other anurans, the pulmocutaneous artery branches to perfuse both the lungs and part of the skin. In aerated water the skin receives 20% of total pulmocutaneous blood flow but when the water is hypoxic, skin flow is reduced to almost nothing. Conversely, a fall in lung P_{O_2} increases skin perfusion (Boutilier *et al.*, 1986).

Regulation of the microcirculation also controls cutaneous gas exchange. Burggren & Moalli (1984) showed that the number of perfused capillaries decreases during periods of air exposure and increases upon return to water, a 'capillary recruitment,' which would augment the gas conductance across the skin. Other studies of frog skin have also shown that the number of perfused capillaries is regulated (Burggren & Moalli, 1984; Malvin & Hlastala, 1986*b*).

Variation in the number of perfused capillaries would alter skin diffusing capacity *if* (and this is where some discrepancy occurs) this parameter includes capillary blood volume, i.e. if the same model that is applied to pulmonary diffusing capacity (membrane diffusing capacity, reaction time with haemoglobin, pulmonary capillary volume) is applied to the skin.

However, most studies do not sort out the relative roles of membrane diffusing capacity, *per se*, and total diffusing capacity including capillary blood volume.

Circulation and blood gas transport

Control of lung and gill perfusion
Amphibians are truly virtuosi of respiration. Some exchange gases by using only their skin (lungless salamanders), others use skin and lungs (adult anurans and urodeles), while larval forms use gills and skin or gills, skin, and lungs (paedomorphic larvae). These multiple breathing modes are supported by rather complex circulations to the gas exchange sites.

How is blood flow to the different gas exchange organs controlled? Major controlling factors are ventilatory state and ambient O_2 availability. As discussed above, most species are periodic air-breathers. Periods of lung ventilation are usually accompanied by a fall in pulmonary vascular resistance so that ventilation/perfusion is optimized (Johansen *et al.*, 1970). Hypoxia, either aquatic or aerial, increases the frequency of air breathing in most amphibians. Hypoxia also affects the distribution of blood flow to gills, skin, and lung (reviewed by Shelton *et al.*, 1986). In tri-modal breathers, the circulation to the gills is both in series and parallel with the lungs and is controlled by water P_{O_2} (Malvin & Heisler, 1988).

Lung perfusion in amphibians and reptiles is determined by cardiac output, the magnitude and direction of intracardiac shunting, and autonomic reflexes. Although there is current debate about the mechanisms of intracardiac shunting (see Ishimatsu *et al.*, 1988), the magnitude and direction of shunts in most species is generally thought to be determined by the resistance of the outflow circuits (see Johansen, 1979; Shelton, 1976, 1985). An important factor controlling resistance of the pulmonary circulation is a vasoactive segment of the pulmonary artery that is under vagal control with tonus linked to the respiratory cycle (Arthaud & Butte, 1890). In addition to peripheral controls (changes in pulmonary and systemic resistances), intracardiac factors influence the degree of distribution of cardiac output (see Johansen, 1979; Ishimatsu *et al.*, 1988 for review and references).

Impact of shunts on blood gases
The impact of cardiovascular shunts on blood gases is an important consideration in three broad areas: (1) pathophysiology of congenital heart disease (Rosoff *et al.*, 1980); (2) mismatch of ventilation and perfusion (Turek & Kreuzer, 1981); and (3) normal physiology of amphibians and reptiles. Amphibians and reptiles can experience hypoxia in three ways: (1)

A low arterial O_2 saturation due to external hypoxia (burrows, high altitude); (2) A low arterial O_2 content, but normal saturation, due to anaemia; and (3) A low arterial saturation due to anatomical or physiological shunts. For amphibians and reptiles the last cause is prevalent and creates a situation, unique to ectothermic vertebrates, in which the P_{O_2} of arterial blood is not directly related to gas exchange or ambient P_{O_2}. Instead, Pa_{O_2} becomes a dependent variable of O_2 saturation and O_2 affinity of blood. This variation of arterial P_{O_2} can be pronounced since the normal fluctuations of body temperature and arterial pH act in concert to alter haemoglobin–oxygen affinity.

As shown in Figure 1, the P_{O_2} difference between lung gas and arterial blood can be enormous at low temperature. For *Pseudemys scripta* at 20 °C this difference is ≈ 100 mmHg. As temperature increases, the difference narrows reaching a value of ≈ 10 mmHg at 40 °C. Lung P_{O_2} declines markedly with rising temperature in *Pseudemys* because this species is one that shows alphastat pH control and this requires that ventilation does not increase as much as O_2 or CO_2 output. Therefore, lung P_{CO_2} rises and lung P_{O_2} falls. Arterial P_{O_2} increases with rising temperature as the oxygen dissociation curve shifts to the right. Arterial O_2 content (Ca_{O_2}) is a function of the shunt fraction and the O_2 contents of venous and pulmonary blood. The increase in Pa_{O_2} with increasing P_{50} reaches a 'breaking point' when the curve becomes so far right shifted that the saturation of pulmonary blood drops sharply. After the breaking point is reached, further decreases in O_2 affinity produce a decrease in Pa_{O_2} (Wood *et al.*, 1987*b*). In comparing *Pseudemys scripta* with *Varanus exanthematicus* (Figure 1) it is clear that temperature at which the 'breaking point' occurs depends on the temperature sensitivity of Hb, the magnitude of the Bohr effect and the degree of R–L shunting. *Varanus exanthematicus* blood is less affected by temperature because its temperature sensitivity is less than that of *Pseudemys scripta* and, more importantly, because its ΔpH$/\Delta T$ is only -0.005, one third that of *Pseudemys*. In general, the breaking point is at or near the normal preferred body temperature, T_b. These theoretical models led to the hypothesis that the normal selected T_b of amphibians and reptiles would be reduced during hypoxia.

Behavioural thermoregulation

Effects of hypoxia

In contrast to homoiotherms, most ectothermic vertebrates tolerate wide fluctuations of T_b. Although tolerant of varying T_b, most species use both behavioural and physiological means to sustain a 'preferred' or

     S.C. Wood and M.L. Glass

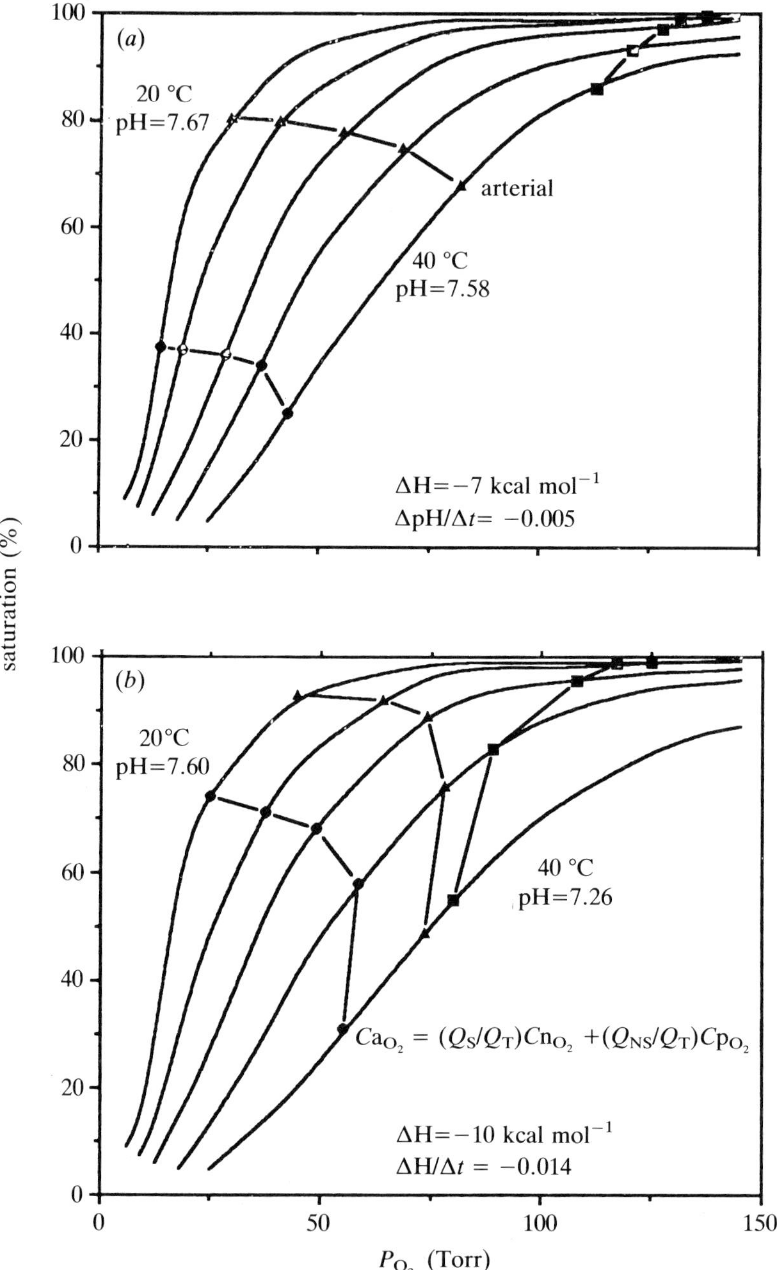

Figure 1. Pulmonary gas and blood P_{O_2} (squares), systemic arterial P_{O_2} (triangles) and mixed venous P_{O_2} (circles) of the lizard *Varanus exan-thematicus* (*a*) and the turtle *Pseudemys scripta* (*b*) as a function of body

'mean selected' T_b. For many species, this selected T_b is surprisingly high, often higher than 37°C. This fact, coupled with the relatively low O_2 affinity of amphibian and reptile blood, led to a hypothesis that hypoxia would elicit a behavioural 'hypothermia' in ectothermic animals (Wood, 1984). This was first confirmed for lizards exposed in a thermal gradient to 7–10% O_2 (Hicks & Wood, 1985) and later for a number of aquatic and terrestrial ectotherms (Wood *et al.*, 1987*b*).

Metabolic depression

The wide variety of responses and adaptations to hypoxia among vertebrates reflects an extremely varied dependence on aerobic metabolism. The effect of rapid exposure to hypoxia (e.g. 8% inspired O_2) ranges from virtually no response in hypoxia-tolerant species (e.g. pond turtles) to stress response and rapid death in hypoxia-intolerant species (e.g. most mammals). With a graded exposure to hypoxia, the adaptations of cardio-pulmonary function that occur are geared to improving the acquisition and distribution of O_2.

Active responses to hypoxia, increased ventilation and cardiac output, are themselves O_2-dependent and therefore potentially deleterious. Alternatively, biochemical depression of metabolism is a potent adaptation to hypoxia (reviewed by Hochachka, 1988; Storey, 1988). Metabolic depression has been documented in invertebrates and vertebrates as a mechanism to survive low O_2 or other harsh conditions. The energetic savings can be impressive. For example, a hibernating ground squirrel realizes an 88% energy saving due to its hypothermia-hypometabolism (Wang, 1978). Similar savings are realized by aestivating lungfish (Lahiri *et al.*, 1970). However, as Storey (1988) points out, this strategy of adaptation to hypoxia is dangerous if sensitive control mechanisms are absent. Mammals, and other O_2 regulators, are generally unable to maintain normal electrochemical gradients and experience a massive stress response when rendered hypothermic (also see Hochachka, 1988). On the other hand, metabolic depression due to behavioural hypothermia has been shown to be reversible in all cases (Wood *et al.*, 1987*b*).

Caption for fig. 1. (cont.).
temperature and pH. (Adapted from Wood, 1984.) The equation in (*b*) applies to any case of venous admixture; i.e. in the presence of a R→L shunt arterial O_2 content (Ca_{O_2}) is a function of the shunt fraction (Q_S/Q_1), mixed venous O_2 content (Cv_{O_2}), and pulmonary venous O_2 content (Cv_{O_2}), and pulmonary venous O_2 content (Cp_{O_2}). Q_{NS}/Q_T is the non-shunt fraction. See text for discussion.

Hypercapnia-induced hypothermia

Physiological thermoregulation of homoiotherms becomes compromised under both hypoxic and hypercapnic conditions. Hypoxia causes increased heat loss; hypercapnia causes a reduced heat production and increased heat loss from sweating and inhibition of shivering (Schaefer *et al.*, 1975).

The established parallel effects of hypoxia on T_b of homoiotherms and ectotherms and the established effect of hypercapnia on T_b of homoiotherms prompted a hypothesis that behavioural thermoregulation would be affected by hypercapnia as well as by hypoxia. A recent study showing behavioral hypothermia in hypercapnic toads supports this hypothesis. Exposure of *Bufo marinus* to 10% CO_2 elicited a significant behavioural reduction of selected T_b whereas 5% had no effect and adding 40% O_2 to the 10% inspired CO_2 reversed the hypothermia (Riedel & Wood, 1988).

Mechanisms of altered behavioural thermoregulation

What triggers the behavioural hypothermia in response to hypoxia and hypercapnia? The answer has proven elusive. The original observations with lizards employed simulated high altitude (low inspired O_2), a condition that reduces both arterial O_2 content and P_{O_2}. Subsequent studies with both lizards (Hicks & Wood, 1985) and toads (Wood *et al.*, 1989) showed that induced anaemia, where arterial O_2 content is reduced but P_{O_2} is normal, elicits behavioural hypothermia. The mechanism for hypercapnia-induced hypothermia is also elusive. Obviously, for ectotherms the mechanism must be independent of sweating or shivering.

A common denominator of hypoxia, anaemia, and hypercapnia could be reduced arterial oxygen. As shown in Figure 2, lizards show no behavioural response to low inspired O_2 until a threshold of $\approx 10\%$ is reached. The shape of these temperature–P_{O_2} curves parallels the oxygen dissociation curve, i.e. there is little drop in arterial saturation until a 'threshold' P_{O_2} is reached. For anaemia (Figure 3), there is a significant behavioural hypothermia with no decrease in arterial P_{O_2}, only a drop in arterial $[O_2]$. It is possible that hypercapnia, by right-shifting the oxyhaemoglobin dissociation curve, would prevent full saturation of blood in the lungs. Hypercapnia is also known to increase pulmonary vascular resistance in some species, including *Bufo marinus* (Smith, 1978), thereby increasing right to left shunt. Both factors could promote low arterial $[O_2]$ and trigger behavioural hypothermia. The study of *Bufo marinus* mentioned above (Riedel & Wood, 1988) tested this hypothesis. As expected from previous studies, *Bufo marinus* exhibits behavioural hypothermia in response to

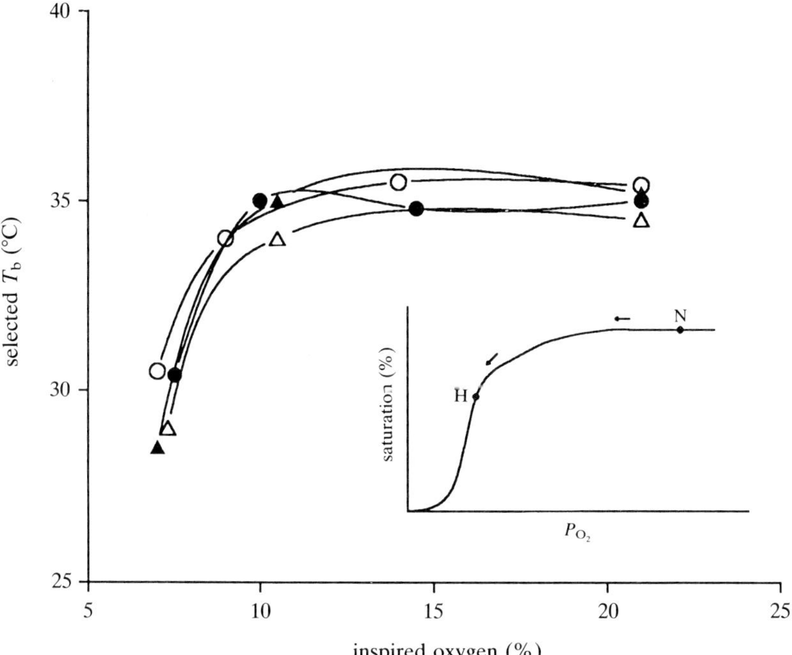

Figure 2. Mean selected body temperature (T_b) of four species of lizard in a temperature gradient chamber to graded reductions of inspired O_2. (Data from Hicks & Wood, 1985.) The inset shows a generic oxyhaemoglobin dissociation curve with approximate values for arterial saturation and P_{O_2} under normoxic (N) and hypoxic (H) conditions.

10% inspired O_2. Figure 4 shows the arterial P_{O_2} of this species at 25 °C breathing various gas mixtures. Behavioural hypothermia is not elicited by 15% inspired O_2 with an arterial P_{O_2} of 64 mmHg and satuaration of 72% (open circle on the dissociation curve) but is elicited by 10% O_2 with an arterial P_{O_2} of 50 mmHg and saturation of 55% (closed circle). *Bufo marinus* also exhibited behavioural hypothermia in response to inspired CO_2. Interestingly, 15% O_2 alone and 5% CO_2 alone did not induce hypothermia but, when combined, they did with an arterial P_{O_2} of 75 mmHg and saturation of 67%. Furthermore, 10% CO_2 combined with 21% O_2 elicited behavioural hypothermia with an arterial P_{O_2} of 120 mmHg and saturation of 80%. These results ruled out arterial oxygen, *per se*, as the factor eliciting the behavioural hypothermia.

The possibility that reduced blood pH triggers the behaviour was tested by combining 10% CO_2 with 40% O_2 in the inspired gas. Under these conditions, arterial blood had a P_{O_2} of 170 mmHg and saturation of 95%.

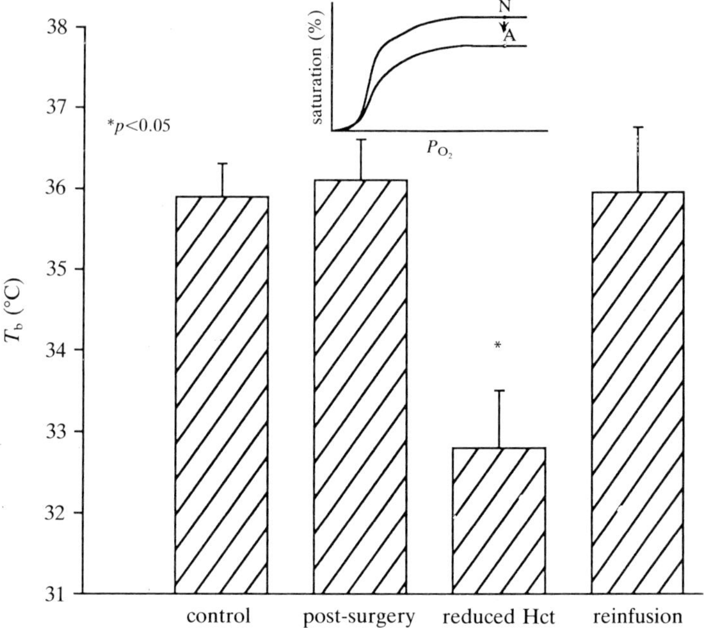

Figure 3. Effect of induced anaemia (blood withdrawn and replaced with saline) on selected body temperature (T_b) of the iguana, *Iguana iguana*, in a temperature gradient. (Adapted from Hicks & Wood, 1985.) Inset shows a generic oxyhaemoglobin dissociation curve with approximate values for arterial saturation and P_{O_2} under normoxic (N) and anaemic (A) conditions. Hct, haematocrit.

No behavioural hypothermia was elicited and the arterial pH was reduced to the same level (7.19) as during 10% CO_2+21% O_2.

In Figure 4, the threshold of arterial oxygen at which behavioural hypothermia is elicited lies somewhere between the open and closed symbols on each curve. If arterial oxygen is triggering the behavioural hypothermia, then the threshold for this response is increased as oxygen affinity of blood is reduced. Current studies are aimed at resolving this issue.

Physiological significance of behavioural hypothermia

Behavioural hypothermia (reducing T_b) is a physiologically significant and adaptive response to hypoxia in four ways. (1) The demand for O_2 is reduced about 11% per K (Q_{10} effect). (2) The O_2 dissociation curve is shifted to the left, providing increased saturation of pulmonary capillary blood. (3) The ventilatory response to hypoxia is reduced or avoided. (4)

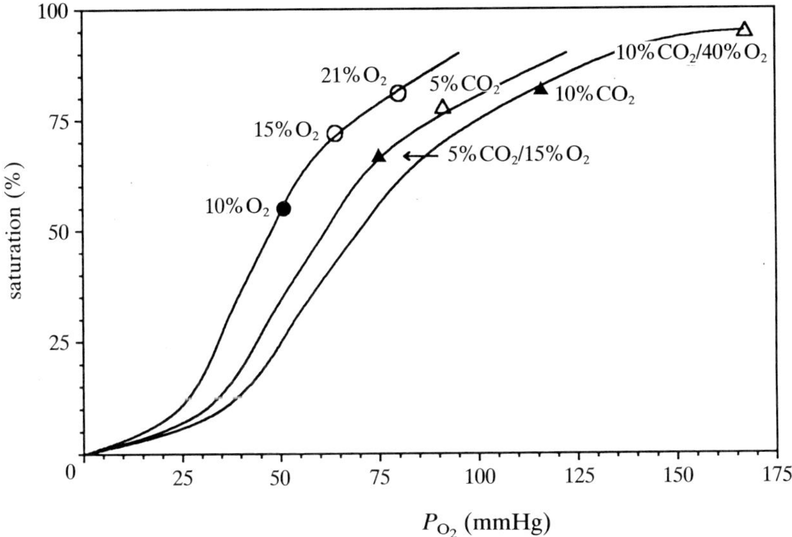

Figure 4. Arterial blood gases and pH of the toad *Bufo marinus*, at 25 °C, during normoxia and during inhalation of 15% O_2, 10% O_2, 5% CO_2, 15% O_2+5% CO_2, 10% CO_2 and 10% CO_2+40% O_2. Open symbols indicate blood gases and pH that did not elicit, and closed symbols blood gases that did elicit, behavioural hypothermia of toads in temperature gradient experiments. (Data from Riedel & Wood, 1988.)

The critical P_{O_2} is reduced and therefore the danger of hypoxic damage to tissues is reduced. The last way deserves additional comment. The P_{O_2} value at which oxidative metabolism cannot be maintained is the 'critical P_{O_2}'. Among ectotherms the critical P_{O_2} may be directly correlated with the O_2 affinity of blood for both inter- and intraspecies comparisons (Hall, 1966; Wood & Hicks, 1985). Therefore, the reduction in arterial P_{O_2} that occurs with lowered T_b in ectotherms is not deleterious as long as Pa_{O_2} remains higher than critical P_{O_2}. Furthermore, the reduction P_{O_2} during hypothermia is consistent with Fick's Law of tissue O_2 uptake by passive diffusion:

$$\dot{V}_{O_2}=Dx(Pcap_{O_2}-Pcell_{O_2}),$$

where lower $\dot{V}_{O_2}$ during hypothermia (Q_{10} effect) can be met by a lower capillary P_{O_2}. A reduction in critical P_{O_2} with a lowering of T_b has been reported for fish (Fry & Hart, 1948), turtles (Glass *et al.*, 1983; White & Somero, 1982); dogs (Warley & Gutierrez, 1984) and pigs (Wilford *et al.*, 1986).

Summary

Amphibians and reptiles share a feature that dominates their lives: a body temperature that is largely uncontrollable by physiological means but is often maintained within a relatively narrow range by behaviour. This feature restricts their distribution to certain habitats, latitudes, and altitudes and becomes an important aspect of control mechanisms for blood pH, ventilation, circulation, and metabolism.

At their preferred body temperature, amphibians and reptiles are usually far more susceptible to hypoxia than are homeotherms because their blood has a relatively lower affinity. By having shunts and a right-shifted O_2 dissociation curve, most species do not have the hypoxic reserve of the upper plateau of the O_2 dissociation curve that allows mammals and birds to tolerate moderate hypoxia with minimum arterial desaturation. Amphibians and reptiles cope with this problem in many ways, including a high capacity and tolerance for anaerobic metabolism. Amphibians and reptiles can also cope with hypoxic stress by taking advantage of their ectothermy to utilize behavioral hypothermia and thereby reduce oxygen demand.

Research supported by grants from the Flinn Foundation, Phoenix, Arizona and NATO (S.C.W.) and the Von Humboldt-Stiftung (FRG) and the Carlsberg Foundation (DK) (M.L.G.)

References

Arthaud, G. & Butte, M. (1890). Action vasomotrice du nerf pneumogastrique sur le poumon. *Comptes rendus Société Biologie, Paris* **42**, 12–13.

Ballam, G.O. (1985). Breathing response of the tegu lizard to 1–4% CO_2 in the mouth or nose or inspired into the lung. *Respiration Physiology* **62**, 375–86.

Boutilier, R.G., Glass, M.L. & Heisler, N. (1986). The relative distribution of pulmocutaneous blood flow in *Rana catesbeiana*: effects of pulmonary or cutaneous hypoxia. *Journal of Experimental Biology* **126**, 33–9.

Boutilier, R.G., Glass, M.L. & Heisler, N. (1987). Blood gases, and extracellular/intracellular acid-base status as a function of temperature in the anuran amphibians *Xenopus laevis* and *Bufo marinus*. *Journal of Experimental Biology* **130**, 13–25.

Boutilier, R.G., Randall, D.J., Shelton, G. & Toews, D.P. (1979). Acid-base relationships in the blood of the toad *Bufo marinus*. I. The effects of environmental CO_2. *Journal of Experimental Biology* **82**, 331–44.

Boutilier, R.G. & Toews, D.P. (1977). The effects of progressive hypoxia on respiration in the toad *Bufo marinus*. *Journal of Experimental Biology* **68**, 99–107.

Burggren, W. & Moalli, R. (1984). 'Active' regulation of cutaneous gas exchange by capillary recruitment in amphibians: experimental evidence and a revised model for skin respiration. *Respiration Physiology* **55**, 379–92.

Davies, D.G. & Sexton, J.A. (1987). Brain ECF pH and central chemical control of ventilation during anoxia in turtles. *American Journal of Physiology* **252**, R848–52.

Feder, M.E. (1985). Effects of thermal acclimation on locomotor energetics and locomotor performance in a lungless salamander. *Physiologist* **28**, 342.

Fry, F.E.J. & Hart, J.S. (1948). The relation of temperature to oxygen consumption in the goldfish. *Biological Bulletin* **94**, 66–77.

Full, R.J. (1985). Exercising without lungs: energetics and endurance in a lungless salamander, *Plethodon jordani*. *Physiologist* **28**, 342.

Gatz, R.N., Glass, M.L. & Wood, S.C. (1987). Pulmonary function of the green sea turtle, *Chelonia mydas*. *Journal of Applied Physiology* **62** 459–63.

Glass, M.L. (1988). Diffusing capacity. In *Comparative Pulmonary Physiology: Current Concepts* (ed. S.C. Wood), pp. 417–38. Marcel Dekker, New York.

Glass, M.L., Abe, A.S. & Johansen, K. (1981*a*). Pulmonary diffusing capacity in reptiles: relations to temperature and O_2-uptake. *Journal of Comparative Physiology* **142**, 509–14.

Glass, M.L., Boutilier, R.G. & Heisler, N. (1983). Ventilatory control of arterial PO_2 in the turtle, *Chrysemys picta bellii*: effects of temperature and hypoxia. *Journal of Comparative Physiology* **151**, 145–53.

Glass, M.L., Boutilier, R.G. & Heisler, N. (1985). Effects of body temperature on respiration, blood gases and acid-base status in the turtle *Chrysemys picta bellii*. *Journal of Experimental Biology* **114**, 37–51.

Glass, M.L., Burggren, W.W. & Johansen, K. (1981*b*). Pulmonary diffusing capacity of the bullfrog (*Rana catesbeiana*). *Acta Physiologica Scandinavica* **113**, 485–90.

Glass, M.L. & Heisler, N. (1986). The effects of hypercapnia on the arterial acid-base status in the tegu lizard, *Tupinambis nigropunctatus*. *Journal of Experimental Biology* **122**, 13–24.

Glass, M.L. & Wood, S.C. (1983). Gas exchange and control of breathing in reptiles. *Physiological Reviews* **63**, 232–55.

Hall, F.R. (1966). Minimal utilizable oxygen and the oxygen dissociation curve of blood of rodents. *Journal of Applied Physiology* **21**, 375–8.

Heisler, N. (1986). Comparative aspects of acid-base regulation. In *Acid-base Regulation in Animals* (ed. N. Heisler), pp. 397–450. Elsevier, Amsterdam.

Hicks, J.W. & Wood, S.C. (1985). Temperature regulation in lizards: effects of hypoxia. *American Journal of Physiology* **248**, R595–600.

Hitzig, B.M. (1982). Temperature-induced changes in turtle CSF pH and central control of ventilation. *Respiration Physiology* **49**, 205–22.

Hitzig, B.M. & Jackson, D.C. (1978). Central chemical control of ventilation in the unanesthetized turtle. *American Journal of Physiology* **235**, R257–64.

Hochachka, P.W. (1988). Metabolic suppression and O_2 availability. *Canadian Journal of Zoology* **66**, 152–8.

Ishii, K., Ishii, K. & Kusakabe, T. (1985*a*). Electrophysiological aspects of reflexogenic area in the chelonian, *Geoclemmys reevessi*. *Respiration Physiology* **59**, 45–54.

Ishii, K., Ishii, K. & Kusakabe, T. (1985*b*). Chemo- and baroreceptor innervation of the aortic trunk of the toad *Bufo vulgaris*. *Respiration Physiology* **60**, 365–75.

Ishimatsu, I., Hicks, J.W. & Heisler, N. (1988). Analysis of intracardiac shunting in the lizard, *Varanus niloticus*: a new model based on blood oxygen levels and microsphere distribution. *Respiration Physiology* **71**, 83–100.

Jackson, D.C. (1973). Ventilatory response to hypoxia in turtles at various temperatures. *Respiration Physiology* **18**, 178–87.

Johansen, K. (1979). Cardiovascular support of metabolic function. In *Evolution of Respiratory Processes* (ed. S.C. Wood & C. Lenfant), pp. 107–92. Marcel Dekker, New York.

Johansen, K., Lenfant, C. & Hansen, D. (1970). Phylogenetic development of the pulmonary circulation. *Federation Proceedings* **29**, 1135–40.

Krogh, A. (1929). Progress of physiology. *American Journal of Physiology* **90**, 243–51.

Kruhøffer, M., Glass, M.L., Abe, A.S. & Johansen, K. (1987). Control of breathing in an amphibian *Bufo paracnemis*: effects of temperature and hypoxia. *Respiration Physiology* **69**, 267–75.

Lahiri, S. & Gelfand, R. (1981). Mechanisms of acute ventilatory responses. In *Regulation of Breathing*, part II (ed T.F. Hornbein), pp. 773–843. Marcel Dekker, New York.

Lahiri, S., Szidon, J.P. & Fishman, A.P. (1970). Potential respiratory and circulatory adjustments to hypoxia in the African lungfish. *Federation Proceedings* **29**, 1141–8.

Lutcavage, M.E., Lutz, P.L. & Baier, H. (1987). Gas exchange in the loggerhead sea turtle, *Caretta*. *Journal of Experimental Biology* **131**, 365–72.

Malvin, G.M. & Heisler, N. (1988). Blood flow patterns in the salamander, *Ambystoma tigrinum*, before, during and after metamorphosis. *Journal of Experimental Biology* **137**, 53–74.

Malvin, G.M. & Hlastala, M.P. (1986*a*). Regulation of cutaneous gas exchange by environmental O_2 and CO_2 in the frog. *Respiratory Physiology* **65**, 99–111.

Malvin, G.M. & Hlastala, M.P. (1986*b*). Effects of lung volume and O_2 and CO_2 content on cutaneous gas exchange in frogs. *American Journal of Physiology* **251**, R941–6.

Piiper, J., Gatz, R.N. & Crawford, E.C., Jr (1976). Gas transport characteristics in an exclusively skin-breathing salamander, *Desmognathus fuscus*. In *Respiration of Amphibious Vertebrates* (ed. G.M. Hughes), pp. 339–56. Academic Press, New York.

Rahn, H. (1967). Gas transport from the environment to the cell. In *Development of the Lung* (ed. A.V.S. de Reuck & R. Porter), pp. 3–23. Churchill, London.

Reeves, R.B. (1972). An imidazole alphastat hypothesis for vertebrate acid-base regultion: tissue carbon dioxide content and body temperature in bullfrogs. *Respiration Physiology* **14**, 219–36.

Riedel, C. & Wood, S.C. (1988). Effects of hypercapnia and hypoxia on temperature selection of the toad, *Bufo marinus*. *Ferderation Proceedings* **2**, A500.

Rosoff, L., Zeldin, R., Hew, E. & Aberman, A. (1980). Changes in blood P_{50}: effects on oxygen delivery when arterial hypoxemia is due to shunting. *Chest* **77**, 142–6.

Sakakibara, Y. (1978). Localization of CO_2 sensor related to the inhibition of the bullfrog respiration. *Japanese Physiology* **28**, 721–35.

Schaefer, K., Messier, A.A., Morgan, C. & Baker, G.T., III (1975). Effect of chronic hypercapnia on body temperature regulation. *Journal of Applied Physiology* **38**, 900–6.

Shelton, G. (1976). Gas exchange, pulmonary blood supply, and the partially divided amphibian heart. In *Perspectives in Experimental Biology*, (ed. P. Spencer Davies), pp. 247–59. Pergamon Press, Oxford.

Shelton, G. (1985). Functional and evolutionary significance of cardiovascular shunts in the Amphibia. In *Cardiovascular Shunts* (Alfred Benzon Symposium 21) (ed. K. Johansen & W. Burggren), pp. 100–20. Munksgaard, Copenhagen.

Shelton, G., Jones, D.R. & Milsom, W.K. (1986). Control of breathing in ectothermic vertebrates. In *Handbook of Physiology*, sect. 3, (*The Respiratory System*), vol. 2 (ed. A.P. Fishman, N.S. Cherniac, J.G. Widdiocombe & S.R. Geiger), pp. 857–909. The American Physiological Society, Washington.

Silver, R.B. & Jackson, D.C. (1986). Ionic compensation with no renal response to chronic hypercapnia in *Chrysemys picta bellii*. *American Journal of Physiology* **252**, R1228–34.

Smith, D.G. (1978). Evidence for pulmonary vasoconstriction during hypercapnia in the toad, *Bufo marinus*. *Canadian Journal of Zoology* **56**, 1530–4.

Storey, K.B. (1988). Suspended animation: the molecular basis of metabolic depression. *Canadian Journal of Zoology* **66**, 124–32.

Turek, Z. & Kreuzer, E. (1981). Effects of shifts of the O_2 dissociation curve upon alveolar-arterial O_2 gradients in computer models of the lung with ventilation-perfusion mismatching. *Respiration Physiology* **45**, 133–9.

Wang, L.C.H. (1978). Energetic and field aspects of mammalian torpor: the Richardson's ground squirrel. In *Strategies in Cold: Natural*

Torpidity and Thermogenesis (ed. L.C.H. Wang & J.W. Hudson), pp. 109–45. Academic Press, New York.

Warley, A. & Gutierrez, G. (1984). Oxygen delivery in hypothermia. *Physiologist* **27**, 251.

White, F.N. & Somero, G. (1982). Acid-base regulation and phospholipid adaptations to temperature: time courses and physiological significance of modifying the milieu for protein function. *Physiological Reviews* **62**, 40–90.

Wilford, D.C., Hill, E.P., White, F.C. & Moores, W.Y. (1986). Decreased critical mixed venous oxygen tension and critical oxygen transport during induced hypothermia in pigs. *Journal of Clinical Monitoring* **2**, 155–68.

Wood, S.C. (1984). Cardiovascular shunts and oxygen transport in lower vertebrates. *American Journal of Physiology* **247**, R3–14.

Wood, S.C., Glass, M.L., Heisler, N. & Andersen, N. (1987*a*). Respiratory properties of blood and arterial blood gases in Tegu lizards: effects of temperature and hypercapnia. *Experimental Biology* **47**, 27–31.

Wood, S.C. & Hicks, J.W. (1985). Oxygen transport in vertebrates with cardiovascular shunts. In *Cardiovascular shunts* (Alfred Benzon Symposium 21) (ed. K. Johansen & W. Burggren), pp. 354–62. Munksgaard, Copenhagen.

Wood, S.C., Hicks, J.W. & Dupre, R.K. (1987*b*). Hypoxic reptiles: blood gases, body temperature, and control of breathing. *American Zoologist* **27**, 21–9.

Wood, S.C., Malvin, G.M. & Riedel, C. (1989). Effect of hematocrit on behavioral temperature regulation of the toad, *Bufo marinus*. *Federation Proceedings* **3**, A234.

J.H. BRACKENBURY

Ventilation, gas exchange and oxygen delivery in flying and flightless birds

Introduction

One of the clearest physiological differences between birds and reptiles is the ability of the former to maintain high metabolic rates for prolonged periods of time irrespective of variations in environmental temperature. Birds have been able to colonize the air because they are warm-blooded, which buffers their working tissues against adverse changes in air temperature, and they possess a cardiopulmonary system that is capable of delivering oxygen to the working tissues at a constant high rate. Both of these factors are tested to their limits in birds that fly in cool, rarefied atmospheres at high altitude. These factors were also prefigured in the putative reptilian ancestors of birds. As Perry & Duncker (1980) speculate, the avian lung air-sac system seems to have had its origin in a multi-chambered reptilian lung of the type found in monitor lizards, which has a large, ventilatable surface area and high compliance. There is evidence that the most immediate ancestral relatives of birds, the dinosaurs and pterosaurs, were warm-blooded and had evolved complex, multi-chambered lungs to increase the ratio of gas exchange area to body mass. For the flying, or for that matter the walking or running, bird, the ability to maintain a high body temperature and to perform strenuous work also necessitates a means for dissipating excess metabolic heat into possibly unfavourable thermal environments. Birds do not possess sweat glands but, like panting dogs, use the respiratory system to thermoregulate. The respiratory control system must therefore take account not only of the prevailing gaseous exchange needs but also of the need to stop overheating. This chapter reviews the metabolic capabilities of birds and the mechanisms that ensure adequate delivery of oxygen to the working tissues in normal conditions and when confronted by environmental challenge such as hypoxia and hyperthermia.

126   *J.H. Brackenbury*

Gas exchange and oxygen delivery during exercise

Respiratory responses of birds to increased metabolic rate have been studied in a variety of physiologically normal activities including flight, running and swimming, but also in experimental situations where oxygen demand has been artificially increased by drug injection. Oxygen consumption data on ten species of bird during wind-tunnel flight have been reviewed by Bernstein (1987). Although there is considerable variation between species, birds in general appear to increase their metabolic rate approximately 15-fold compared with rest, and this relative increase is independent of body size.

Some non-flying birds that have been trained to run on treadmills can achieve maximum metabolic rates that come within the range of those measured during flight. The South American rhea *Rhea americana* (Taylor *et al.*, 1971), domestic cockerel (Brackenbury & Avery, 1980) and emu *Dromiceius novae-hollandiae* (Grubb *et al.*, 1983) all increased their metabolic rates by 10–12 times resting. Although this order of increase is to be considered maximal in the fowl, it is unlikely to represent the best performance of the two ratites, which are highly specialized for plains running. There is evidence of a strong sex difference in aerobic scope: female domestic fowl can achieve metabolic rate increases of only approximately five times during treadmill exercise (Brackenbury *et al.*, 1981; Brackenbury & El-Sayed, 1985) and the onset of egg laying further reduces this scope. Running ability is also certainly influenced by body morphology and it seems likely that the 3–3.5-fold increases in metabolic rate measured in running ducks *Anas platyrhynchos* (Bech & Nomoto, 1982; Grubb, 1982; Kiley *et al.*, 1985) and *Anas superciliosa* (Baudinette & Gill, 1985) and penguin *Eudyptula minor* (Baudinette & Gill, 1985), which are not natural runners, are close to their maximum potential. On the other hand, the greylag goose *Anser anser* (Fedak *et al.*, 1974) and the Marabou stork *Leptoptilos crumeniferus* (Bamford & Maloiy, 1980), neither of which would seem well adapted for treadmill exercise, both achieved increases of approximately five times resting oxygen consumption. Running pigeons, *Columba livia*, also increased metabolic rate by five times (Grubb, 1982) although, not surprisingly, this is much less than the 10–12-fold increase measured in this species during wind-tunnel flight (Butler *et al.*, 1977). The Japanese quail *Coturnix coturnix japonica* is also by disposition a runner rather than a flier and almost certainly capable of a larger metabolic rate increment than the 2.5-fold increase measured by Nomoto *et al.* (1983). In all these comparisons it must be borne in mind that the definition of 'resting' metabolic rate may not have been identical in the different studies: in very few instances are the 'basal' metabolic rate figures given. With this proviso, it appears that, apart from a few exceptions, running birds are capable of only half or less of the proportional elevation of metabolic rate seen in flying species.

The leg muscles are not only used for terrestrial locomotion but also for surface swimming. Tufted ducks, *Aythya fuligula*, showed no significant increase in oxygen consumption at swimming speeds up to about 0.5 m s^{-1}, but thereafter it increased rapidly, attaining a maximum value of 3.8 times resting at 0.78 m s^{-1} (Woakes & Butler, 1983). Similarly, the oxygen consumption rates of the penguin *Eudyptula minor* and the black duck *Anas superciliosa* were independent of swimming speeds up to 0.5 ms^{-1}, then increased to maximum values of approximately two and three times resting respectively, at a swimming speed of 0.72 ms^{-1} (Baudinette & Gill, 1985).

Pharmacological stimulation of hypermetabolism has been investigated in duck (Geiser *et al.*, 1984) and domestic fowl (Gleeson, 1986) after injection of 2,4-dinitrophenol. This drug induced 3–4-fold increments in oxygen consumption and the ventilatory responses were similar to those observed in normally exercised birds (see next section). Although the technique can be useful for examining the effects of increased metabolic rate on respiration, its value as a tool for investigating normal exercise is strictly limited. The two situations are not equivalent biomechanically, or in many instances, physiologically.

All flightless birds have evolved secondarily from flying ancestors and seem to have conserved the cardiopulmonary fitness of the latter. Although only limited morphometric data are available it seems probable that the relative metabolic capabilities of flying, running and swimming birds are limited, not by oxygen delivery, but by the relative masses and aerobic capacities of the working muscles. In a recent study Turner & Butler (1988) compared the aerobic capacity of total hindlimb and total flight muscle in the tufted duck *Aythya fuligula* on the basis of anatomically measured mitochondrial density. The total mitochondrial volume of the flight muscle, and therefore its potential for oxidation, was approximately six times that of the hindlimb muscle. This was roughly in proportion to the physiologically measured oxygen consumption during maximal swimming and flight in this species, when due account was taken of the increased cardiac work.

Increased oxygen delivery to the working muscles can be achieved by increased cardiac output or increased O_2 extraction from the blood or by a combination of both. Enhanced O_2 extraction is reflected in a rise in the O_2

content difference between arterial and mixed venous blood. Cardiac output can be raised either as a result of increased stroke volume or heart rate, or both. All flying birds that have so far been investigated show increases in heart rate of the order of 2–4 times compared with resting (Berger & Hart, 1974), although increases of 6–7 times have been monitored during wind-tunnel flight in the pigeon (Butler *et al.*, 1977) and open-air flight in the Barnacle goose *Branta leucopsis* (Butler & Woakes, 1980). According to the latter authors this increase is probably maximal in the Barnacle goose, and in both this species and the kestrel (Johnson & Gessaman, 1973) heart rate remained constant over a range of flight speeds. Butler & Woakes (1980) suggest that this constancy of heart rate may be related to the independence of power input from flight speed. Although theoretical calculations had predicted that the power input of flying birds should vary in a U-shaped manner with air speed (Pennycuick, 1969; Tucker, 1973) most experimental studies have in contrast revealed an almost flat relationship (Tucker, 1972; Bernstein *et al.*, 1973; Torre-Bueno & Larochelle, 1978; Hudson & Bernstein, 1983).

In running birds a linear relation between oxygen consumption and heart rate has been observed over a range of work loads in the Marabou stork (Bamford & Maloiy, 1980), pigeon and duck (Grubb, 1982) and emu (Grubb *et al.*, 1983). This was also found in surface swimming Tufted duck (Woakes & Butler, 1983). Thus, as Woakes & Butler (1983) suggest, heart rate can be used as a rough practical guide to metabolic rate but this does not apply to naturally diving ducks as in this case a degree of diving bradycardia reduces heart rate by roughly 30% compared with its value at the same work rate in a swimming duck.

There is much less information available on cardiac output and stroke volume than on heart rate in exercising birds. By using the Fick equation, Butler *et al.* (1977) estimated that the 12-fold increase in oxygen consumption in flying pigeons was met by a 6-fold rise in cardiac output combined with a doubling of the arteriovenous O_2 difference. Cardiac stroke volume, however, did not change. In running Pekin duck, oxygen delivery was increased 2.5 times as a result of 60% and 50% rises in cardiac output and arteriovenous O_2 difference respectively (Bech & Nomoto, 1982). Similar responses were observed in the surface swimming Tufted duck *Aythya fuligula* (Butler *et al.*, 1988): oxygen delivery approximately doubled as a result of a 70% increase in cardiac output and a 20% increase in arteriovenous O_2 content difference. Beck and Nomoto (1982) stress the important fact that the high resting venous oxygen content of birds confers an innate advantage in raising the scope for oxygen extraction during exercise.

As in the flying pigeon, cardiac stroke volume varied little in running ducks (Bech & Nomoto, 1982; Grubb, 1982, pigeon (Grubb, 1982) and emu (Grubb *et al.*, 1983). Grubb (1982) highlighted the point that exercising birds appear to have a higher cardiac output for a given rate of oxygen consumption than mammals of the same mass. She speculated that the large cardiac output response of birds to exercise could be an important pre-adaptation to increased oxygen delivery during flight at high altitude, where a high oxygen flux must be maintained against a very low environmental P_{O_2}.

Ventilation during normoxic exercise

Some of the earliest and most valuable studies on respiration in flying birds were carried out by Hart & Roy (1966, 1967). They used pigeons fitted with transducers and telemetric equipment, which relayed information about respiratory airflow and heart rate during brief periods of free flight. Minute ventilation increased 20-fold compared to rest and as respiratory rate also rose by the same amount, tidal volume remained unaltered. Lefebvre (1964) had shown that during flight the metabolic rate of pigeons rose 8-fold, so Hart & Roy reasoned that ventilation must rise 250% in proportion to metabolic rate. This agreed with the early prediction by Zeuthen (1942) that a pigeon flying at $50\,\mathrm{km\,h^{-1}}$ would double its ventilation in proportion to metabolic rate in order to lose body heat by respiratory evaporation.

Later studies of flight in wind tunnels (Torre-Bueno, 1978; Tucker, 1968; Bernstein, 1976; Hudson & Bernstein, 1978; Butler *et al.*, 1977) or in the open air (Butler & Woakes, 1980) showed that birds increase respiratory rate but changes in tidal volume appear more limited, varying from zero in the pigeon to a fourfold increase in the starling *Sturnus vulgaris* (Figure 1). It is difficult to assess the true pattern of ventilatory response to exercise in birds without taking into account effects of increased body temperature. In treadmill-exercised fowl it has been possible to separate the primary effect of increased metabolic rate from secondary effects of hyperthermia by comparing ventilatory responses to the same work loads in isothermic and hyperthermic conditions. When body temperature is held in check by running at reduced environmental temperatures lung ventilation increases in proportion to gas exchange and there is little change in intrapulmonary or arterial P_{CO_2} over the full range of work loads of which the birds are capable (Figure 2). This precise matching of ventilation to gas exchange is accomplished by a ventilatory strategy that relies principally on increasing respiratory rate to meet demand for ventilation and that permits only small

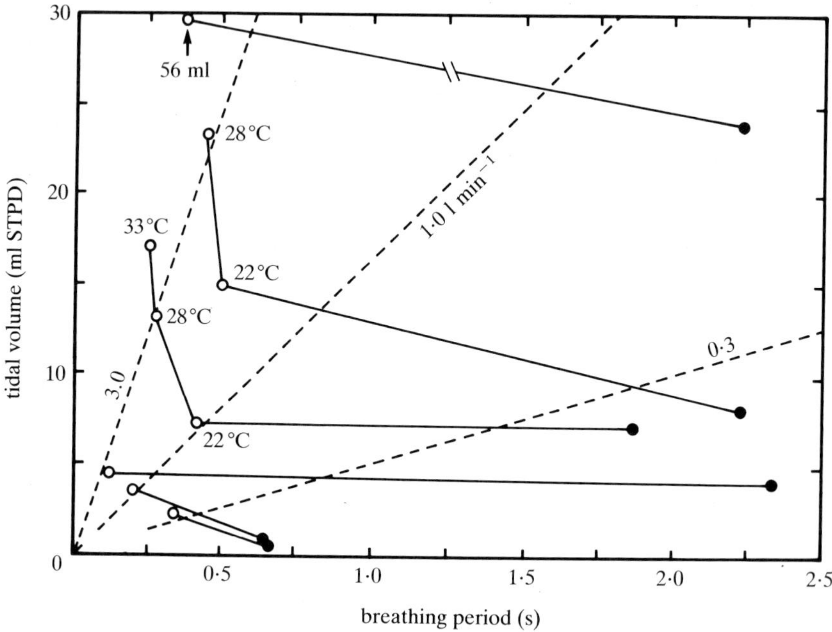

Figure 1. Ventilatory characteristics measured in six species of bird during wind-tunnel flight. Shaded circles designate resting birds, open circles flying birds. Ambient temperatures indicated for selected data. Solid lines connect resting and flight data for each species; dashed lines indicate minute volume. (From Bernstein, 1987; identification of species given in original reference.)

increments in tidal volume. When exercise is performed in hyperthermic conditions by running the birds at raised environmental temperatures the ventilatory strategy must be adjusted to cater for the needs of thermoregulation as well as gas exchange and acid–base balance. The bird hyperventilates but, at least in the fowl, the breathing pattern is adjusted towards a polypneic mode, which has the effect of limiting the thermoregulatory component of the ventilation to the dead space. This kind of ventilatory strategy has been well documented in resting hyperthermic birds (Figure 3). In the running fowl the strategy is imperfect, as it cannot prevent a significant drop in arterial P_{CO_2} due to parabronchial hyperventilation. A similar trend towards polypneic breathing and hypocapnia was also found when Pekin duck performed treadmill exercise in hyperthermic conditions (Kiley *et al.*, 1979).

In contrast to running birds, flying ravens increased tidal volume as well as respiratory rate in response to elevated environmental temperature (Hudson & Bernstein, 1981) (Figure 1). Bernstein (1987) has recently

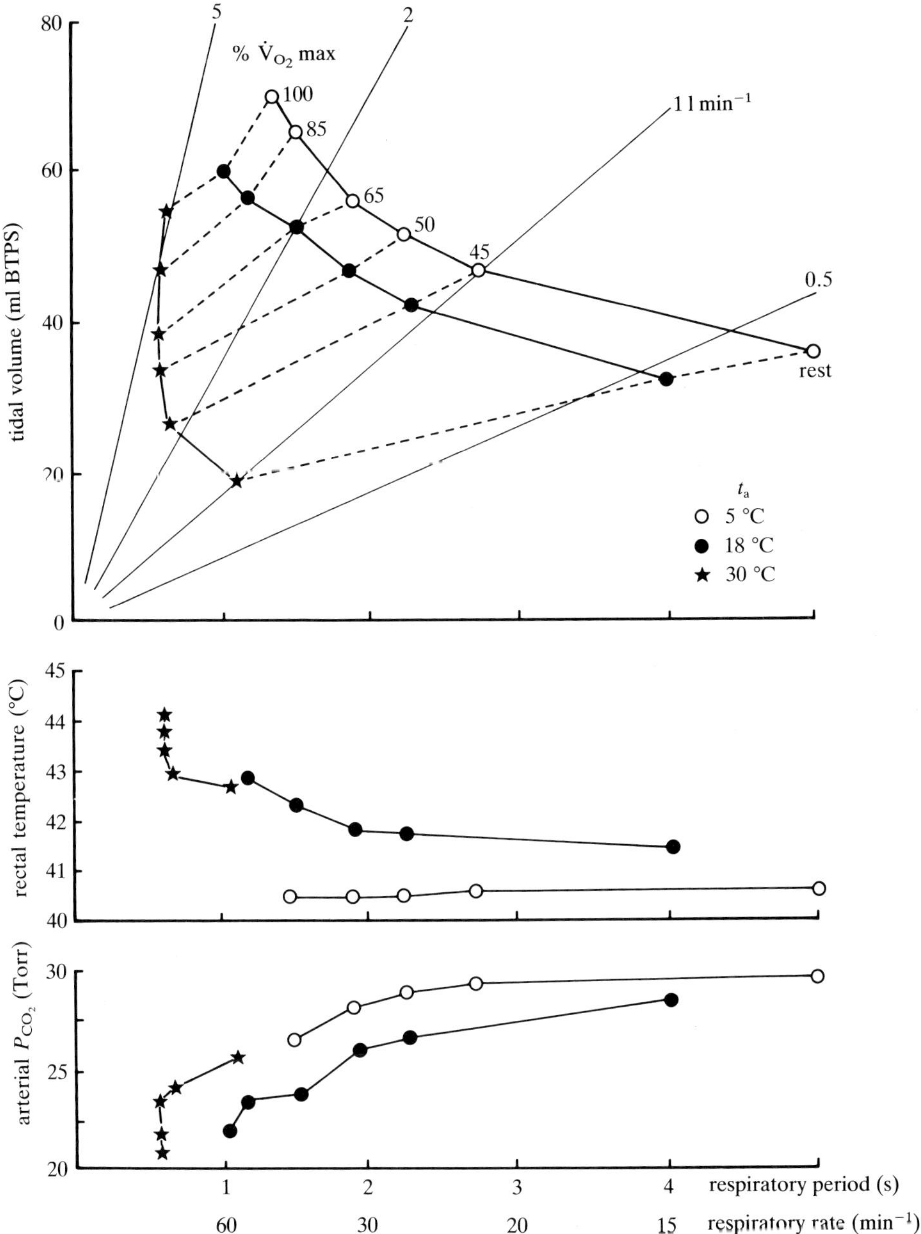

Figure 2. Ventilatory characteristics, rectal temperatures, and arterial P_{CO_2} in domestic fowl following 10 minutes of treadmill exercise at graded work loads and in environmental temperatures (t_a) of 5, 18 or 30 °C. Dotted lines in the upper graph join points of equal work loads and show how increased environmental temperature produced increased respiratory rate but decreased tidal volume. Thermal hyperventilation also led to decreased arterial P_{CO_2}. (From Brackenbury & Gleeson, 1983.)

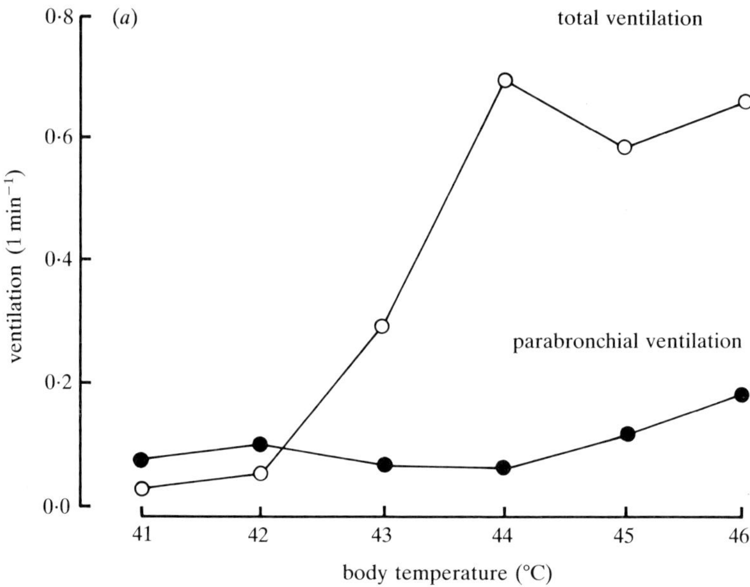

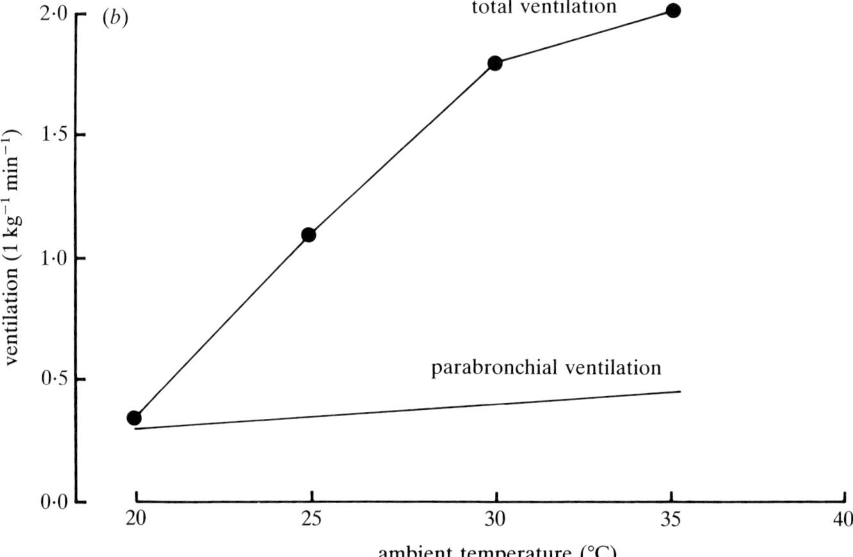

Figure 3. Comparison of total and estimated parabronchial ventilation at different body temperatures in resting pigeon (*a*) (from Bernstein & Samaniego, 1981) and at different environmental temperatures in the Pekin duck (*b*) (from Bouverot *et al.*, 1976).

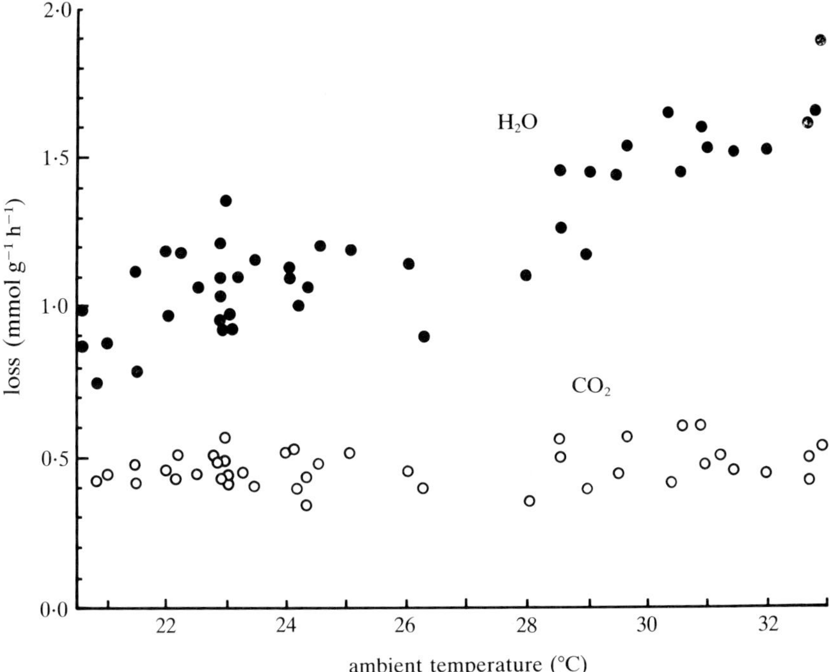

Figure 4. Respiratory evaporative water loss (shaded circles) and CO$_2$ loss (filled circles) at different environmental temperatures during wind-tunnel flight in the white-necked raven (from Bernstein, 1987, after Hudson, 1978).

stressed the importance of tidal volume, rather than respiratory rate, as the principal variable in the thermoregulatory response to flight. It is not known whether flying birds avoid alkalosis during hyperthermic hyperventilation or whether they tolerate it as an acceptable trade-off against overheating. Data presented by Bernstein (1987) suggest that in the White-necked raven hypocapnia is avoided: despite large increases in respiratory evaporative loss at increased air temperatures, the rate of expired CO$_2$ loss remained constant (Figure 4). Apparently this was not due to shallow breathing. Without offering an explanation of the mechanism, Bernstein (1987) suggests that during hyperthermia increased tidal volume leads to increased dead-space rebreathing with the result that the caudal air sacs accumulate CO$_2$. This in turn reduces the P_{CO_2} gradient in the parabronchi and limits CO$_2$ loss. Caudal sac P_{CO_2} certainly rises during post-exercise hyperthermia in domestic fowl (Brackenbury *et al.*, 1981) and this is directly related to increased dead-space rebreathing, but as a result of a marked decrease in tidal volume, not an increase. Similarly, Powell & Hempleman (1985) were able to raise caudal sac P_{CO_2} in the penguin by lowering tidal volume. Unless flying birds are able to valve intrapulmonary airflow in some way that has not yet been identified in birds at rest, it is difficult to see how increased tidal volume would have the effect that Bernstein (1987) proposes.

There are only few data available on changes in intrapulmonary or arterial P_{CO_2} during exercise in birds. Air sac P_{CO_2} fell during wind-tunnel flight in starlings (Torre-Bueno, 1978) and during treadmill exercise in Pekin duck (Kiley *et al.*, 1979) and domestic fowl (Brackenbury *et al.*, 1981); arterial P_{CO_2} dropped by 11 Torr during wind-tunnel flight in pigeons (Butler *et al.*, 1977). In all these cases it is likely that lung hyperventilation in response to increased body temperature was the cause of CO_2 washout; in support of this conclusion, the drop in arterial P_{CO_2} was found to be less when Pekin ducks were run at lowered air temperature (Kiley *et al.*, 1982). According to Bernstein's (1987) analysis on ten species, flying birds increase their ventilation approximately 17 times compared with basal rates, whereas oxygen consumption increases only 15 times. This analysis did not take account of differences in environmental factors, which may have influenced the responses in individual studies. It is likely that in many cases hyperventilation was caused in part by thermal stress, although metabolic acidosis may also have contributed.

The possible role of spinal thermosensitive neurons in the control of ventilation during exercise has been investigated in cockerels (Gleeson *et al.*, 1985). When these animals ran on a treadmill at an environmental temperature of 9 °C, experimental cooling of the spinal cord resulted in increments in oxygen consumption and ventilation. Minute volume increases were due to increased respiratory rate and tidal volume and the effect was identical to that of an increase in exercise intensity. In contrast, spinal cooling at 34 °C reduced both the minute volume and the respiratory rate, implying that cooling breaks the normal linkage between the thermoregulatory and respiratory control centres. It appears that spinal thermoreceptors may exert direct effects on motor output from the respiratory centre, as distinct from secondary influences on respiratory pattern resulting from altered arterial P_{CO_2}.

Breathing pattern in treadmill-exercised Pekin ducks, which are not natural runners, appears to be controlled in a somewhat different manner from chickens. In this species, increased respiratory rate alone accounted for the hyperpnea of exercise and tidal volume dropped even when running took place in isothermic conditions (Kiley *et al.*, 1979, 1982, 1985). Gleeson *et al.* (1985) have commented that the tracheal cannulation technique employed in the running duck experiments might alter the pattern of breathing by inhibiting the normal changes in tidal volume. When Pekin ducks were exercised in conditions where arterial P_{CO_2} was preserved at its normal level, both the rate and the depth of breathing increased (Kiley & Fedde, 1983*a*,*b*). Recently, respiratory responses of Tufted ducks, *Aythya fuligula*, to swimming exercise have been studied by using a non-invasive mask technique (Woakes & Butler, 1986). During surface swimming most of the ventilatory increase was due to accelerated rhythm but at the highest work loads investigated tidal volume also increased. In Tufted duck, running Pekin duck (Kiley *et al.*, 1982) and running chicken (Gleeson & Brackenbury, 1983; Brackenbury & El-Sayed, 1985) blood lactate increased substantially at higher work loads, implying that enhancement of tidal volume, with a consequent increase in parabronchial ventilation, may represent a respiratory compensation of metabolic acidosis.

Additional evidence for the importance of tidal volume as well as respiratory rate in the ventilatory response to increased activity has come from experiments in anaesthetized Muscovy ducks (*Airina moschata*) in which metabolic rate was artificially raised by injection of 2,4-dinitrophenol (Geiser *et al.*, 1984). Increased ventilatory demand was met by coupled increases in rate and depth of breathing. Gleeson (1986) also examined effects of 2,4-dinitrophenol, as well as cold exposure, on breathing in anaesthetized chickens. In these experiments, significantly, arterial P_{CO_2} remained at its control value and the observed increases in respiratory rate and tidal volume were identical to those that would have been occasioned by the same level of metabolic increase in a conscious chicken running in isothermic, isocapnic conditions.

The fact that the respiratory pattern generator is sensitive to alterations in arterial and intrapulmonary P_{CO_2} during exercise is not surprising in view of the role of the specifically CO_2-sensitive intrapulmonary chemoreceptors in the control of avian respiration (Bouverot, 1978). These receptors, however, do not appear to be directly involved in the promotion of exercise ventilation as both Pekin duck (Kiley & Fedde, 1983*a*,*b*) and chicken (Brackenbury & Gleeson, 1986) showed the same increase in minute ventilation during exercise when they were breathing air or CO_2-enriched air (Figure 5).

The demand for air during exercise is therefore set by factors other than P_{CO_2}, although the precise combination of rate and depth of breathing used

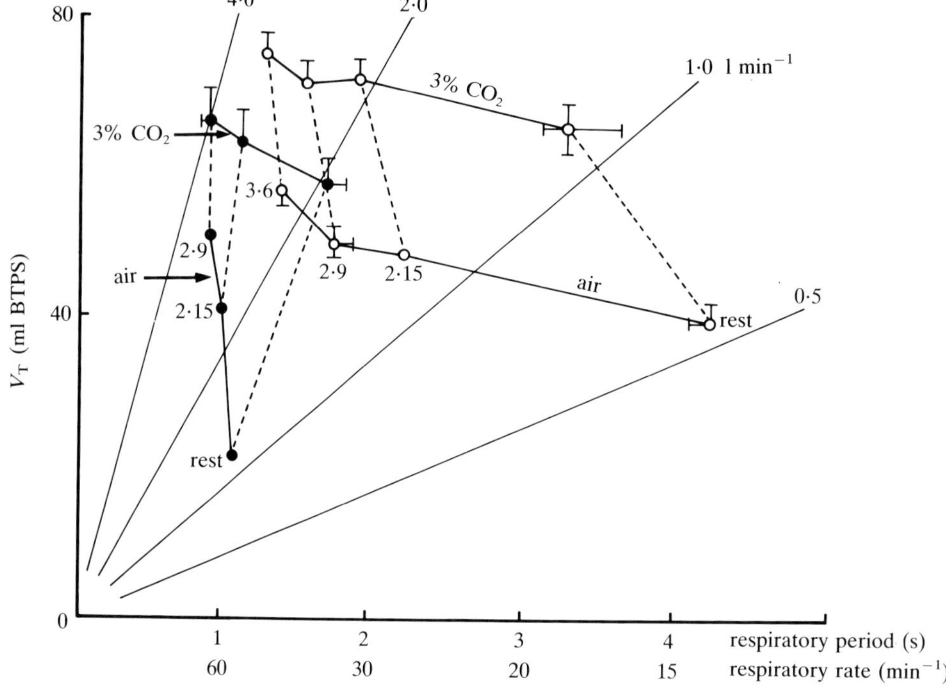

Figure 5. Ventilatory characteristics of domestic fowl measured during treadmill exercise at different speeds when the animals were breathing air or 3% CO_2 in air. Exercise was performed at air temperatures of 3–5 °C (open circles) or 28–30 °C (filled circles). Solid lines join points of equal inspired gas composition. Broken lines join points of equal exercise intensity; numbers beside each point represent treadmill speed in kilometres per hour. Data from duplicate runs in five birds. Note how administration of CO_2 in the hyperthermic birds reverses hypocapnia and results in a pattern of ventilation similar to that displayed in the euthermic isocapnic birds breathing air (J.H. Brackenbury & M. Gleeson, unpublished data).

to satisfy this demand is highly dependent on P_{CO_2}. When chickens are exercised at raised air temperatures, respiration becomes fast and shallow compared with controls exercising euthermically (Figure 5) (see also Brackenbury & Gleeson, 1983). After inhalation of CO_2-enriched air to restore normal arterial and intrapulmonary P_{CO_2}, respiratory characteristics become identical to those of controls. In chickens, therefore, regardless of the type of respiratory drive, hyperthermia or exercise, there appears to be a unique combination of tidal volume and respiratory rate to satisfy each given value of minute ventilation, so long as P_{CO_2} remains unaltered.

Respiration in hypoxic conditions

Many birds are capable of sustained flight at high altitudes and the combination of high aerobic work loads in the presence of very low environmental oxygen pressures imposes extreme demands on the oxygen delivery system. Unfortunately, very few studies have been carried out on exercising birds and most information is based on inferences from experiments on resting animals. Tucker (1968) was one of the first workers to examine hypoxic responses in relation to high-altitude flight; he concluded that the ability to withstand low oxygen saturation of the blood, plus a relatively high cardiac output compared with mammals, were important features in the hypoxia tolerance of birds. A high tolerance to blood hypocapnia resulting from hypoxic hyperpnea was also implicated. An increase in ventilation in response to low environmental oxygen pressures has been observed in the Bar-headed goose *Anser indicus* (Black *et al.*, 1978; Black & Tenney, 1980), pigeon (Bouverot *et al.*, 1976) and duck (Bouverot & Hildwein, 1978). Increased ventilation is mainly due to an elevation of respiratory rate and the lack of a tidal volume response may represent an attempt to limit excessive CO_2 loss (Bernstein, 1987). Hypoxic hypernea usually results in reduced arterial P_{CO_2} and elevated pH (Lutz & Schmidt-Nielsen, 1977; Faraci *et al.*, 1984; Kiley *et al.*, 1985; Weinstein *et al.*, 1985; Faraci, 1986; Shams & Scheid, 1987). In ducks, mild hypoxia produces alkalosis, then lactacidosis, which compensates the respiratory alkalosis (Shams & Scheid, 1987); at the deepest hypoxic levels tolerated there was a distinct acidosis despite very low arterial P_{CO_2} (5.7 Torr at 11 580 m). Hypoxic hyperpnea obviously maximizes oxygen loading, as shown by the diminution of the P_{O_2} gradient between arterial and inspired gas, but, according to Shams & Scheid (1987) this achievement is not due to the particularly high efficiency of the gas exchange system but to a very high parabronchial ventilation relative to oxygen uptake. In the severely hypoxic resting duck ventilation is not a limiting factor since the further increments to the ventilatory drive produced by CO_2 inhalation had little advantageous effect on the arterial–inspired P_{O_2} difference. Tolerance to hypoxic hypocapnia may also confer a direct advantage with respect to oxygen loading. In the duck, hypoxic hypocapnia evoked a Bohr effect, which raised the oxygen content of the blood at a given P_{O_2} (Grubb *et al.*, 1979). Hypoxic ducks, in contrast to mammals, do not seem to suffer a decrease in cerebral blood flow (Grubb *et al.*, 1977); consequently, the hypocapnia-induced rise in blood O_2 content, together with maintained blood flow, was sufficient to maintain oxygen delivery to the brain. As

Bernstein (1987) indicated, the repeated finding by different authors of hypoxic hyperpnea leading to hypocapnia and alkalosis contributes directly to brain oxygenation; this ventilatory response does not simply represent an attempt to elevate parabronchial P_{O_2}.

The above considerations based on resting respiratory responses suggest that birds may be pre-adapted to high-altitude breathing, although resting experiments take no account of the greatly increased oxygen demands of the active animal. The only data on true in-flight respiratory responses to altitude comes from Berger (1978) who found that, based on measurements of expired water loss, ventilation in the hummingbird *Colibri coruscans* increased by 30% at a simulated altitude of 4000 m. Running cockerels exposed to 10% O_2 in nitrogen increased their ventilation by 35% (Brackenbury, 1986), entirely as a result of increased respiratory rate. The hypoxic hyperpnea of resting domestic fowl is also due almost exclusively to increased rate of breathing (Brackenbury & Gleeson, 1985). The absence of a tidal volume response appears to be due at least in part to hypocapnia, which inhibits the normal drive from the intrapulmonary chemoreceptors; as inhalation of CO_2 in running hypoxic fowl increased tidal volume (Brackenbury, 1986).

Considerations based on gas-exchange models in resting birds suggest that even in normoxic conditions physical exercise could be limited by the oxygen diffusion capacity of the lung (Piiper & Scheid, 1977). Kiley *et al.* (1985) examined this possibility in treadmill-exercised ducks but found that the oxygen diffusion capacity increased by 150% over resting levels during air breathing and by 220% when the birds breathed 12% oxygen. An increased oxygen diffusion capacity was also observed when the oxygen consumption of ducks was artificially raised by injection of 2,4-dinitrophenol (Geiser *et al.*, 1984). Part of the increased oxygen diffusion capacity during exercise appears to be due to the elevation of cardiac output, which automatically reduces the disadvantageous effects of ventilation–perfusion inhomogeneities in the lung. A further factor may be the increase in available area for gas exchange as a result of the distension and recruitment of blood capillaries under the increased perfusion pressure in the lung.

Arterial chemoreceptors are known to be involved in hypoxic responses of resting birds (Jones & Purves, 1970; Bouverot & Sébert, 1979) and presumably also mediate the hyperpneic response of high-altitude fliers. Treadmill-exercised fowl failed to elicit either a transient or a steady-state ventilatory response to oxygen inhalation; this result suggests that, at least during exercise in air at sea level, the tonic drive from arterial chemoreceptors is weak or non-existent (Brackenbury *et al.*, 1982). Arterial P_{O_2} is higher than normal in flying pigeons (Butler *et al.*, 1977) and running ducks (Kiley *et al.*, 1979, 1982) and fowl (Brackenbury & Gleeson, 1983) and probably exceeds the threshold for discharge of the arterial O_2 receptors. At high altitude, however, arterial P_{O_2} must fall to exceedingly low levels, presumably triggering a powerful response from the systemic receptors.

Intrapulmonary airflow during exercise

Although it has been possible to monitor intrapulmonary airflow directly in resting (albeit anaesthetized) birds, the technical difficulties involved in making comparable measurements in exercising birds would be formidable. It is generally assumed that the unidirectional airflow pattern through the palaeopulmo is preserved during exercise; indeed, there is growing evidence that the so-called inspiratory aerodynamic valve, which prevents entry of air into the ventrobronchi, may be even more efficient during exercise hyperpnea. Recent work by Banzett *et al.* (1987) and Wang *et al.* (1988) has shown that aerodynamic valving in the avian lung is heavily dependent on gas velocity and density. The 'convective inertia' hypothesis presented by the latter authors is essentially a restatement of the familiar idea that inspiratory valving is due to the momentum of the inspired gas which after it has entered the mesobronchus, carries it past the ventrobronchial openings. Such a valve must inevitably become more efficient at increased flow velocities, as both Wang *et al.* (1988) and Kuethe (1988) have demonstrated in models of the lung. Wang *et al.* (1988) also pointed out that, because the valve was dependent on gas density, its efficiency might be expected to drop during high-altitude breathing. This would be partly compensated for, however, by the increased gas velocity resulting from hypoxic hyperpnea.

In an attempt to assess indirectly the relative importance of different air sacs in maintaining aerodynamic valving during exercise we recently investigated respiratory responses of domestic fowl to selective blocking of individual pairs of air sacs (Figure 6). Exercise took place on a treadmill for 10–15 min at work loads of 3.5 times resting metabolic rate. Ventilation, blood gases and clavicular air-sac gas composition were compared in control birds and birds in which either the cranial thoracic sacs alone or the cranial and caudal thoracic sacs together had been completely blocked by cotton wool. Additional measurements were made after combined blockage of the caudal thoracic and abdominal sacs, although these were less successful owing to the very high distensibility of the abdominal sac. The main results of these experiments are shown in Table 1. All experimental

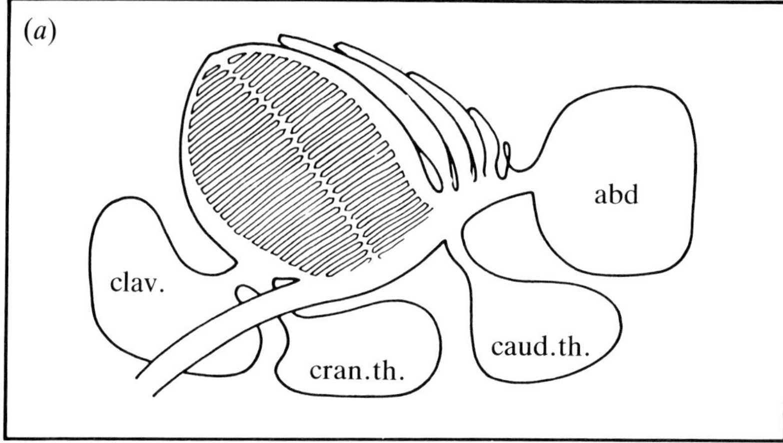

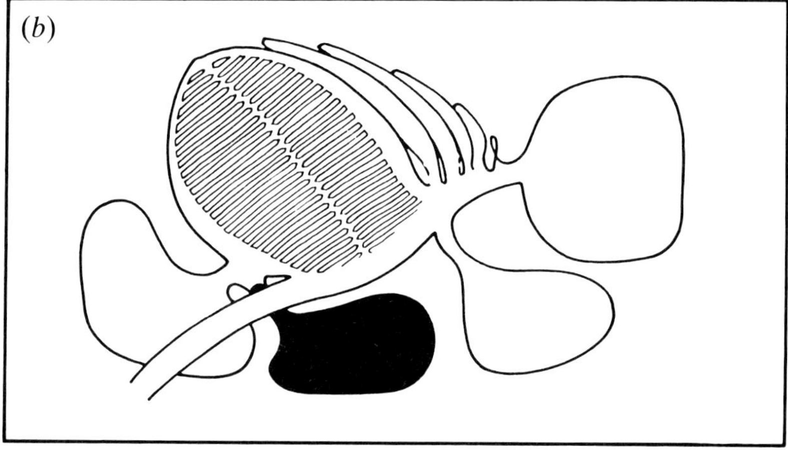

Figure 6. Scheme of air sac blocks carried out in domestic fowl. Blocked sacs are shaded. Respiratory data for control (*a*), cranial thoracic-blocked (*b*) and cranial–caudal thoracic-blocked (*c*) groups are given in Table 1;

groups, including that in which the combined caudal thoracic and abdominal sacs had been blocked, were capable of maintaining normal or even slightly elevated ventilation compared with controls. In addition all groups, control and experimental, showed an increase in clavicular sac P_{O_2} and a decrease in P_{CO_2} during exercise. Blood gases were maintained at control values in the cranial thoracic and caudal thoracic–abdominal blocked groups but there was a small but significant hypoxaemia–hypercapnaemia in the group in which both pairs of thoracic sacs had been blocked.

We view with caution the data on the combined caudal thoracic–abdomi-

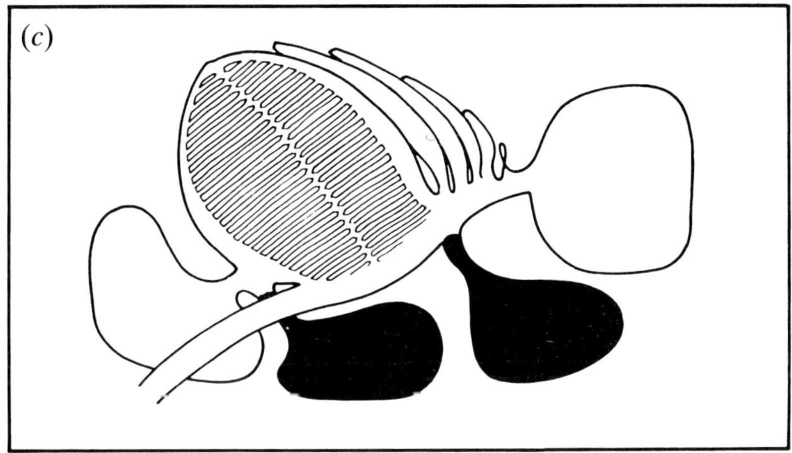

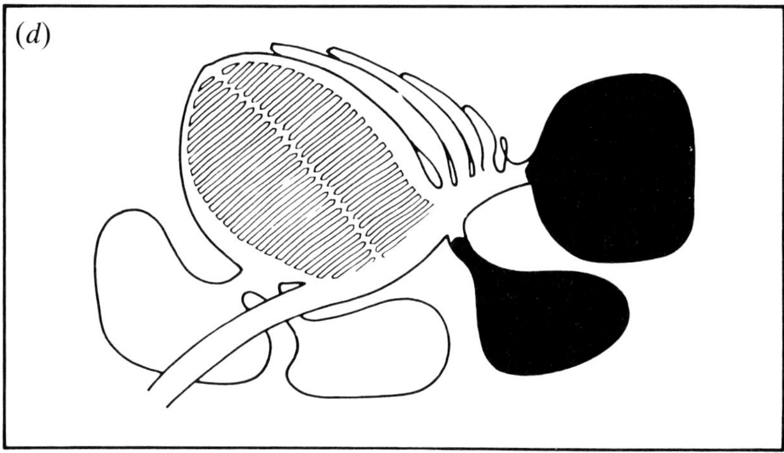

Caption for fig. 6. (cont.).
preliminary data from the caudal thoracic–abdominal-blocked group (*d*)
are discussed in the text. (In part from Scheid, 1979.)

nal blocked birds, owing to uncertainties as to whether complete blockage
of the abdominal sac had been achieved. With this proviso, it is clear that
there is a considerable redundancy within the lung air-sac system in terms
of its ability to adapt to removal of functional capacity, both at rest and
during exercise. This adaptation presumably takes place through increased
excursion of the rib-cage. The second feature to emerge from these experiments is that the combined thoracic air sacs appear to be necessary for
efficient valving of intrapulmonary airflow. The coexistence in the combined cranial–caudal thoracic-blocked birds of a mild hyperoxia–hypo-

Table 1. *Respiratory characteristics of domestic fowl at rest and during exercise after blockage of the cranial thoracic air sac alone* (CRT) *or the cranial and caudal thoracic air sacs together* (CRT & CT)

Mean values (± 1 s.e.m). except for tidal volume, which was calculated from f and $\dot{V}_E$. N, number of birds; $\dot{V}_E$, minute ventilation; V_T, tidal volume; f, respiratory frequency; Pa, arterial gas partial pressure; Pc_s, clavicular air sac gas partial pressure. (Unpublished data from J.H. Brackenbury, C. Darby & M.S. El-Sayed.)

	Control ($N=6$)		CRT ($N=5$)		CRT & CT ($N=5$)	
	Rest	Exercise	Rest	Exercise	Rest	Exercise
$\dot{V}_E$	319	976	367	1231	335	1110
(ml kg^{-1} min^{-1})	(32)	(39)	(24)	(42)	(16)	(59)
V_T	13.3	16.0	14.7	18.6	14.0	15.4
(ml kg^{-1})						
f	24	61	25	66	24	72
(min^{-1})	(0)	(3)	(1)	(1.5)	(1)	(5)
Pa_{CO_2}	30.4	28.5	30.7	30.2	34.3*	32.6*
(Torr)	(0.7)	(0.6)	(0.4)	(0.4)	(0.7)	(0.6)
Pa_{O_2}	82.9	86.0	79.6	79.4*	73.7 *	76.2*
(Torr)	(1.3)	(2.0)	(1.5)	(1.2)	(1.1)	(1.1)
Pc_{sCO_2}	40.2	37.1	39.4	36.3	41.7	35.0
(Torr)	(0.6)	(0.7)	(0.8)	(1.1)	(0.5)	(0.7)
Pc_{sO_2}	99.4	107.9	104.1*	112.2*	101.5	113.8*
(Torr)	(1.0)	(0.9)	(1.2)	(0.9)	(1.0)	(0.9)

*Significantly different from control values ($p<0.05$).

capnia in the clavicular air sac and a mild hypoxia–hypercapnia in the arterial blood during exercise implies that gas reaching the clavicular air sac had not all passed across the parabronchi but that some may have leaked from the mesobronchus into the ventrobronchial connections to the cranial air sacs. This leakage presumably occurred at the expense of the parabronchi, resulting in parabronchial hypoventilation. Although the picture is still by no means clear, the results of these experiments suggest that selective air-sac blocking may be a simple but promising tool for the investigation of airflow within the lung air sac system during exercise hyperpnea.

I thank Sue Insole and Ian Edgar for the art work, and Jane Seymour-Shove for typing the manuscript.

References

Bamford, O.S. & Maloiy, G.M.O. (1980). Energy metabolism and heart rate during treadmill exercise in the Marabou stork. *Journal of Applied Physiology: Respiratory, Environmental and Exercise Physiology* **49**, 491–6.

Banzett, R.B., Butler, J.P., Nations, C.S., Barnas, G.M., Lehr, J.L. & Jones, J.H. (1987). Inspiratory aerodynamic valving in goose lungs depends on gas density and velocity. *Respiration Physiology* **70**, 287–300.

Baudinette, R.V. & Gill, P. (1985). The energetics of 'flying' and 'paddling' in water: locomotion in penguins and ducks. *Journal of Comparative Physiology* **B155**, 373–80.

Bech, C. & Nomoto, S. (1982). Cardiovascular changes associated with treadmill running in the Pekin duck. *Journal of Experimental Biology* **97**, 345–58.

Berger, M. (1978). Ventilation in the humming bird *Colibri coruscans* during altitude hovering. In *Respiratory Function in Birds, Adult and Embryonic* (ed. J. Piiper), pp. 85–8. Springer-Verlag, Berlin.

Berger, M. & Hart, J.S. (1974). Physiology and energetics of flight. In *Avian Biology* (ed. D.S. Farner & J.R. King), vol. 4, pp. 415–77. Academic Press, New York.

Bernstein, M. (1976). Ventilation and respiratory evaporation in the flying crow, *Corvus ossifragus*. *Respiration Physiology* **26**, 371–82.

Bernstein, M. (1987). Respiration in flying birds. In *Bird Respiration* (ed. T.J. Seller), vol. 2, pp. 43–73. CRC Press, Florida.

Bernstein, M. & Samaniego, F.C. (1981). Ventilation and acid-base status during thermal panting in pigeons (*Columa livia*). *Physiological Zoology* **54**, 303–15.

Bernstein, M., Thomas, S.P. & Schmidt-Nielsen, K. (1973). Power input during flight of the fish crow, *Corvus ossifragus*. *Journal of Experimental Biology* **58**, 401–10.

Black, C.P. & Tenney, S.M. (1980). Oxygen transport during progressive hypoxia in high-altitude and sea-level waterfowl. *Respiration Physiology* **39**, 217–39.

Black, C.P., Tenney, S.M. & Kroonenburg, Van M. (1978). Oxygen transport during progressive hypoxia in Bar-headed geese (*Anser indicus*) acclimatized to sea-level and 5,600 metres. In *Respiratory Function in Birds, Adult and Embryonic* (ed. J. Piiper), pp. 79–83. Springer-Verlag, Berlin.

Bouverot, P. (1978). Control of breathing in birds as compared with mammals. *Physiological Reviews* **58**, 604–55.

Bouverot, P. & Hildwein, G. (1978). Combined effects of hypoxia and moderate heat load on ventilation in awake Pekin ducks. *Respiration Physiology* **33**, 378–84.

Bouverot, P., Hildwein, G. & Oulhen, P. (1976). Ventilatory and circulatory O_2 convection at 4000 m in pigeons at neutral or cold temperature. *Respiration Physiology* **28**, 371–85.

Bouverot, P. & Sébert, Ph. (1979). O_2-chemoreflex drive of ventilation in awake birds at rest. *Respiration Physiology* **37**, 201–18.

Brackenbury, J.H. (1986). Blood gases and respiratory pattern in exercising fowl: comparison in normoxic and hypoxic conditions. *Journal of Experimental Biology* **126**, 423–31.

Brackenbury, J.H. & Avery, P. (1980). Energy consumption and ventilatory mechanisms in exercising fowl. *Comparative Biochemistry and Physiology* **66A**, 439–45.

Brackenbury, J.H., Avery, P. & Gleeson, M. (1981). Respiration in exercising fowl. I. Oxygen consumption, respiratory rate and respired gases. *Journal of Experimental Biology* **93**, 317–25.

Brackenbury, J.H. & El-Sayed, M.S. (1985). Comparison of running energetics in male and female domestic fowl. *Journal of Experimental Biology* **117**, 349–55.

Brackenbury, J.H. & Gleeson, M. (1983). Effects of P_{CO_2} on respiratory pattern during thermal and exercise hyperventilation in domestic fowl. *Respiration Physiology* **54**, 109–19.

Brackenbury, J.H. & Gleeson, M. (1985). Separate and combined effects of temperature and hypoxia on breathing pattern in domestic fowl. *Journal of Comparative Physiology* **156**, 109–13.

Brackenbury, J.H. & Gleeson, M. (1986). Exercise hyperpnea in birds: evidence against a primary role for p_{CO_2}. *Comparative Biochemistry and Physiology* **83A**, 337–339.

Brackenbury, J.H., Gleeson, M. & Avery, P. (1982). Control of ventilation in running birds: effects of hypoxia, hyperoxia, and CO_2. *Journal of Applied Physiology: Respiratory, Environmental and Exercise Physiology* **53**, 1397–404.

Butler, P.J., Turner, D.L., Al-Wassia, A. & Bevan, R.M. (1988). Regional distribution of blood flow during swimming in the Tufted duck (*Aythya fuligula*). *Journal of Experimental Biology* **135**, 461–72.

Butler, P.J. & Woakes, A.J. (1980). Heart-rate, respiratory frequency and wing beat frequency of free flying Barnacle geese, *Branta leucopsis*. *Journal of Experimental Biology* **85**, 213–26.

Butler, P.J., West, N.G. & Jones, D. R. (1977). Respiratory and cardiovascular responses of the pigeon to sustained level flight in a windtunnel. *Journal of Experimental Biology* **71**, 7–26.

Faraci, F.M. (1986). Circulation during hypoxia in birds. *Comparative Biochemistry and Physiology* **85A**, 613–20.

Faraci, F.M., Kilgore, D.L. Jr & Fedde, M.R. (1984). Oxygen delivery to the heart and brain during hypoxia: Pekin duck vs Bar-headed goose. *American Journal of Physiology* **247**, R69–75.

Fedak, M.A., Pinshow, B. & Schmidt-Nielsen, K. (1974). Energy cost of bipedal running. *American Journal of Physiology* **227**, 1038–44.

Geiser, J., Gratz, R.K., Hiramoto, T. & Scheid, P. (1984). Effects of increasing metabolism by 2,4–dinitrophenol on respiration and pulmonary gas exchange in the duck. *Respiration Physiology* **57**, 1–14.

Gleeson, M. (1986). Respiratory adjustments of the anaesthetized chicken, *Gallus domesticus*, to elevated metabolism elicited by 2,4-dinitrophenol or cold exposure. *Comparative Biochemistry and Physiology* **83A**, 283–9.

Gleeson, M., Barnas, G.M. & Rautenberg, W. (1985). Respiratory and cardiovascular responses of the exercising chicken to spinal cord cooling at different ambient temperatures. II. Respiratory responses. *Journal of Experimental Biology* **114**, 427–41.

Gleeson, M. & Brackenbury, J.H. (1983). Respiratory and blood gas responses in exercising birds. *Comparative Biochemistry and Physiology* **76A**, 211–16.

Grubb, B.R. (1982). Cardiac output and stroke volume in exercising ducks and pigeons. *Journal of Applied Physiology: Respiratory, Environmental and Exercise Physiology* **53**, 207–11.

Grubb, B.R., Jones, J.H. & Schmidt-Nielsen, K. (1979). Avian cerebral blood flow: influence of the Bohr effect on oxygen supply. *American Journal of Physiology* **236**, H744–9.

Grubb, B., Jorgensen, D.D. & Conner, M. (1983). Cardiovascular changes in the exercising emu. *Journal of Experimental Biology* **104**, 193–201.

Grubb, B.R., Mills, C.D., Colacino, J.M. & Schmidt-Nielsen, K. (1977). Effects of arterial carbon dioxide on cerebral blood flow in ducks. *American Journal of Physiology* **232**, H596–601.

Hart, J.S. & Roy, O.Z. (1966). Respiratory and cardiovascular responses to flight in pigeons. *Physiological Zoology* **39**, 291–306.

Hart, J.S. & Roy, O.Z. (1967). Temperature regulation during flight in pigeons. *American Journal of Physiology* **213**, 1311–16.

Hudson, D.M. (1978). Power input, ventilation and thermoregulation during steady-state flight in the White-necked raven, *Corvus cryptoleucus*. Ph.D. thesis, New Mexico State University, Las Cruces.

Hudson, D.M. & Bernstein, M.H. (1978). Respiratory ventilation during steady-state flight in the White-necked raven *Corvus cryptoleucus*. *Federation Proceedings* **37**, 472.

Hudson, D.M. & Bernstein, M.H. (1981). Temperature regulation and heat balance in flying White-necked ravens, *Corvus cryptoleucus*. *Journal of Experimental Biology* **90**, 267–81.

Hudson, D.M. & Bernstein, M.H. (1983). Gas exchange and energy cost of flight in the White-necked raven, *Corvus cryptoleucus*. *Journal of Experimental Biology* **103**, 121–30.

Jones, D.R. & Purves, M.J. (1970). The effect of carotid body denervation upon the respiratory response to hypoxia and hypercapnia in the duck. *Journal of Physiology (London)* **211**, 295–309.

Johnson, S.F. & Gessaman, J.A. (1973). An evaluation of heart rate as an indirect monitor of free-living energy metabolism. In *Ecological Energetics of Homeotherms* (ed. J.A. Gessaman), pp. 44–54. Utah State University Press, Utah.

Kiley, J.P., Faraci, F.M. & Fedde, M.R. (1985). Gas exchange during exercise in hypoxic ducks. *Respiration Physiology* **59**, 105–15.

Kiley, J.P. & Fedde, M.R. (1983*a*). Cardiovascular control during exercise in the duck. *Journal of Applied Physiology: Respiratory, Environmental and Exercise Physiology* **55**, 1514–81.

Kiley, J.P. & Fedde, M.R. (1983*b*). Exercise hyperpnea in the duck without intrapulmonary chemoreceptor involvement. *Respiration Physiology* **53**, 355–65.

Kiley, J.P., Kuhlmann, W.D. & Fedde, M.R. (1979). Respiratory and cardiovascular responses to exercise in the duck. *Journal of Applied Physiology: Respiratory, Environmental and Exercise Physiology* **47**, 827–33.

Kiley, J.P., Kuhlmann, W.D. & Fedde, M.R. (1982). Ventilatory and blood gas adjustments in exercising isothermic ducks. *Journal of Comparative Physiology* **147**, 107–12.

Kuethe, D.O. (1988). Fluid mechanical valving of air flow in bird lungs. *Journal of Experimental Biology* **136**, 1–12.

Lefebvre, E. (1964). The use of D_2O^{18} for measuring the energy metabolism in *Columba livia*, at rest and in flight. *Auk* **81**, 403–16.

Lutz, P.L. & Schmidt-Nielsen, K. (1977). Effect of simulated altitude on blood gas transport in the pigeon. *Respiration Physiology* **30**, 383–8.

Nomoto, S., Rautenberg, W. & Iriki, M. (1983). Temperature regulation during exercise in the Japanese Quail (*Coturnix coturnix japonica*). *Journal of Comparative Physiology* **149**, 519–25.

Piiper, J. & Scheid, P. (1977). Diffusion within parabronchial air capillaries as a limiting factor for pulmonary gas exchange in birds. *Journal of Physiology (London)* **270**, 66–7.

Pennycuick, C.J. (1969). The mechanisms of bird migration. *Ibis* **3**, 525–56.

Perry, S.F. & Duncker, H.R. (1980). Interrelationship of static mechanical factors and anatomical structure in lung evolution. *Journal of Comparative Physiology* **138**, 321–34.

Powell, F.L. & Hempleman, S.C. (1985). Sources of carbon dioxide in penguin air sacs. *American Journal of Physiology* **17**, R748–52.

Shams, H. & Scheid, P. (1987). Respiration and blood gases in the duck exposed to normocapnic and hypercapnic hypoxia. *Respiration Physiology* **67**, 1–12.

Taylor, C.R., D'miel, R., Fedak, M. & Schmidt-Nielsen, K. (1971). Energetic cost of running and heat balance in a large bird, the rhea. *American Journal of Physiology* **221**, 597–601.

Torre-Bueno, J.R. (1978). Respiration during flight in birds. In *Respiratory Function in Birds, Adult and Embryonic* (ed. J. Piiper), pp. 89–94. Springer-Verlag, Berlin.

Torre-Bueno, J.R. & Larochelle, J. (1978). The metabolic cost of flight in unrestrained birds. *Journal of Experimental Biology* **75**, 223–9.

Tucker, V.A. (1968). Respiratory physiology of House sparrows in relation to high-altitude flight. *Journal of Experimental Biology* **48**, 55–66.

Tucker, V.A. (1972). Metabolism during flight in the laughing gull, *Larus atricilla*. *American Journal of Physiology* **222**, 237–45.

Tucker, V.A. (1973). Bird metabolism during flight: Evaluation of a theory. *Journal of Experimental Biology* **58**, 689–709.

Turner, D.L. & Butler, P.J. (1988). The aerobic capacity of locomotory muscles in the Tufted duck *Aythya fuligula*. *Journal of Experimental Biology* **135**, 445–60.

Wang, N., Banzett, R.B., Butler, J.P. & Fredberg, J.J. (1988). Bird lung models show that convective inertia effects inspiratory aerodynamic valving. *Respiration Physiology* **73**, 109–24.

Weinstein, Y., Bernstein, M., Bickler, P.E., Gonzales, D.V., Samaniego, F.C. & Escobedo, M.A. (1985). Blood respiratory properties in pigeons at high altitudes: effects of acclimation. *American Journal of Physiology* **249**, R765–75.

Woakes, A.J. & Butler, P.J. (1983). Swimming and diving in Tufted ducks, *Aythya fuligula*, with particular reference to heart rate and gas exchange. *Journal of Experimental Biology* **107**, 311–29.

Woakes, A.J. & Butler, P.J. (1986). Respiratory, circulatory and metabolic adjustments during swimming in the Tufted duck, *Aythya fuligula*. *Journal of Experimental Biology* **120**, 215–31.

Zeuthen, E. (1942). The ventilation of the respiratory tract in birds. *Kongelige Danske Videnskabernes Selskabs Skrifter* **17**, 1–50.

ERICH GNAIGER

Animal energetics at very low oxygen: information from calorimetry and respirometry

Introduction: 'facultative' anaerobiosis?

For more than two decades the topic of anaerobiosis has been the subject of intensive investigation in invertebrate physiology and biochemistry (Hammen, 1969; Hochachka & Mustafa, 1972; Saz, 1971; Simpson & Awapara, 1966; De Zwaan & Zandee, 1972; see also von Brand, 1946). Still, mammalian biochemists are insufficiently aware of the importance of anaerobic mitochondrial energy transformation in metazoans, although ATP formation coupled to anaerobic mitochondrial electron transport occurs not only in invertebrates (Saz, 1981) but in mammalian myocyte mitochondria (Wiesner *et al.*, 1988).

The term invertebrate facultative anaerobiosis, adopted from the microbiological literature, implies the 'unsubstantiated assumption that there are facultative anaerobes among metazoans' (Hammen, 1976). It has been applied with reference to animals that do not tolerate anoxia for more than several days. Growth is arrested in such euryoxic animals as oxygen is depleted and other energy-demanding activities are greatly reduced, whereas by definition facultative microorganisms grow under anoxia. The most tolerant animals enter a state of dormancy for long-term anoxic survival, and metabolic heat flux is lowered to as little as 2% of the aerobic level (Hand & Gnaiger, 1988). To date it is not known if any animal is capable of completing its life cycle under strict anoxia, with the possible exception of meiofauna living in the anoxic sulphide system (Meyers *et al.*, 1987; Powell *et al.*, 1980; Wieser *et al.*, 1974). A reported example of facultative anaerobiosis among aquatic oligochaetes has been critically discussed by Gnaiger & Staudigl (1987).

The previous critique of the term invertebrate facultative anaerobiosis focused on the duration of anoxia. An additional, important, yet much neglected aspect is discussed here, namely the extent and quantification of the 'anaerobic' condition. The actual measurement and control of oxygen is not simple under conditions when oxygen is supposedly not available to

A
B
C
30 min
P_{O_2} (kPa)
0.8
0.4
0.0
0.4
0
diffusion (pmol O$_2$ s^{-1})
60
40
20
0
20
0
A_m
A_c
A_c
A_m
B_m
B_c
B_m
C_m
B_c
B_c
B_m
B_c
C_c
B_m

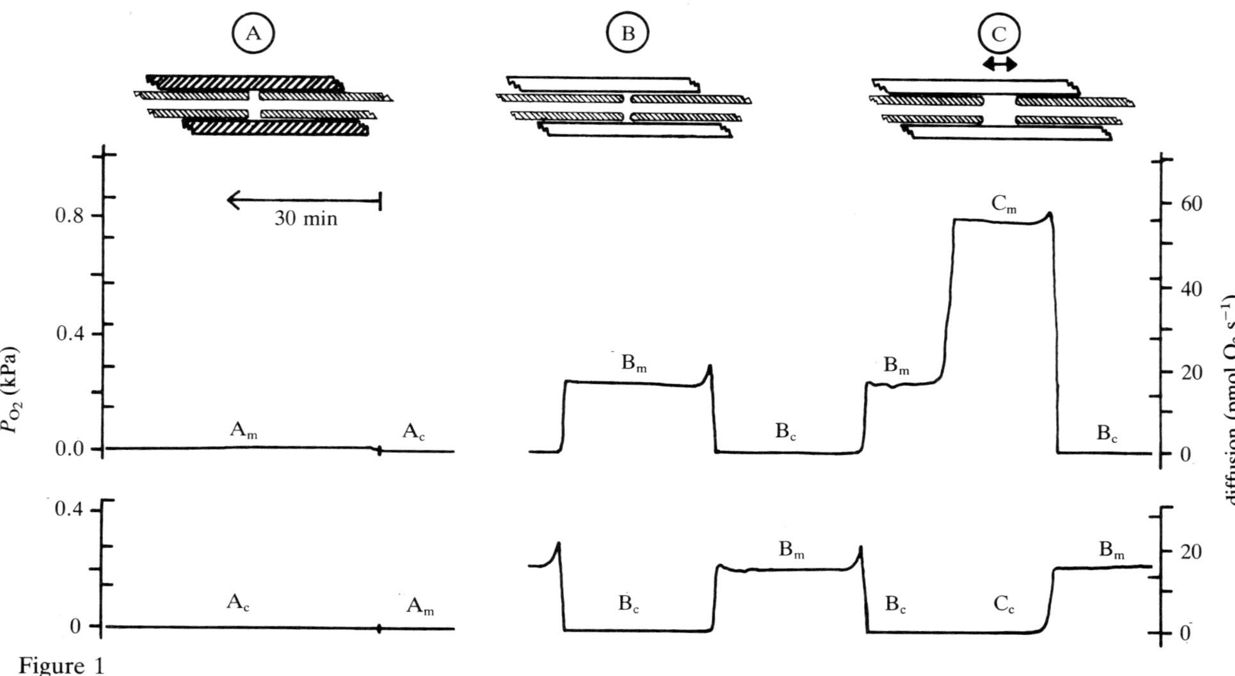

Figure 1

the animal. Special methods are required to detect oxygen uptake at very low oxygen concentrations. Whereas specific anaerobic aspects of intermediary metabolism are quantified by biochemical analysis of accumulated and excreted organic end products, direct calorimetry detects indiscriminately both aerobic and anaerobic sources of metabolic heat at any oxygen regime (Pamatmat, 1978; Gnaiger, 1983*a*). Therefore, direct calorimetry should complement respirometric and biochemical approaches in the study of energy metabolism at very low oxygen.

Anaerobic metabolism

In zoophysiology, 'anaerobic' (without air) is rarely defined in terms of controlled measurements of the actual extent of anaerobic conditions. By comparison, oxygen exclusion techniques are highly advanced in anaerobic microbiology (Holland *et al.*, 1987) and in studies of isolated microxic cells and mitochondria (see below). Despite careful precautions to exclude molecular oxygen, this is not always achieved completely owing to insufficient oxygen removal and diffusion of oxygen through the materials of experimental vessels, seals and tubes (Figure 1). If no measurements of the actual oxygen level in 'anaerobic' media are available, caution is required when interpreting the biological results. When strictly *anoxic* conditions are not achieved, *anaerobic* metabolism proceeds simultaneously with oxygen consumption (Gnaiger & Staudigl, 1987).

For instance, in one of the few biochemical publications where the upper level of oxygen in 'anaerobic' experiments is reported, the value 'less than

Figure 1. Oxygen diffusion through silicone tubing connecting two stainless steel capillaries (1.1 mm inner diameter) in an open-flow calorespirometry system. A, Viton tubing without detectable diffusion; B and C, silicone tubing with narrow spacing and a 3 mm gap between the inner steel capillaries, respectively, resulting in elevated oxygen of (B) 0.24 kPa (1.8 mmHg; 3.0 μmol dm^{-3}; 1.2% air saturation) and (C) 0.80 kPa (6.0 mmHg; 10.0 μmol dm^{-3}; 3.9% air saturation). Recorder traces are shown of the two polarographic oxygen sensors (POS) of the Twin-Flow respirometer (Cyclobios, Austria). Water is equilibrated with nitrogen in a glass reservoir, connected to the Twin-Flow valve system (Gnaiger, 1983*b*) by a stainless steel capillary. While one POS is in *calibration* position directly connected to the water reservoir (subscript c) the other POS is in *measuring* position after the water has passed the capillary connection (subscript m). The trace in the upper panel shows, from right to left: B_c, the calibration at 0 kPa; C_m, the switch to measuring position while the capillary gap is simultaneously increased to 3 mm; B_m, the response to a narrowing of the gap; and B_c, back to calibration. Perfusion with deionized water is kept constant at 5.56 mm^3 s^{-1} (20.0 cm^3 h^{-1}) at 20 °C.

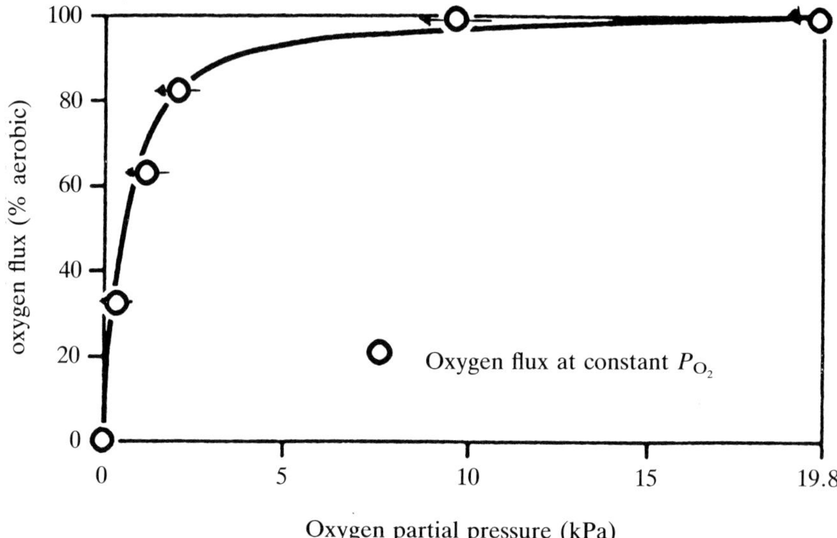

Oxygen partial pressure (kPa)

Figure 2. Oxygen flux–pressure relation in the planktonic copepod *Cyclops abyssorum*. Oxygen flux was measured at 6 °C in a 0.5 cm^3 glass chamber of a Cyclobios Twin-Flow respirometer at 5 steady-state P_{O_2} levels. The P_{O_2} of the inflow and outflow water is shown by arrows (after Gnaiger 1983*b*).

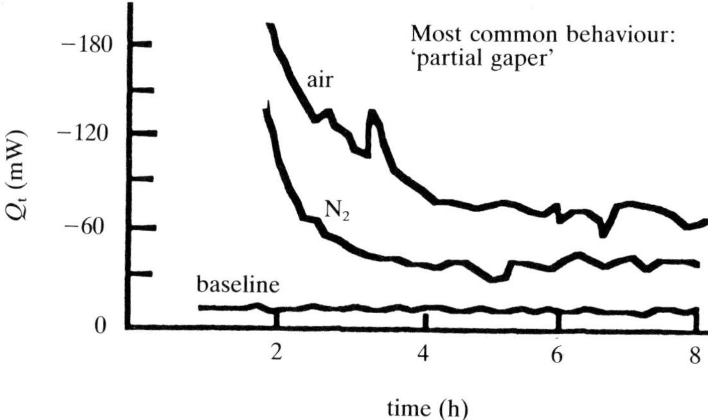

time (h)

Figure 3. Heat flux of the intertidal mussel *Mytilus edulis* during exposure to air and nitrogen gas. From the difference between the traces at steady state it is concluded that 40% of total heat flux in air is aerobic. Heat flux was measured in a 25 ml stainless steel chamber of a Thermal Activity Monitor (ThermoMetric, Sweden) at 15 °C (from Shick *et al.*, 1986).

0.5 ml O_2' per litre is specified for anaerobic energy metabolism in an isopod (De Zwaan & Skjoldal, 1979). At experimental conditions (13 °C; seawater) the oxygen concentration at 100% air saturation is 264 µmol dm^{-3}; therefore the oxygen concentration of 0.5 ml l^{-1} (22.3 µmol dm^{-3}) corresponds to 8.4% air saturation (1.7 kPa). Although the actual oxygen content may have been less and this particular isopod species is probably not an oxygen regulator, it is important to note that the oxygen consumption of some Crustacea is virtually unaffected down to oxygen pressures of 1.7 kPa (Figure 2).

Another example illustrates the relative meaning of 'anaerobic'. During exposure to air, the intertidal mussel *Mytilus edulis* maintains an aerobic metabolism responsible for up to 40% of the total (anaerobic and aerobic) heat flux (Figure 3). Thus, the simultaneous operation of aerobic and anaerobic pathways of *M. edulis* exposed to air is important, and the ratio depends on season and nutritional status (Widdows & Shick, 1985; Shick *et al.*, 1988). Obviously, total metabolism of mussels exposed to air cannot be referred to as anaerobic without testing for aerobic contributions to total ATP turnover. Too little attention has been paid to the quantitative significance of residual oxygen consumption relative to anaerobic ATP generation under 'anaerobic' conditions.

Methods at low oxygen

Most studies on the oxygen dependence of respiration have been carried out in closed respirometric chambers under conditions of continuously declining oxygen pressure. In this approach limited time is available for observation of hypoxic respiration. Transient conditions may elicit a different response compared with steady-state conditions, which can be maintained for long periods of time in open- or intermittent-flow respirometers (Bayne & Livingstone, 1977). Oxygen consumption of *Cyclops abyssorum* (Figure 2) and of the aquatic oligochaete *Lumbriculus variegatus* was measured in a Cyclobios Twin-Flow respirometer at various constant P_{O_2} levels of the inflow (Gnaiger, 1983*b*). The inflow–outflow P_{O_2} difference varies according to the experimental oxygen uptake at a constant perfusion flow maintained through the system (Figure 4, arrows).

If the inflow water is immediately mixed with the bulk water in a well-stirred animal chamber, then the outflow P_{O_2} is the relevant partial pressure experienced by the animal. In an unstirred chamber the turbulence generated by the animals is irregular and undefined. Then the average P_{O_2} in the gradient between inflow and outflow is a more accurate choice than

Table 1. *Amperometric and redox titration in chemical Winkler analysis of dissolved oxygen in air-equilibrated and N_2-equilibrated water samples*
Values are means±standard deviations; n=4.

	Amperometric (μmol O_2 dm^{-3})	% Air	Redox (μmol O_2 dm^{-3})	% Air
Air	261.6±1.06	98.56	263.8±0. 75	99.36
Air	264.7±0.31	99.06	256.3±0.28	99.34
N_2 techn.	1.8±0.09	0.67	1.9±0.06	0.71
N_2 pure	1.5±0.19	0.58	1.5±0.16	0.57

From Gnaiger (1983*c*).

reference to the inflow oxygen pressure (Gnaiger, 1983*b*; Hughes *et al.*, 1983). At low perfusion flow and low oxygen levels, respiration may be limited by the total oxygen delivery into the system rather than by P_{O_2}, particularly when the outflow P_{O_2} remains near zero irrespective of changes in the inflow P_{O_2}. The oxygen delivery is the inflow oxygen concentration multiplied by the perfusion flow. Different problems are associated with either perfusion or closed-chamber respirometry when defining the dependence of oxygen flux on oxygen pressure, and a systematic investigation of these methodological aspects is required.

Calorespirometric studies show that at a P_{O_2} of 0.1 kPa (0.5% air saturation), aerobic metabolism may amount to as much as 33% of total heat flux (see Figure 6). It is worrying that this oxygen pressure is near the limit of detection in some routine applications of polarographic oxygen sensors with small cathodes (Hitchman, 1983). This is overcome in the Twin-Flow respirometer by frequent, automatic calibration of the oxygen sensors with large cathodes (Figure 1). Moreover, 0.5% air saturation is at the limit of detection of the Winkler method in N_2 equilibrated water (Table 1).

The sensitivity of a polarographic oxygen sensor, with a guard cathode for scavenging the oxygen that diffuses from the electrolyte and sensor body (Orbisphere, Switzerland) (Hitchman, 1983), is extended to oxygen concentrations of 0.003 μmol dm^{-3} (*ca.* 0.001% air saturation, 0.2 Pa or 0.002 mmHg). This sensor was elegantly used to study the respiration of isolated myocytes (Wittenberg & Wittenberg, 1985). Optical methods, particularly for studies of tissue P_{O_2}, have been reviewed by Tamura *et al.* (1989).

Critical and limiting oxygen pressure

Oxygen regulators are characterized by a critical P_{O_2} (Pc), above which the resting, routine or standard metabolic energy flux is relatively independent of oxygen pressure. Below the Pc the regulation of oxygen flux becomes less efficient (Bayne & Livingstone, 1977). The Pc of regulators is significantly below normoxia. Metabolic hypoxia is indicated as a reduced oxygen flux below the critical oxygen pressure (Figure 2) and is either fully or partly anaerobic. The metabolism of the planktonic copepod *Cyclops abyssorum* is completely aerobic at 2 kPa (15 mmHg) as measured by simultaneous direct and indirect calorimetry (calorespirometry; Gnaiger, 1983*b*, and unpublished results). In the lugworm *Arenicola marina*, the Pc is near air saturation at 16 kPa (120 mmHg) (Toulmond, 1975) but anaerobic processes are not switched on until partial oxygen pressure reaches 6 kPa (45 mmHg) (Schöttler *et al.*, 1983). Above this 'second critical' (Toulmond, 1975) or 'limiting' oxygen pressure (Schöttler *et al.*, 1983) there is an extended phase of fully aerobic hypoxia. Anaerobiosis compensates progressively for the declining oxidative energy supply below the limiting oxygen pressure.

Oxygen flux of *L. variegatus* is nearly independent of oxygen pressure between 50 and 100% air saturation (94% of the normoxic respiration observed at 8 kPa) and decreases non-linearly under hypoxic conditions (Figure 4; Figure 5*a*, heavy line). A hyperbolic fit (Figures 2 and 4) does not imply that the theory of Michaelis–Menten kinetics is applicable to a complex animal system. Interpretation of a calculated K_M as the whole organism's oxygen affinity is not meaningful (50% of normoxic respiration is reached at 0.6 and 1.2 kPa; Figures 2 and 4). Furthermore, a 50% drop in oxygen flux does not reflect the functional aspect of hypoxia. Neither can an unambiguous definition of the Pc be applied when a sharp discontinuity does not occur in the oxygen flux-pressure relation (Rosenthal *et al.*, 1976). If the critical P_{O_2} is arbitrarily defined as the P_{O_2} at which respiration declines below 75% of the normoxic value, then *C. abyssorum* has a Pc of *ca.* 1.5 kPa (Figure 2), and in *L. variegatus* the Pc is at 3 kPa (Figure 4). As an oxygen regulator, *L. variegatus* shows a steep decline of oxygen consumption in the microxic region towards anoxia. A significant oxygen uptake is observed at the lowest P_{O_2} of 0.08 kPa (0.6 mmHg) (see also Figure 6).

The distinction between the critical and limiting P_{O_2} is also important at the cellular level. Resting myocytes isolated from heart muscle regulate oxygen flux down to 0.05 kPa (0.4 mmHg). The limiting P_{O_2} of these cells,

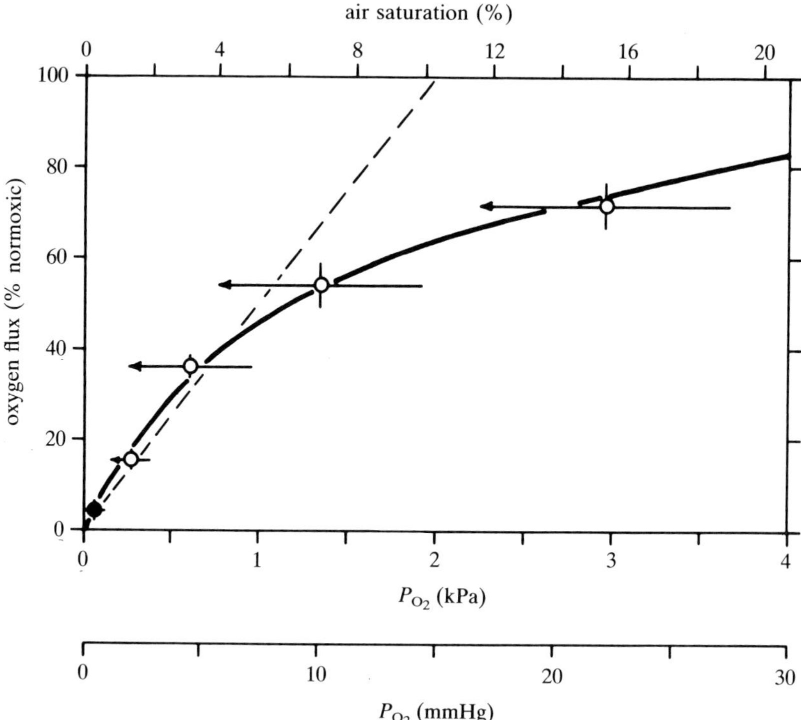

Figure 4. Oxygen flux–pressure relation in the aquatic oligochaete *Lumbriculus variegatus*. Oxygen flux was measured at 20 °C in a 0.5 cm³ glass chamber of a prototype Twin-Flow respirometer at 7 steady-state P_{O_2} levels. The P_{O_2} of the inflow and outflow water is shown by horizontal arrows. The vertical bars show the ranges of oxygen flux of 2 (open circles) or 4 experiments (full circle) normalized relative to the oxygen consumption at normoxia (11.8 ± 4.6 nmol O_2 s^{-1} g^{-1} dry mass, $n=6$). At the lowest P_{O_2} (full circle) respiration was calculated as the difference of the oxygen diffusion into the system in the absence and presence of animals (Gnaiger & Staudigl, 1987; see also Figure 6). Steady-state respiration was observed for at least two hours at each P_{O_2}. (The broken line and heavy line are those of Figure 5a.)

observed as a steep increase in lactate production, is about 5 times lower (Wittenberg & Wittenberg, 1985) (Figure 7).

Microxic regulators versus oxygen conformers

In contrast to oxygen regulators, oxygen conformers show a linear decline in oxygen flux as P_{O_2} decreases from the normoxic level. Conformity and regulation are a matter of degree, depending on various environmental and physiological variables (Bayne, 1971; Mangum & van

Winkle, 1973). The oxygen flux–pressure relation reflects the operation of complex regulatory feedback mechanisms of oxygen supply and ATP demand, described as a dynamic balance of drive-coupled and load-coupled metabolism (Gnaiger, 1987).

Various types of conforming pattern are distinguished depending on the intercept of the linear slope between oxygen flux and P_{O_2} (Herreid, 1980). In a perfect conformer the relation between oxygen flux and pressure is not only linear but proportional (for a discussion of generalized flux–pressure relations, see Gnaiger (1989)). A steeper slope with a negative intercept results in zero or very low oxygen flux before completely anoxic conditions are reached. Thus, uptake of oxygen can be diminished owing to cessation of blood circulation, to retraction behaviour or to valve closure.

A linear slope is commonly observed between oxygen flux and partial pressure over an extended hypoxic range, with a positive intercept when extrapolated to zero oxygen. Examples are found in polychaetes (Petersen & Johansen, 1967), nematodes (Atkinson, 1973), sipunculids (Pörtner *et al.*, 1985), crustaceans (reviewed by Herreid, 1980), fish (Hughes *et al.*, 1983; Pullin *et al.*, 1980; Ultsch *et al.*, 1981) and snakes (Abe & Mendes, 1980). The extrapolated intercept can be as high as 40% of the oxygen flux at normoxia, with a linear slope from a P_{O_2} of 20 kPa down to 4–6 kPa (150 to 30–45 mmHg). Although classically described as conformers with emphasis on the hypoxic region, these animals behave just like regulators from the perspective of the microxic region around 1% air saturation (0.2 kPa or 1.5 mmHg) (Figure 5*a*). Microxic regulation, therefore, effectively increases the slope of the flux–pressure relation in the microxic region. The slope of an ideal conformer is equivalent to a 5% increase of oxygen flux per kPa (Figure 5*a*).

Irrespective of the position of the *Pc* below or above normoxia in a regulator or conformer, respectively, the onset of anaerobic processes at the limiting P_{O_2} can be defined experimentally by specific biochemical indices (Livingstone & Bayne, 1977; Burton & Heath, 1980; Pörtner *et al.*, 1985; Schöttler *et al.*, 1983), by more general acid–base balance studies (Toulmond, 1975), and most generally by calorespirometry (Gnaiger *et al.*, 1989) (see below).

The importance of 'microxic regulation' (Figure 5*a*) for total energy metabolism at very low oxygen is illustrated by a simplified model of physiological energetics at anoxic and various microxic steady states. Based on a stoichiometric and thermodynamic analysis (Gnaiger, 1983*a*), the energy saving on account of residual oxygen consumption under 'anaerobic' conditions can be calculated. It is assumed (Figure 5*b–d*) that

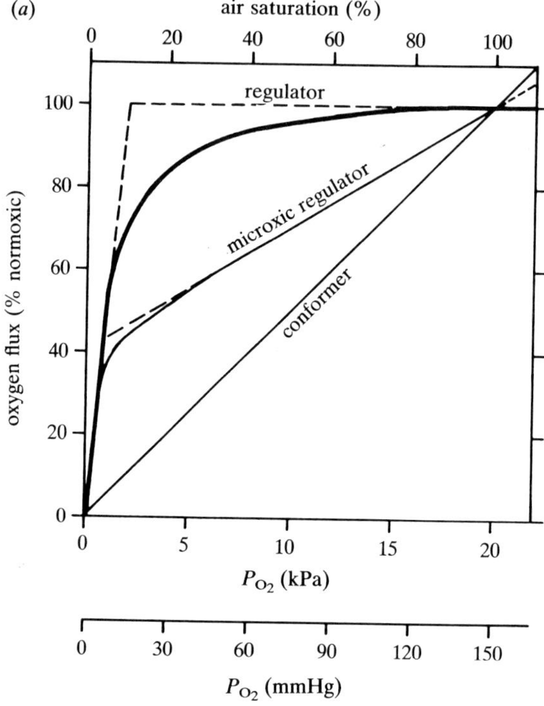

Figure 5. Energetics at low oxygen: Comparison of an ideal conformer with an ideal regulator and a microxic regulator. (*a*) Oxygen flux–pressure relations, normalized relative to the flux under normoxia. The thick hyperbolic line for the regulator is the experimental observation in *Lumbriculus variegatus* (Figure 4). (*b*) Contribution of aerobic metabolism to total heat flux under microxic conditions, assuming that anoxic heat flux is constant in the microxic region at 10% of heat flux under normoxia (see

the animal utilizes glycogen exclusively; the propionate–acetate pathway is the only anoxic process, with excretion of the acids; anoxic heat flux is 10% of normoxic heat flux, and the corresponding propionate and acetate production is maintained at a constant level at P_{O_2} from 0 to 0.5 kPa. In this region, the metabolism of the regulator and microxic regulator is identical, with a slope 10 times that of the perfect conformer (Figure 8.5*a*).

At 1% air saturation (0.2 kPa or 1.5 mmHg), 50% of the total heat flux is aerobic in the regulators. Even the conformer is 9% aerobic (Figure 5*b*). Few respiratory measurements have been made with animals at microxic conditions.

The calculated ATP turnover increases steeply in the regulators, reaching nearly 150% of the anoxic ATP turnover at 0.2 kPa (1% air saturation),

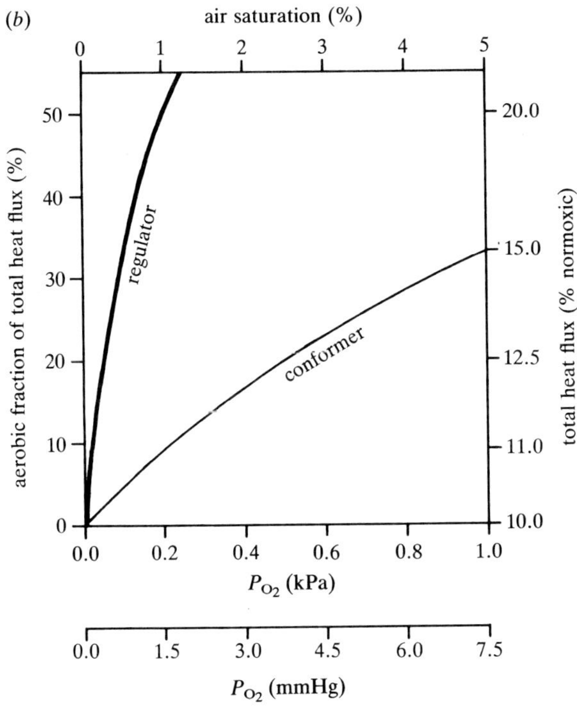

Caption for fig. 5. (cont.)

text). (*c*) Glycogen utilization (solid lines) and ATP turnover (broken lines) in the microxic range, normalized relative to the fluxes under anoxia. The hatched area indicates the range of deviation between the regulator (upper boundary line) and the microxic regulator (lower boundary line). Heat flux (*b*) is converted into the flux of glycogen utilization by division by the catabolic enthalpy per mol glycosyl-unit, $\Delta_k H_{Glyc}$,

while glycogen consumption is only increased to 108% of the anoxic demand (Figure 5*c*). This difference is due to the high ATP coupling stoichiometry of aerobic metabolism, 37 mol ATP per mol glycosyl-unit compared with 6.4 mol ATP per mol glycosyl-unit in the propionate–acetate pathway (Gnaiger, 1983*a*). Thus, microxic regulation is an effective mechanism to increase simultaneously the flux of ATP turnover and the ATP stoichiometry per unit of substrate (Figure 5*d*).

Various quantitative aspects of microxic regulation require modification for analysis of specific cases. (i) Glycogen is rarely the exclusive or even the predominant substrate of aerobic catabolism under normoxia (see, for example, Hawkins *et al.*, 1985). Information on the substrate of aerobic catabolism under hypoxic and microxic conditions is rarely available.

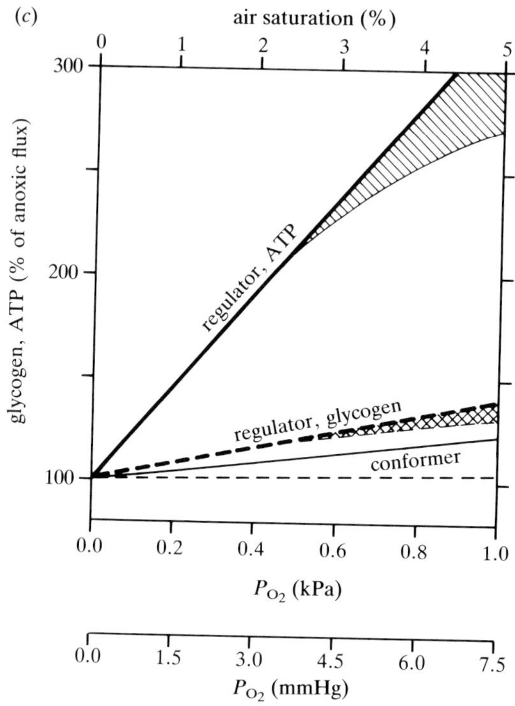

Caption for fig. 5. (cont.)
which is -2868 and $-226\,$kJ mol^{-1} in aerobic oxidation and anaerobic propionate–acetate excretion, respectively (Gnaiger, 1983a). At an anoxic heat flux of 10% of the normoxic level, therefore, anoxic glycogen utilization (100% in c) is $10\times -2868/-226=127\%$ of the normoxic glycogen utilization. Heat flux is converted into the catabolic flux expressed as ATP-turnover by division by the catabolic enthalpy per mol ATP

Therefore, the small increase of glycogen utilization (Figure 5c) merely illustrates the demand for organic carbon. (ii) Accumulation and excretion of different anaerobic end-products occurs, depending on the P_{O_2}, temperature and time of exposure (Livingstone, 1978). Accordingly, the appropriate thermochemical constants have to be used, ranging from -160 to $-260\,$kJ mol^{-1} glycosyl-unit, or -35 to $-72\,$kJ mol^{-1} ATP turnover with glycogen as the substrate (Gnaiger, 1983a). (iii) Anaerobic metabolism is constant in the microxic range up to 2 kPa (15 mmHg) in *Arenicola marina* (Schöttler *et al.*, 1983). Even if aerobic ATP turnover is simply added to a constant anaerobic flux, microxic regulation is a highly significant strategy (Figure 5). Microxic regulation, however, is energetically even more

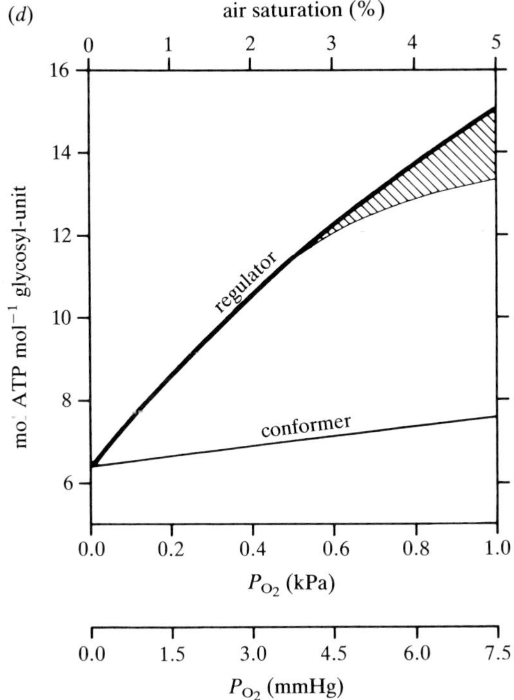

Caption for fig. 5. (cont.)
turnover, $\Delta_k H_{\infty \text{ATP}}$, which is -77.5 and $-35.3\,\text{kJ mol}^{-1}$, respectively. At an anoxic heat flux of 10% of the normoxic level, therefore, anoxic ATP turnover (100% in *c*) is $10 \times -77.5/-35.3 = 22\%$ of the normoxic ATP turnover. (*d*) Stoichiometric coefficient of ATP production per unit glycogen utilization, increasing with an increasing aerobic contribution to total heat flux.

important if anaerobic catabolism is progressively compensated by increasing aerobic ATP turnover in the microxic P_{O_2} range.

Aerobic compensation is actually observed in the microxic regulator *Sipunculus nudus*, where anaerobic metabolism is reduced from a P_{O_2} of *ca.* 0 to 0.4 kPa (3 mmHg) (Pörtner *et al.*, 1985), and in the regulator *Lumbriculus variegatus*, where anaerobic ATP turnover at 0.2 kPa is *ca.* 60% of ATP turnover under anoxia (Putzer, 1985). The stoichiometric coefficient of ATP production increases from 6.4 to 9.6 mol ATP mol^{-1} glycosyl-unit with an increase in the P_{O_2} from 0 to 0.2 kPa in the 'additive' microxic regulator (Figure 5*d*), whereas aerobic compensation at the extent observed in *L. variegatus* increases the stoichiometry to 10.0 mol

ATP mol^{-1} glycosyl-unit at 0.2 kPa under otherwise identical metabolic conditions.

In addition to these important energetic aspects of microxic regulation, residual oxygen uptake contributes to the redox balance of intermediary metabolism. Under strictly anoxic conditions the molar acetate:(succinate+propionate) ratio has a theoretical maximum value of 0.5 in metabolic pathways of glycogen utilization known in animals (Gnaiger, 1983*a*). If microbial contamination is excluded, therefore, an acetate ratio of more than 0.5 indicates residual oxygen uptake (Gnaiger & Staudigl, 1987). Residual oxygen uptake at low oxygen and exhaustion of internal oxygen stores under anoxia may be related to the partial oxidation of protein and amino acids. This might explain some of the variability observed in the stoichiometric patterns of 'anaerobic' aspartate utilization and provide an alternative explanation to anaerobic ammonia fixation in the case of excess alanine and glutamate accumulation.

Calorimetry and residual oxygen consumption

Owing to the low substrate demand of aerobic metabolism, the measurement of glycogen contents for the construction of biochemical carbon balances is not precise enough (De Zwaan & Wijsman, 1976) for the detection of residual oxygen uptake. This is true even if the other potential aerobic substrates, protein and lipid, are not utilized. The problem arises owing to the interindividual variability of organic biomass and composition, which affects the estimation of biochemical change by the destructive whole-body analysis. Since calorimetry is a non-invasive method which monitors the enthalpy change of metabolic processes directly, a higher accuracy is obtained.

The caloric quotient of heat production to glycogen consumption is -2868 kJ mol^{-1} glycosyl-unit in aerobic dissipative catabolism, but only -226 kJ mol^{-1} when propionate and acetate are excreted (Gnaiger, 1983*a*). If only 1% of the total anaerobic glycogen consumption is due to residual oxygen uptake, then the calorimetric signal of heat flux increases by 12% relative to completely anoxic catabolism. Therefore, calorimetry is a particularly sensitive method for the detection of such residual oxygen uptake (Figure 6). The corresponding P_{O_2} may be as low as 0.02 kPa (0.15 mmHg or 0.1% air saturation) in a microxic regulator (Figure 5*b*); this P_{O_2} is below the limit of detection of conventional analytical methods (Figure 7).

Open-flow calorespirometry was applied by Widdows *et al.* (1989) to study the energetics of the oyster *Crassostrea virginica* at low oxygen and

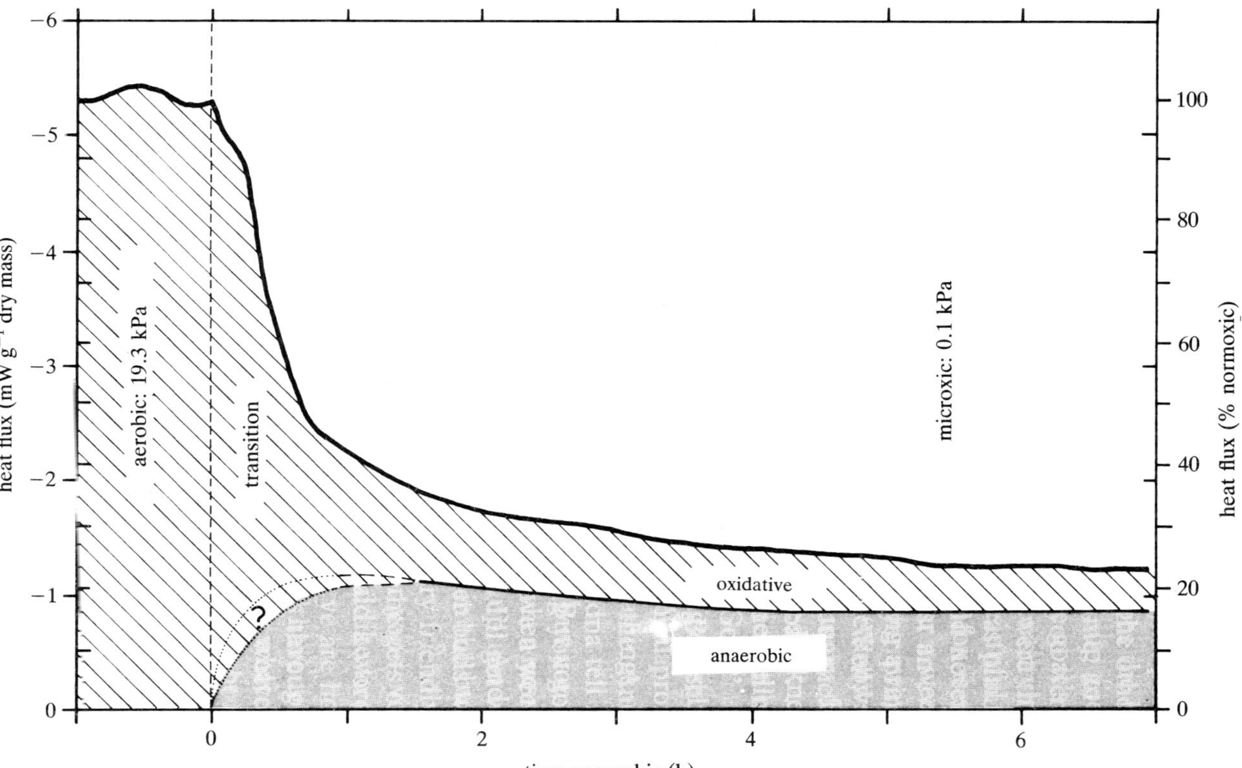

Figure 6. Heat flux of *Lumbriculus variegatus* at 20 °C under normoxic conditions ($P_{O_2} = 19.3$ kPa, fully aerobic) and microxic conditions ('anaerobic', $P_{O_2} = 0.1$ kPa, 0.75 mmHg, 0.5% air saturation), measured in a calorespirometer system. The low time resolution of the Twin-Flow respirometer during the aerobic–anaerobic transition does not allow for an accurate separation of aerobic and anoxic sources of heat during the transition period (? on graph). At 0.1 kPa a significant contribution to total heat flux is aerobic (oxidative, 30%) owing to residual oxygen consumption (modified after Gnaiger & Staudigl, 1987).

different developmental stages (see also Widdows, 1989). There is no major anaerobic component under hypoxia down to limiting P_{O_2} levels of 1.3 kPa, which is below the *Pc*. The flux–oxygen pressure plots are complicated by the fact that a percentage of the oysters close the shell valves at low oxygen and generate an internally reduced P_{O_2}. The early larval stages are active under severe hypoxia and anoxia, a mechanism to avoid long-term oxygen depletion. These larvae show particularly pronounced microxic regulation of oxygen flux and some aerobic compensation of anaerobic metabolism at a P_{O_2} of 0.67 kPa (Widdows *et al.*, 1989) or presumably even lower if the outflow P_{O_2} or an average P_{O_2} in the perfusion system were considered.

The energetic counterpart of microxic regulation and residual oxygen uptake is *partial* anaerobiosis in aerobic metabolism. If the calorimetrically measured heat flux under aerobic conditions is increased by 5% owing to anoxic catabolism, the glycogen demand rises by 63% and the energetic cost of ATP production increases by 44%, assuming excretion of propionate and acetate (Gnaiger & Staudigl, 1987). Free-living aerobic animals, therefore, are under strong selective pressure to utilize a limited energy supply fully aerobically. This expectation agrees with experimental findings of balanced aerobic energy budgets under aerobic conditions in whole animals (Gnaiger & Staudigl, 1987).

Mitochondria: the terminal oxygen acceptor

Ultimately, the oxygen-dependence of respiration is a function of oxygen pressure at the mitochondria and of the response of mitochondrial oxygen flux to the local oxygen pressure. Mitochondria are protected from high oxygen by extracellular oxygen diffusion barriers and by clustering. The critical P_{O_2} in vasculature of the heart is tenfold higher than that in isolated mitochondria (Tamura *et al.*, 1989) (Figure 7). In the working heart, the oxygen pressure difference between capillary lumen and mitochondria is 2.7 kPa (20 mmHg) and is mostly extracellular. The drop in P_{O_2} between cystosol and mitochondrion is small, less than 0.03 kPa (Clark *et al.*, 1987; Wittenberg & Wittenberg, 1985). In myocytes, low oxygen pressure is buffered by myoglobin (Wittenberg & Wittenberg, 1985). Connett *et al.* (1985*a*) argue that in red muscle cells the *Pc* must be 0.07 kPa (0.5 mmHg; 0.3% air saturation) and oxygen is not limiting even at maximally stimulated oxygen flux. Respiration of isolated mitochondria is not affected by decreasing oxygen pressure until below $1\,\mu\text{mol dm}^{-3}$ (0.01 to 0.1 kPa, depending on metabolic state) (Oshino *et al.*, 1974; Sugano *et al.*, 1974; Cole *et al.*, 1982).

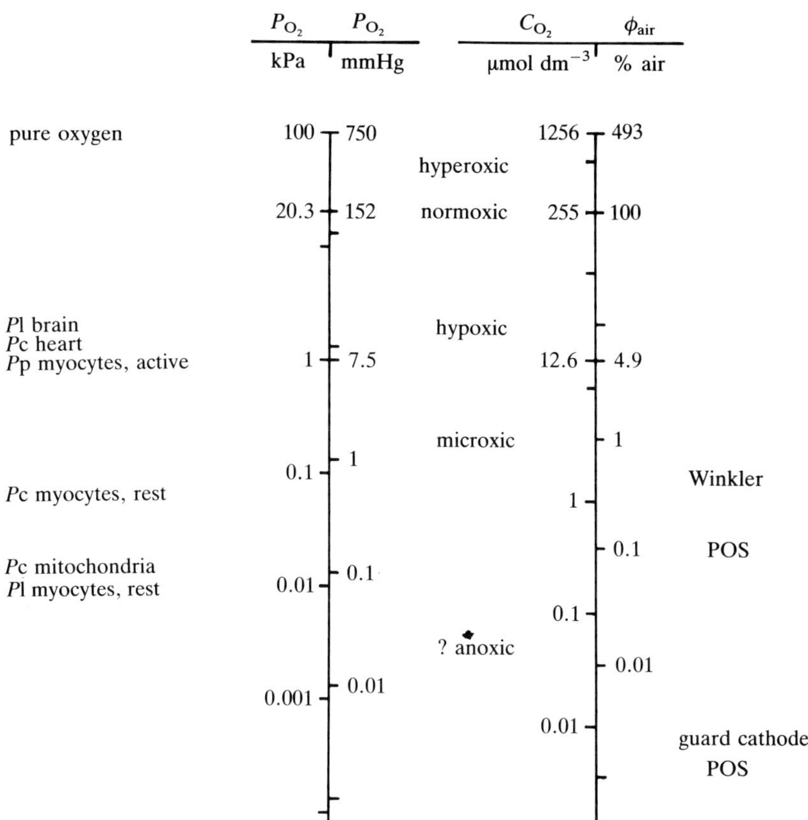

Figure 7. Logarithmic presentation of partial pressure of oxygen, P_{O_2} (1 kPa=0.133322 mmHg). The corresponding oxygen concentration, C_{O_2} (µmol dm^{-3}), is given for 25 °C in pure water. The volume–percentage air saturation values, ϕ_{air}, are for the standard barometric pressure of 100 kPa and corrected for a 3.2 kPa water vapour saturation pressure at 25 °C. Left: some critical and limiting partial oxygen pressure values, Pc and Pl, for mammalian systems (see text for references). Right: limit of detection of some routine methods (POS, polarographic oxygen sensor).

Owing to clustering of mitochondria in sites of high ATP demand, the oxygen dependencies of isolated liver mitochondria and mitochondria in intact hepatocytes are different (Jones, 1986). In the liver, glycolysis is known to play an important role in overall energy-yielding metabolism in the activated state of gluconeogenesis (Soboll *et al.*, 1978), possibly owing to local hypoxia (Baumgärtl & Lübbers, 1983). The respiratory flux in intact hypoxic cells is controlled at constant levels by adjustments of the phosphorylation potential, resulting in independence of cellular oxygen consumption despite the effect of lowered oxygen on the mitochondrial redox state (Wilson *et al.*, 1977, 1979*a,b*). Aerobic glycolysis plays a regulatory role in muscular rest–work transitions without occurrence of anoxia at any time or location (Connett *et al.*, 1985*b*).

Importantly, severe hypoxia occurs under pathological conditions due to impaired oxygen delivery, and little is known about the role of mitochondrial energy metabolism under these conditions. In tissues sensitive or intolerant to hypoxia, mitochondria suffer from a decay in structure, the degree of coupling is reduced, and membrane potentials decrease (Zimmer *et al.*, 1985). Rapidly growing tumour cells have high glycolytic ATP production correlated with a deficiency in mitochondria, as is well known from cancer tissues. However, even in these neoplastic cells a significant fraction of total ATP production stems from mitochondrial oxidative phosphorylation, although tumours *in vivo* may be less oxygenated than the isolated cells *in vitro* (Pedersen, 1978). In contrast to the fully aerobic energy budget of whole animals under normoxia, even well-aerated mammalian cell cultures display a partly anaerobic metabolism. Highly negative calorimetric–respirometric ratios (to $-1100\,\text{kJ mol}^{-1}\,O_2$, compared with the oxycaloric equivalent of $-450\,\text{kJ mol}^{-1}$) can be explained on the basis of simultaneous oxygen consumption and lactate production, with molar lactate:O_2 ratios up to 14 (Gnaiger & Kemp, 1990).

Summary

Below the *critical* oxygen pressure, the aerobic ATP production decreases, and below the *limiting* oxygen pressure anaerobic processes compensate increasingly for the diminished aerobic flux. The respiration of cells, tissues, organs and whole organisms is more dependent on external p_{O_2} than that of isolated mitochondria. The capacity of organisms to utilize traces of oxygen at reduced metabolic flux must be studied with great experimental care, and calorespirometry provides a sensitive approach.

Microxic regulation enables a steep increase of oxygen flux with P_{O_2} at very low oxygen levels. This is a metabolic adaptation to environments in the boundary to anoxic conditions, offering an energetic advantage to active organisms.

The difficulties involved in defining an absolute limit between microxic and anoxic conditions are best illustrated by a logarithmic P_{O_2} scale (Figure 7). Ideally the term 'anoxic' (without oxygen) should be restricted to any situation where molecular oxygen is strictly absent. Absence of oxygen, however, has to be practically defined in terms of tested oxygen removal techniques and relative to the sensitivity limit of analytical methods (Figure 7). Effective anoxia is obtained when a further decrease of P_{O_2} does not elicit any physiological or biochemical response.

The capacity for energy assimilation and growth under anoxia should be considered as the ultimate criterion for 'facultatively anaerobic' animals. Very little direct evidence exists documenting the persistence of animal populations under strict anoxia, except for members of the meiofauna in the sulphide system.

Supported by FWF Austria, project no. P7162-BIO. I thank Dr J.M. Shick for reading the manuscript.

References

Abe, A.S. & Mendes, E.G. (1980). The effect of inspired oxygen tension on oxygen uptake and tolerance to anoxia in the water snakes *Helicops modestus* and *Liophis miliaris* (Colubridae). *Comparative Biochemistry and Physiology* **65A**, 459–63.

Atkinson, H.J. (1973). The respiratory physiology of the marine nematodes *Enoplus brevis* (Bastian) and *E. communis* (Bastian). I. The influence of oxygen tension and body size. *Journal of Experimental Biology* **59**, 255–66.

Baumgärtl, H. & Lübbers, D.W. (1983). Microcoaxial needle sensor for polarographic measurement of local O_2 pressure in the cellular range of living tissue. Its construction and properties. In *Polarographic oxygen sensors: Aquatic and Physiological Applications* (ed. E. Gnaiger & H. Forstner), pp. 37–65. Springer, Berlin.

Bayne, B.L. (1971). Oxygen consumption by three species of lamellibranch mollusc in declining ambient oxygen tension. *Comparative Biochemistry and Physiology* **40A**, 955–70.

Bayne, B.L. & Livingstone, D.R. (1977). Responses of *Mytilus edulis* L. to low oxygen tension: Acclimation of the rate of oxygen consumption. *Journal of Comparative Physiology* **114**, 129–42.

Brand, Th. von (1946). Anaerobiosis in invertebrates. *Biodynamica Monographs* **4**, 1–328.

Burton, D.T. & Heath, A.G. (1980). Ambient oxygen tension (p_{O_2}) and transition to anaerobic metabolism in three species of freshwater fish. *Canadian Journal of Fisheries and Aquatic Sciences* **37**, 1216–24.

Clark, A. Jr, Clark, P.A.A., Connett, R.J., Gayeski, T.E.J. & Honig, C.R. (1987) How large is the drop in p_{O_2} between cytosol and mitochondrion? *American Journal of Physiology* **252**, C583–7.

Cole, R.P., Sukanek, P.S., Wittenberg, J.B. & Wittenberg, B.A. (1982). Mitochondrial function in the presence of myoglobin. *Journal of Applied Physiology* **53**, 1116–24.

Connett, R.J., Gayeski, T.E.J. & Honig, C.R. (1985a). An upper bound on the minimum p_{O_2} for O_2 consumption in red muscle. In *Oxygen transport to tissue*, vol. 7 (ed. F. Kreuzer, S.M. Cain, Z. Turek & T.K. Goldstick), pp. 291–300. Plenum Publishing Corporation, New York.

Connett, R.J., Gayeski, T.E.J. & Honig, C.R. (1985*b*). Energy sources in fully aerobic rest–work transitions: a new role for glycolysis. *American Journal of Physiology* **248**, H922–9.

De Zwaan, A. & Skjoldal, H.R. (1979). Anaerobic energy metabolism of the scavenging isopod *Cirolana borealis* (Lilljeborg). *Journal of Comparative Physiology* **129**, 327–31.

De Zwaan, A. & Wijsman, T.C.M. (1976). Anaerobic metabolism in bivalvia (Mollusca). Characteristics of anaerobic metabolism. *Comparative Biochemistry and Physiology* **54B**, 313–24.

De Zwaan, A. & Zandee, D.I. (1972). The utilization of glycogen and accumulation of some intermediates during anaerobiosis in *Mytilus edulis* L. *Comparative Biochemistry and Physiology* **43B**, 47–54.

Gnaiger, E. (1983*a*). Heat dissipation and energetic efficiency in animal anoxibiosis. *Journal of Experimental Biology* **228**, 471–90.

Gnaiger, E. (1983*b*). The twin-flow microrespirometer and simultaneous calorimetry. In *Polarographic Oxygen Sensors: Aquatic and Physiological Applications* (ed. E. Gnaiger & H. Forstner), pp. 134–66. Springer, Berlin.

Gnaiger, E. (1983*c*). In situ measurement of oxygen profiles in lakes: Microstratification, oscillations and the limits of comparison with chemical methods. In *Polarographic Oxygen Sensors: Aquatic and Physiological Applications* (ed. E. Gnaiger & H. Forstner), pp. 245–64. Springer, Berlin.

Gnaiger, E. (1987). Optimum efficiencies of energy transformation in anoxic metabolism. The strategies of power and economy. In *Evolutionary Physiological Ecology* (ed. P. Calow), pp. 7–36. Cambridge University Press.

Gnaiger, E. (1989). Mitochondrial respiratory control: Energetics, kinetics and efficiency. In *Energy Transformations in Cells and Animals* (ed. W. Wieser & E. Gnaiger), pp. 6–17. Thieme, Stuttgart.

Gnaiger, E. & Kemp, R.B. (1990). Anaerobic metabolism in aerobic mammalian cells. Information from the ratio of calorimetric heat flux and respirometric oxygen flux. *Biochimica et Biophysica Acta* **1016**, 328–32.

Gnaiger, E., Shick, J.M. & Widdows, J. (1989). Metabolic microcalorimetry and respirometry of aquatic animals. In *Techniques in Comparative Respiratory Physiology: An Experimental Approach* (ed. C.R. Bridges & P.J. Butler), pp. 113–35. Cambridge University Press.

Gnaiger, E. & Staudigl, I. (1987). Aerobic metabolism and physiological responses of aquatic oligochaetes to environmental anoxia. Heat dissipation, oxygen consumption, feeding and defecation. *Physiological Zoology* **60**, 659–77.

Hammen, C.S. (1969). Metabolism of the oyster, *Crassostrea virginica. American Zoologist* **9**, 309–18.

Hammen, C.S. (1976). Respiratory adaptations: Invertebrates. In *Estuarine Processes* (ed. M. Wiley), pp. 347–55. Academic Press, New York.

Hand, S.C. & Gnaiger, E. (1988). Anaerobic dormancy quantified in *Artemia* embryos: A calorimetric test of the control mechanism. *Science* **239**, 1425–7.

Hawkins, A.J.S., Salkeld, P.N., Bayne, B.L., Gnaiger, E. & Lowe, D.M. (1985). Feeding and resource allocation in the mussel *Mytilus edulis*: Evidence for time-averaged optimization. *Marine Ecology Progress Series* **20**, 273–87.

Herreid, C.F. II (1980). Hypoxia in invertebrates. *Comparative Biochemistry and Physiology* **67A**, 311–20.

Hitchman, M.L. (1983). Calibration and accuracy of polarographic oxygen sensors. In *Polarographic Oxygen Sensors: Aquatic and Physiological Applications* (ed. E. Gnaiger & H. Forstner), pp. 18–30. Springer, Berlin.

Hochachka, P.W. & Mustafa, T. (1972). Invertebrate facultative anaerobiosis. *Science* **178**, 1056–60.

Holland, K.T., Knapp, J.S. & Shoesmith, J.G. (1987). *Anaerobic bacteria*. Blackie, Glasgow and London.

Hughes, G.M., Albers, C., Muster, D. & Goetz, K.H. (1983). Respiration of the carp, *Cyprinus carpio* L., at 10 and 20 °C and the effects of hypoxia. *Journal of Fish Biology* **22**, 613–28.

Jones, D.P. (1986). Intracellular diffusion gradients of O_2 and ATP. *American Journal of Physiology* **250**, C663–75.

Livingstone, D.R. (1978). Anaerobic metabolism in the posterior adductor muscle of the common mussel *Mytilus edulis* L. in response to altered oxygen tension and temperature. *Physiological Zoology* **51**, 131–9.

Livingstone, D.R. & Bayne, B.L. (1977). Responses of *Mytilus edulis* L. to low oxygen tensions: Anaerobic metabolism of the posterior adductor muscle and mantle tissues. *Journal of Comparative Physiology* **114**, 143–55.

Mangum, C. & Van Winkle, W. (1973). Responses of aquatic invertebrates to declining oxygen conditions. *American Zoologist* **13**, 529–41.

Meyers, M.B., Fossing, H. & Powell, E.N. (1987). Microdistribution of interstitial meiofauna, oxygen and sulfide gradients, and the tubes of macro-infauna. *Marine Ecology Progress Series* **35**, 223–41.

Oshino, N., Sugano, T., Oshino, R. & Chance, B. (1974). Mitochondrial function under hypoxic conditions: The steady states of cytochrome a+a3 and their relation to mitochondrial energy states. *Biochimica et Biophysica Acta* **368**, 298–310.

Pamatmat, M.M. (1978). Oxygen uptake and heat production in a metabolic conformer (*Littorina irrorata*) and a metabolic regulator (*Uca pugnax*). *Marine Biology* **48**, 317–25.

Pedersen, P.L. (1978). Tumor mitochondria and the bioenergetics of cancer cells. *Progress in Experimental Tumor Research* **22**, 190–274.

Petersen, J.A. & Johansen, K. (1967). Aspects of oxygen uptake in *Mesochaetopterus taylori*, a tube-dwelling polychaete. *Biological Bulletin* **3**, 600–5.

Pörtner, H.O., Heisler, N. & Grieshaber, M.K. (1985). Oxygen consumption and mode of energy production in the intertidal worm *Sipunculus nudus* L.: Definition and characterization of the critical p_{O_2} for an oxyconformer. *Respiration Physiology* **59**, 361–77.

Powell, E.N., Crenshaw, M.A. & Rieger, R.M. (1980). Adaptations to sulfide in sulfide-system meiofauna. Endproducts of sulfide detoxification in three turbellarians and a gastrotrich. *Marine Ecology Progress Series* **2**, 169–77.

Pullin, R.S.V., Morris, D.J., Bridges, C.R. & Atkinson, R.J.A. (1980). Aspects of the respiratory physiology of the burrowing fish *Cepola rubescens* L. *Comparative Biochemistry and Physiology* **66A**, 35–42.

Putzer, V. (1985). Ph.D. thesis, University of Innsbruck.

Rosenthal, M., Lamanna, J.C., Joebsis, F.F., Levasseur, J.E., Kontos, H.A. & Patterson, J.L. (1976). Effects of respiratory gases on cytochrome a in intact cerebral cortex: Is there a critical P_{O_2}? *Brain Research* **108**, 143–54.

Saz, H.J. (1971). Facultative anaerobiosis in the invertebrates: pathways and control systems. *American Zoologist* **11**, 125–35.

Saz, H.J. (1981). Energy metabolism of parasitic helminths: Adaptations to parasitism. *Annual Review of Physiology* **43**, 323–41.

Schöttler, U., Wienhausen, G. & Zebe, E. (1983). The mode of energy production in the lugworm *Arenicola marina* at different oxygen concentrations. *Journal of Comparative Physiology* **149**, 547–55.

Shick, J.M., Gnaiger, E., Widdows, J., Bayne, B.L. & De Zwaan, A. (1986). Activity and metabolism in the mussel *Mytilus edulis* L. during intertidal hypoxia and aerobic recovery. *Physiological Zoology* **59**, 627–42.

Shick, J.M., Widdows, J. & Gnaiger, E. (1988). Calorimetric studies of behavior, metabolism and energetics of sessile intertidal animals. *American Zoologist* **28**, 161–81.

Simpson, J.W. & Awapara, J. (1966). The pathway of glucose degradation in some invertebrates. *Comparative Biochemistry and Physiology* **18**, 537–48.

Soboll, S., Scholz, R. & Heldt, H.W. (1978). Subcellular metabolite concentrations. Dependence of mitochondrial and cytosolic ATP systems on the metabolic state of perfused rat liver. *European Journal of Biochemistry* **87**, 377–90.

Sugano, T., Oshino, N. & Chance, B. (1974). Mitochondrial functions under hypoxic conditions. The steady states of cytochrome *c* reduction and of energy metabolism. *Biochimica et Biophysica Acta* **347**, 340–58.

Tamura, M., Hazeki, O., Nioka, S. & Chance, B. (1989). In vivo study of tissue oxygen metabolism using optical and nuclear magnetic resonance spectroscopies. *Annual Review of Physiology* **51**, 813–34.

Toulmond, A. (1975). Blood oxygen transport and metabolism of the confined lugworm *Arenicola marina* (L.). *Journal of Experimental Biology* **63**, 647–60.

Ultsch, G.R., Jackson, D.C. & Moalli, R. (1981). Metabolic oxygen conformity among lower vertebrates: The toadfish revisited. *Journal of Comparative Physiology* **142**, 439–43.

Widdows, J. (1989). Calorimetric and energetic studies of marine bivalves. In *Energy Transformations in Cells and Animals* (ed. W. Wieser & E. Gnaiger), pp. 145–54. Thieme, Stuttgart.

Widdows, J., Newell, R.I.E. & Mann, R. (1989). Effects of hypoxia and anoxia on survival, energy metabolism, and feeding of oyster larvae (*Crassostrea virginica*, Gmelin). *Biological Bulletin* **177**, 154–66.

Widdows, J. & Shick, J.M. (1985). Physiological responses of *Mytilus edulis* and *Cardium edule* to aerial exposure. *Marine Biology* **85**, 217–32.

Wieser, W., Ott, J., Schiemer, F. & Gnaiger, E. (1974). An ecophysiological study of some meiofauna species inhabiting a sandy beach at Bermuda. *Marine Biology* **26**, 235–48.

Wiesner, R.J., Kreutzer, U., Roesen, P. & Grieshaber, M.K. (1988). Subcellular distribution of malate-aspartate cycle intermediates during normoxia and anoxia in the heart. *Biochimica et Biophysica Acta* **936**, 114–23.

Wilson, D.F., Erecinska, M., Drown, C. & Silver, I.A. (1977). Effect of oxygen tension on cellular energetics. *American Journal of Physiology* **233**, C135–40.

Wilson, D.F., Erecinska, M., Drown, C. & Silver, I.A. (1979a). The oxygen dependence of cellular energy metabolism. *Archives of Biochemistry and Biophysics* **195**, 485–93.

Wilson, D.F., Owen, C.S. & Erecinska, M. (1979b). Quantitative dependence of mitochondrial oxidative phosphorylation on oxygen concentration: A mathematical model. *Archives of Biochemistry and Biophysics* **195**, 491–504.

Wittenberg, B.A. & Wittenberg, J.B. (1985). Oxygen pressure gradients in isolated cardiac myocytes. *Journal of Biological Chemistry* **260**, 6548–54.

Zimmer, G., Beyersdorf, F. & Fuchs, J. (1985). Decay of structure and function of heart mitochondria during hypoxia and related stress and its treatment. *Molecular Physiology* **8**, 495–513.

GUIDO VAN DEN THILLART & AREN VAN
WAARDE

pH changes in fish during environmental anoxia and recovery: the advantages of the ethanol pathway

Abstract

This review compares data on blood acid–base balance during environmental anoxia and recovery in canulated individuals of 3 fish species (goldfish, *Carassius auratus*; crucian carp, *Carassius carassius*; and common carp, *Cyprinus carpio*) with measurements of muscle high-energy phosphate compounds and muscle pH_i by *in vivo* ^{31}P-NMR spectroscopy. Common carp can survive for short periods in the absence of oxygen by a classical Embden–Meyerhof glycolysis, whereas crucian carp and goldfish have a modified anaerobic metabolism (low flux type) with ethanol, carbon dioxide and ammonia as endproducts. Common carp show a rapid depletion of creatine phosphate stores and a steep decline of the pH of muscle and blood, but in the other species, high-energy phosphate compounds reach a steady-state and the development of acidosis is retarded. The regulation of the pyruvate dehydrogenase complex in the *Carassius* genus is an important aspect of the anaerobic metabolism, which requires further investigation.

Introduction

Anaerobic metabolism is generally characterized by the degradation of phosphorylated compounds (mainly creatine phosphate and ATP) and by the accumulation of lactate. The acid–base balance is disturbed by net acid production, which is partly neutralized by hydrolysis of creatine phosphate, since the depletion of phosphagen stores is accompanied by consumption of H^+ ions at physiological pH (Hochachka & Mommsen, 1983; Pörtner *et al.*, 1984). The acidotic effect of anaerobiosis is reflected by low bicarbonate and high lactate levels in the blood, and is potentially harmful when blood pH becomes too low.

A low intracellular pH is strongly correlated with a general suppression of metabolic activity (Nuccitelli & Deamer, 1982). Many enzymes are extremely pH-sensitive. In particular, phosphofructokinase (PFK), the key enzyme of glycolysis, can be completely switched off by a pH decrease (Leaver & Burt, 1981; Nuccitelli & Deamer, 1982). The acidotic effect of anaerobic metabolism will therefore exert a negative feedback on glycolysis.

Although the tissue pH decreases in almost all anoxia-resistant animals in a situation of oxygen lack, there appears to be a difference with non-resistant species: the pH decrease is only minor and the pH stabilizes at a lower value. The reasons for pH-stabilization are either a modified metabolism that takes care of the acidic end-products, or the presence of a buffer such as $CaCO_3$, which neutralizes the acids. Of both strategies, we can find many examples in the animal kingdom (Van den Thillart & Van Waarde, 1985).

In the past, there have been discussions about the efficiency of the anaerobic pathways of invertebrates with respect to ATP yield. The fact that the pH_i may also be an important parameter from a thermodynamic point of view has had considerably less attention. Since the ΔG_o of ATP is pH-dependent, a high ATP yield does not necessarily mean a higher ΔG value for the summated reactions. An efficient anaerobic metabolism can only continue if acidosis is avoided. It is not surprising that animals that are unable to stabilize pH_i and [ATP] either die or remain in a state of suspended animation. Turtles, hibernating amphibians, frozen insects and dormant cysts of *Artemia* are examples of the latter kind; crucian carp and goldfish, but also invertebrates such as flukes and *Tubifex*, represent the former more efficient category (see Hochachka & Guppy, 1987).

Anoxia-resistant animals that show signs of activity during anaerobiosis must have means to stabilize pH and [ATP] and will thus be different from species that fall into torpor or enter a dormant state. The genus *Carassius* (goldfish, crucian carp) is an interesting object of study, since these fish are not only able to survive anoxia for long periods (depending on temperature) but, in addition, they remain conscious. Although goldfish usually rest at the bottom of a closed respirometer, they move at regular intervals and respond immediately to the inlet of oxygenated water (authors' personal observation). In ice-covered Finnish lakes, crucian carp cannot be netted even after several months of anoxia (Holopainen & Hyvärinen, 1985).

Since Blažka (1958) found that goldfish and crucian carp were very anoxia-resistant and showed some peculiar metabolic phenomena such as high CO_2 excretion rates under anoxic conditions, several other studies have been published in which the anaerobic metabolism of these fishes is examined. Anderson (1975) found indications from changes in the internal temperature gradient that the rate of heat production by goldfish is reduced during anoxia. Van den Thillart (1977) observed many metabolic features in the anoxic goldfish, such as a stable energy charge in muscle tissues and glucose–CO_2 conversion. Shoubridge (1980) showed that ethanol is produced as an anaerobic endproduct. Mourik (1982), analysing the ethanol pathway, found a modified pyruvate dehydrogenase in muscle mitochondria, capable of the conversion of pyruvate to acetaldehyde under anaerobic conditions. Acetaldehyde is rapidly reduced to ethanol by a cytoplasmic alcohol dehydrogenase. Johnston & Bernard (1983) confirmed the existence of the ethanol route in crucian carp. An important aspect of this pathway has been described in an ecological study by Holopainen & Hyvärinen (1985). They showed that crucian carp is the only fish species that survives the long, severe Finnish winters under conditions of complete anoxia, probably owing to their modified metabolism. In a comparative study where a large number of Northern European fish species were examined, Wissing & Zebe (1988) found a third fish species with an ethanol pathway: the bitterling, *Rhodeus amarus*. Despite the possession of a modified metabolism, the bitterling survives anoxia for only 2 hours at 12 °C; this is comparable to the anoxia survival of the common carp, which does not have the ethanol pathway.

The mechanism of ethanol formation is not completely resolved, but it appears to involve a modified pyruvate dehydrogenase (PDH), which produces acetaldehyde instead of acetyl coenzyme A (Mourik *et al.*, 1982). Acetaldehyde is reduced in the cytoplasm by alcohol dehydrogenase (ADH). The system is in redox-balance, because glyceraldehyde 3-phosphate dehydrogenase and ADH form a redox couple (van den Thillart, 1982).

The production and excretion of ethanol and CO_2 by goldfish and crucian carp cannot be coupled to energy conservation (Van den Thillart, 1982), but the modified metabolism will probably retard the development of metabolic acidosis. Ecological studies suggest that the pathway is of vital importance for *Carassius carassius*, which is often exposed to long periods of anoxia (Holopainen & Hyvärinen, 1985). So, at this point two questions arise:

1. Is the acid–base balance indeed stabilized by lactate–ethanol conversion?

2. How is this pathway regulated?

Changes in blood pH of common and crucian carp during anoxia

Measurements of changes in the acid–base balance of carp blood during hypoxia, anoxia and recovery from anoxia have been performed by G. Van den Thillart & N. Heisler (unpublished). The results of typical experiments are shown in Figure 1.

Changes in the blood of common carp were as expected: on anoxia exposure there was a sudden pH decrease, which was obviously caused by a sharp increase in the lactate concentration. The exposure of 2 hours was about maximal for this species, since the animals eventually lost equilibrium and were lying on their sides, completely exhausted and ventilating heavily. During the hypoxic period, there was little change in blood pH. At 30% air saturation, the pH showed a small rise because of moderate hyperventilation, but at 10% air saturation it went down, owing to a combination of respiratory alkalosis (hyperventilation) and metabolic acidosis (lactate formation). Owing to exhaustion of the buffer capacity, a further increase in lactate concentration caused a dramatic fall in blood pH (in some cases even to pH 7.0).

The recovery of the normoxic acid–base balance is not as fast as casual examination of Figure 1 might reveal. It took about 15 h for blood lactate to reach levels below 1 mM. Lactate is thought to be the prime combustible substrate and is quickly oxidized in trout during both rest and active swimming (Van den Thillart, 1986). Assuming that all accumulated lactate is oxidized, we can estimate the lactate clearance rate. The oxygen consumption of carp at 15 °C is about 60 mg O_2 h^{-1} kg^{-1}; the O_2:lactate molar ratio for complete oxidation is 3, so the maximal clearance rate by oxidation will be 0.67 mmol h^{-1} kg^{-1}. Assuming a mean lactate concentration of 10 mM after 1 h of anoxic exposure, it will take about 15 h before blood levels are back to normal. This estimation corroborates well the observed decline of the blood lactate concentration during recovery.

Eventual changes in the blood parameters of crucian carp and common carp are rather different, although the initial effects of oxygen deficiency are similar in both species. During hypoxia and early anoxia, a respiratory alkalosis due to hyperventilation is followed by the development of metabolic acidosis due to the production of lactate. However, after about 1.5 h of anoxia, lactate accumulation in crucian carp ceases and the lactate concentration reaches a plateau (Figure 1*b*). At the same time, blood pH stabilizes at a value of about 7.4. In a similar experiment with crucian carp, the duration of the anoxic exposure was about 26 h. Simultaneously, we measured accumulation of ethanol in the blood and the environmental

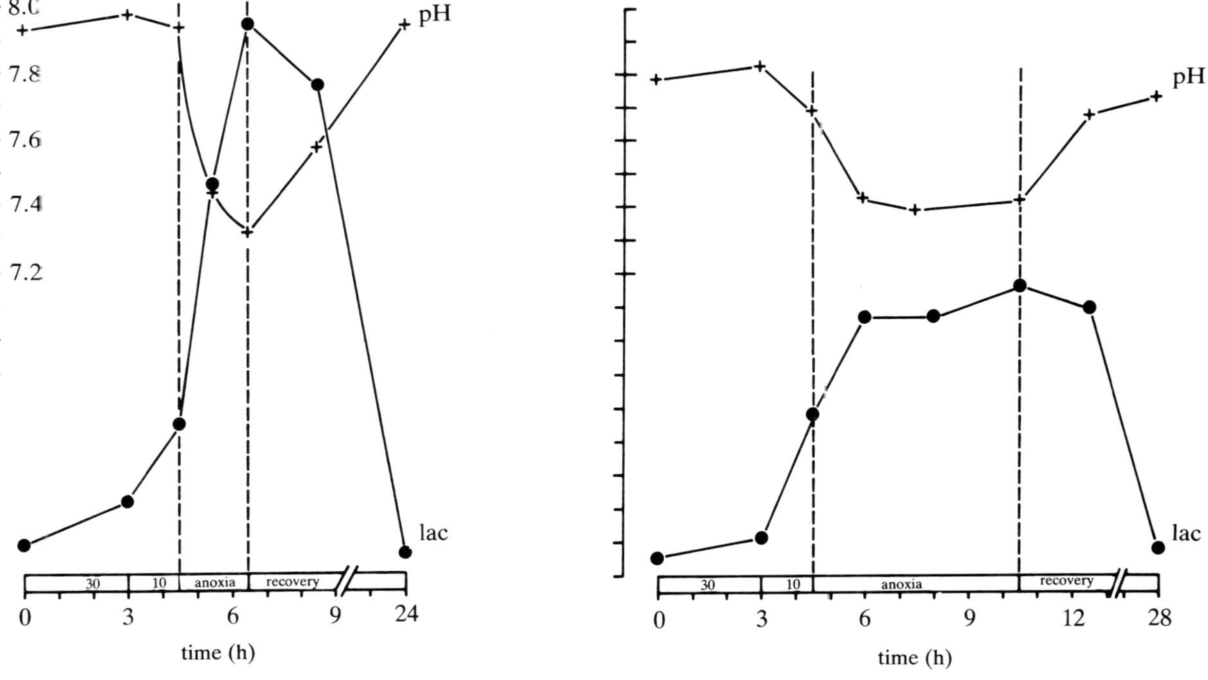

Figure 1. Blood pH and blood lactate content of (*a*) common carp (*Cyprinus carpio*) and (*b*) crucian carp (*Carassius carassius*) exposed to different P_{O_2} levels. The fish had a body mass of 0.75–1.5 kg. After fitting of a PE 50 cannula into the dorsal aorta, they were allowed to recover in a respirometer for 48 h at 15 °C. The lactate level and pH were measured in small blood samples, which were taken at regular intervals during normoxia, after 3 h at 30% air saturation (P_{O_2} *ca.* 50 Torr), after 1.5 h at 10% air saturation (P_{O_2} *ca.* 15 Torr), during anoxia and during subsequent recovery.

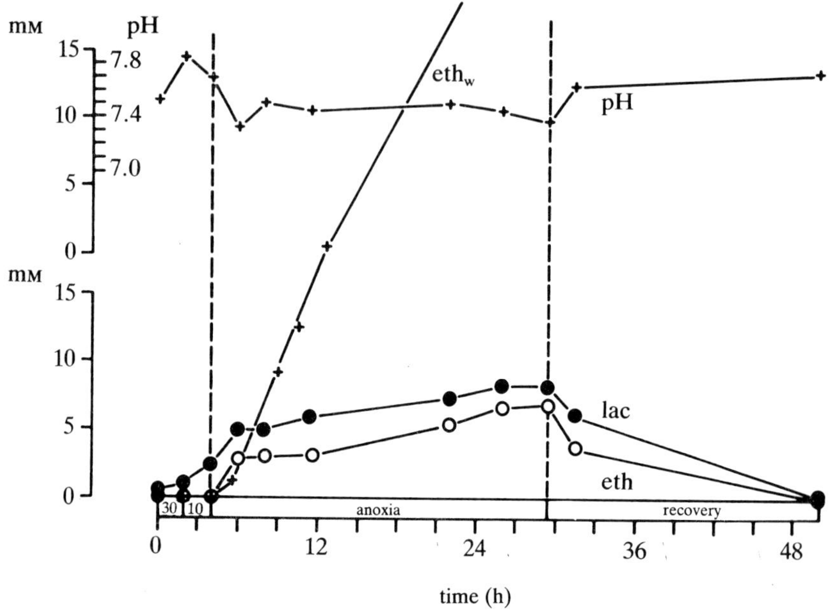

Figure 2. Effect of long-term anoxia exposure (26 h) on blood pH, blood lactate and blood ethanol content of *Carassius carassius*. The total amount of excreted ethanol (eth$_w$) is presented as micromoles per kilogram of body mass.

water (Figure 2). No ethanol was observed in the blood of hypoxic animals; ethanol formation started only after the onset of anoxia. The ethanol excretion took place at about the same rate as the initial appearance of lactic acid (Figure 2). The moment of stabilization of the lactate concentration coincided with the establishment of a rapid alcohol production.

The cannulation experiments described above demonstrate that a metabolic switch is activated after about 1 h of anoxia (15 °C). This switch puts the accumulation of lactate to a halt and initiates the formation of ethanol. The development of metabolic acidosis is suppressed, despite the maintenance of a relatively high glycolytic rate.

Changes in muscle pH and direct energy reserves of common carp and goldfish during anoxia

Recent observations on muscle metabolism by *in vivo* ^{31}P-NMR produced new insights into the anaerobic metabolism of fish and corroborated the results obtained in the cannulation experiments (Van den Thillart *et al.*, 1989). Common carp and goldfish (50 g body mass) were mounted in a modified bioprobe of a Bruker MSL-400 NMR spectrometer. The signal from the lateral muscle was picked up with an 18 mm surface coil, which was double-tuned to the ^{1}H and ^{31}P frequencies. ^{31}P-NMR spectra (Figure 3) were obtained at 10 min intervals during normoxia, anoxia and recovery from anoxia. The duration of anoxic exposure was 1 h for carp and 4 h for goldfish, at a temperature of 20 °C. Anoxia was introduced by switching the gas phase from air to nitrogen. The intracellular pH was calculated from the difference in chemical shift between phosphocreatine and inorganic phosphate in the NMR spectra. The pH measurements were calibrated using two different model solutions (Van den Thillart *et al.*, 1989).

Typical results are presented in Figures 4 and 5. Anoxia caused an immediate decline in creatine phosphate and an increase in inorganic phosphate in all experiments. The initial rates of creatine phosphate breakdown were similar in both species. Only after about half of the direct energy reserves had disappeared did a striking difference become apparent. Creatine phosphate degradation in carp continued for 1 h until the phosphagen store had been exhausted by more than 90%. In goldfish, the net hydrolysis of creatine phosphate came to an end after 20 min of anoxia and a new steady state was reached in which creatine phosphate was maintained at 50% of the control level. The change in the concentration of inorganic phosphate seemed to be directly related to the extent of creatine phosphate breakdown. Inorganic phosphate thus showed a continuous increase in carp, whereas a stable plateau was reached in goldfish.

Environmental anoxia caused a development of muscle acidosis in both species (Figure 5), but the rate of the pH decline in carp was four times greater than in goldfish. During 10 min after reoxygenation, creatine phosphate and inorganic phosphate did not change. Then, creatine phosphate rose sharply and inorganic phosphate declined (Figure 4). The rate of net creatine phosphorylation was three times greater in carp than in goldfish. In both species, the pH decline continued for 60 min after reoxygenation, probably because of the rapid recovery of creatine phosphate. Phosphorylation of creatine produces one acid equivalent (Hochachka & Mommsen, 1983; Pörtner *et al.*, 1984).

The ^{31}P-NMR experiments with goldfish suggest that a metabolic switch is activated after about 20 min of anoxia (20 °C). The switch stops the depletion of the myotomal creatine phosphate stores. Muscle pH_i continues to decrease, albeit slowly. The cannulation experiments show that blood pH in crucian carp stabilizes at the onset of ethanol excretion. In the goldfish NMR experiments, however, muscle pH_i does not reach a plateau,

     G. van den Thillart and A. van Waarde

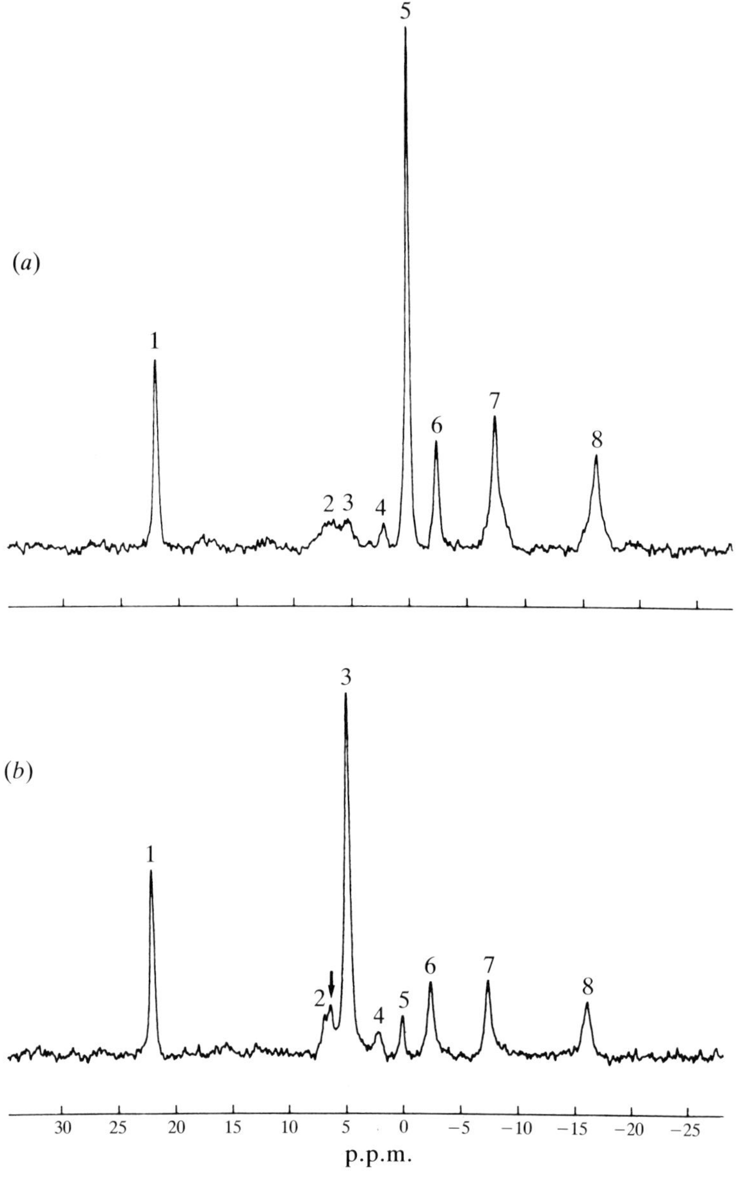

Figure 3. ^{31}P-NMR spectra from the lateral musculature of carp (*Cyprinus carpio*) before (*a*) and after (*b*) a 1 h period of anoxia. The following compounds are indicated: 1, methylene diphosphonate (external reference); 2, sugar phosphates; 3, inorganic phosphate; 4, phosphodiesters; 5, creatine phosphate; 6–8, ATP (γ, α, β-phosphate resonances). The arrow in (*b*) indicates the position of IMP, which accumulates during anoxia as the degradation product of ATP.

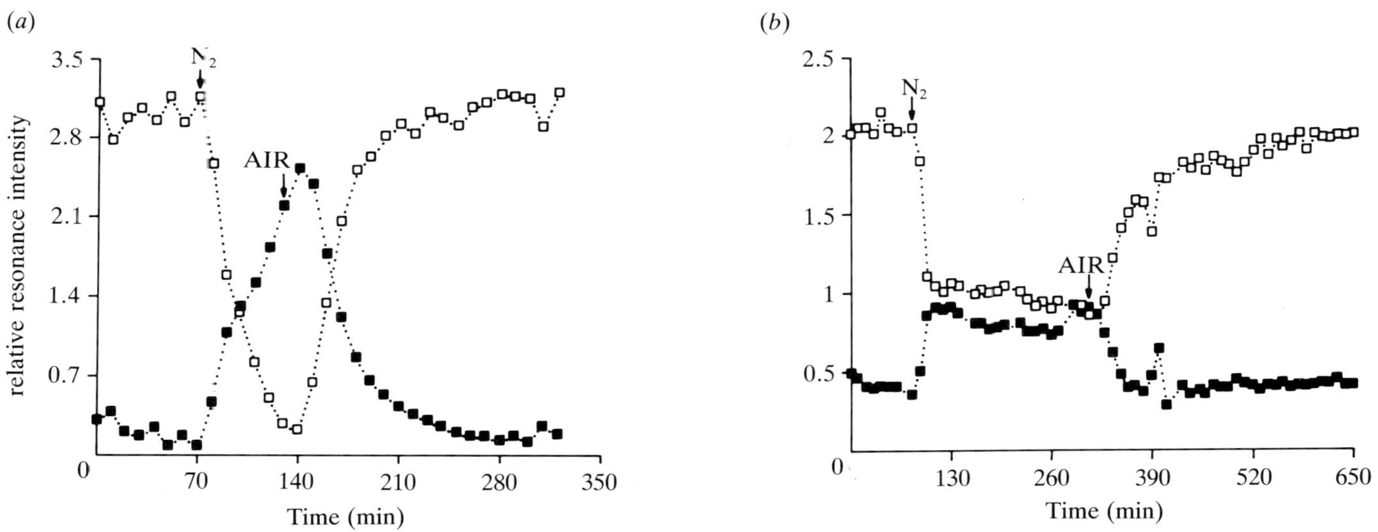

Figure 4. Time course changes of CrP and inorganic phosphate content of fish muscle before, during and after a period of anoxia. (*a*) *Cyprinus carpio* (1 h of anoxia); (*b*) *Carassius auratus* (4 h of anoxia). The arrows indicate the time when the P_{O_2} was changed. The intensity of the peaks was expressed relative to the external standard.

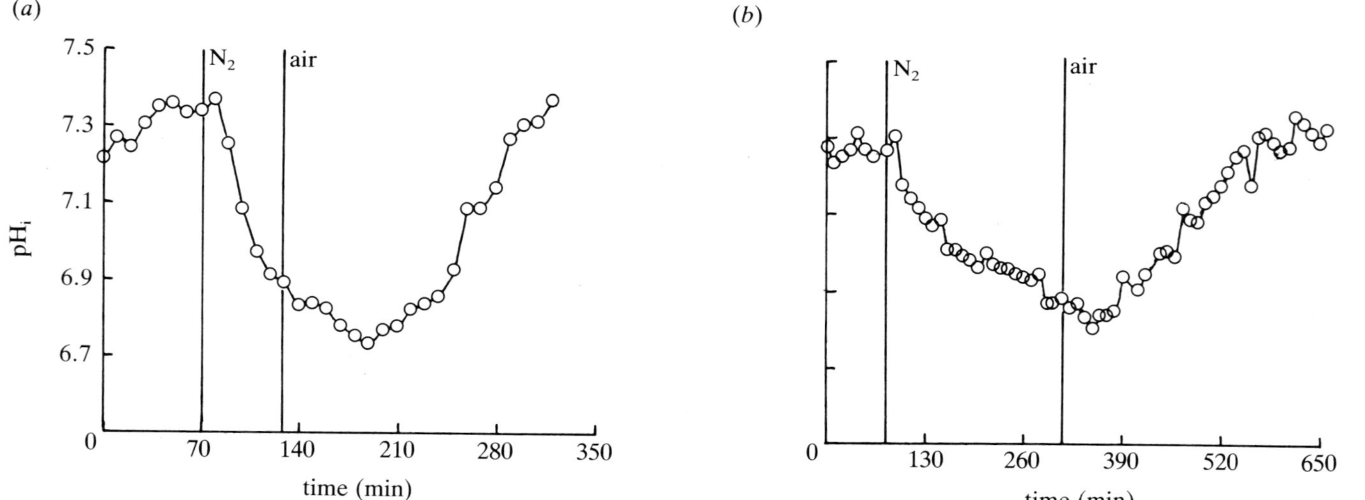

Figure 5. Changes of pH in fish muscle before, during and after a period of anoxia. (a) *Cyprinus carpio* (1 h of anoxia); (b) *Carassius auratus* (4 h of anoxia).

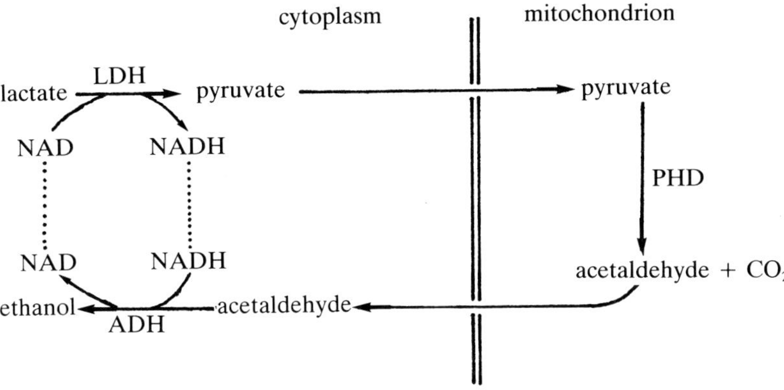

Figure 6. Metabolic scheme for the conversion of glucose to ethanol.

although the development of acidosis slows down at the same time when phosphagen utilization comes to a halt. The cannulation experiments require large animals (0.75–1.5 kg) and were carried out at 15 °C, while for NMR studies we used small goldfish (50 g) at 20 °C. The small goldfish will therefore have had a higher metabolic rate; under these conditions, the capacity for conversion of lactate to ethanol may be insufficient to deal with the glycolytic flux. Nevertheless, a transition to a lower rate of acidification was observed in both species.

Pyruvate dehydrogenase and the control of the ethanol pathway

Conversion of glucose to ethanol involves decarboxylation of pyruvate by the mitochondrial PDH complex (Figure 6). PDH is probably the key enzyme in the ethanol pathway, since alcohol dehydrogenase (ADH) is a simple enzyme which lacks regulatory sites (Mourik, 1982). PDH in contrast is an allosteric enzyme complex, which is also subject to phosphorylation (Denton *et al.*, 1975). Since acetaldehyde is a very reactive compound, which binds to proteins, high acetaldehyde levels cannot be tolerated *in situ*. An active ADH must therefore be combined with a well-controlled PDH to keep acetaldehyde levels low. The structure of the PDH complex is illustrated in Figure 7. The enzyme consists of three different subunits: pyruvate decarboxylase (E$_1$–TPP), dihydrolipoate transacetylase (E$_2$–S–S) and dihydrolipoate dehydrogenase (E$_3$–FAD). The reaction proceeds as follows.

$$E_1-TPP+CH_3-CO-COOH \rightarrow E_1-TPP-CHO-CH_3+CO_2. \qquad (1)$$

$$E_1-TPP-C\,HO-CH_3+E_2-S-S \rightarrow E_1-TPP+E_2-S \quad SH. \tag{2}$$
$$\underset{\underset{O}{\overset{\parallel}{C}-CH_3}}{}$$

$$E_2-S \quad SH+coASH \rightarrow E_2-S \quad S+coA-S-CO-CH_3. \tag{3}$$

$$E_2-S \quad S +E_3-FAD \rightarrow E_2-S-S+E_3-FADH_2. \tag{4}$$

$$E_3-FADH_2+NAD^+ \rightarrow E_3-FAD+NADH^+H^+. \tag{5}$$

Net reaction:

$$Pyruvate+NAD^++CoASH \rightarrow Acetyl\ CoA+NADH+H^++CO_2. \tag{6}$$

During anoxia, oxidation of mitochondrial NADH is difficult owing to lack of the terminal electron acceptor. Reaction (5) can therefore no longer proceed rapidly. This will lead to accumulation of NADH and acetyl CoA, which are known to be strong inhibitors of PDH (Garland & Randle, 1964).

It could therefore be anticipated that oxygen lack will block pyruvate decarboxylation. However, Mourik *et al.* (1982) found that anoxic mitochondria isolated from goldfish red muscle converted pyruvate to acetaldehyde and CO_2. Reaction (1) can apparently proceed under anoxia. Acetaldehyde is probably released from E_1–TPP, since conversion of the enzyme-bound hydroxyethyl derivative to acetyl CoA is impaired.

J. Warnaar & G. van den Thillart (unpublished) performed experiments with goldfish muscle mitochondria in which the lactate–ethanol shuttle proposed by Shoubridge & Hochachka (1981, 1983) was reconstituted. When the mitochondria were incubated with pyruvate, a relatively slow formation of acetaldehyde and CO_2 occurred. However, when lactate, lactate dehydrogenase, NAD and ADH were added instead of pyruvate, much higher fluxes were reached and ethanol replaced acetaldehyde as the endproduct. A lactate–ethanol shuttle seems therefore indeed to be operative in anoxic goldfish. It is probable that little pyruvate catabolism occurs in the absence of LDH and ADH, because the toxicity of acetaldehyde causes product inhibition of the enzymatic reaction. The conversion of

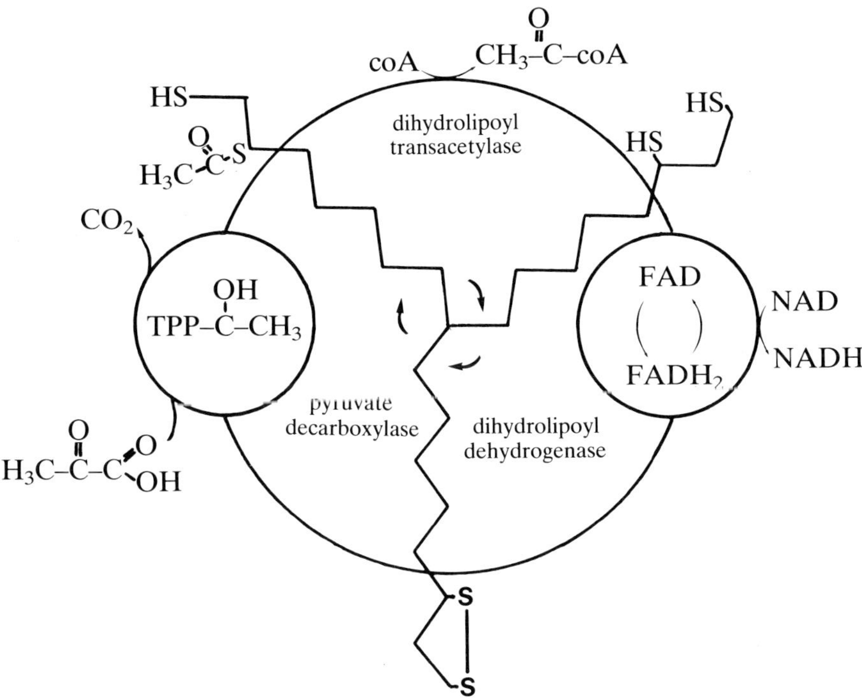

Figure 7. Schematic drawing of the structure of the pyruvate dehydrogenase complex (see text for explanation).

pyruvate to acetaldehyde, or lactate to ethanol, was strongly inhibited when arsenite was included in the incubation medium. This suggests that both processes are dependent on the activity of pyruvate dehydrogenase.

It is likely that production of acetaldehyde by E_1–TPP requires modification of the PDH complex. When muscles of bitterling become anoxic owing to strenuous exercise, no acetaldehyde or ethanol are produced (Wissing & Zebe, 1988). In contrast, ethanol is excreted by bitterling, crucian carp and goldfish after the fish have been exposed to 1–2 h of environmental anoxia (Johnston & Bernard, 1983; van den Thillart *et al.*, 1983; Holopainen & Hyvärinen, 1985; Wissing & Zebe, 1988). Acetaldehyde production is therefore a low-flux type of anaerobic metabolism (De Zwaan & van den Thillart, 1985), which is activated after a short anoxic interval.

In higher vertebrates, PDH is known to exist as an inactive phosphorylated form and an active non-phosphorylated form (Denton *et al.*, 1975). Interconversion of the two forms is regulated by a Mg–ATP-dependent protein kinase and a specific protein phosphatase (Linn *et al.*, 1969). Exercise leads to PDH dephosphorylation and activation of the PDH complex (Ward *et al.*, 1982). The enzyme is also under hormonal control by insulin (Seals & Jarett, 1980).

The mechanisms by which goldfish PDH is regulated have not yet been resolved. However, Rahman & Storey (1988) and Storey (1988) have shown that several enzymes involved in goldfish carbohydrate metabolism are phosphorylated during environmental anoxia. We propose therefore that PDH is also phosphorylated, resulting in a decoupling of the decarboxylase part from the rest of the complex.

Which mechanism activates the ethanol pathway?

We have discussed (above) the pH changes in blood and muscle tissues of carp, goldfish and crucian carp during and after anoxia exposure. Although crucian carp and goldfish have the ethanol pathway to prevent metabolic acidosis, we observed a significant decrease in blood and muscle pH in the early phase of anoxia. All initial metabolic changes in these fishes are quite similar to those in the common carp. However, after about 1 h of oxygen lack, pH stabilizes simultaneously with blood lactate and muscle creatine phosphate levels. Also at this point the ethanol excretion reaches a steady state. So it appears that some metabolic switch is activated not immediately, but after an interval of about 1 h of anoxia.

It has been said many times that the goldfish is able to suppress its metabolic rate (Hochachka, 1982, 1985). This statement was based on data from Anderson (1975) and on changes in metabolite concentrations. Thus far, the evidence had to be considered as only circumstantial, since no direct measurements of metabolic rate could be made. Recently, however, by means of new technology, heat flux measurements of goldfish (20 °C) under steady-state conditions were carried out (Van Waversveld *et al.*, 1988). It was found that the heat flux was depressed from about 710 to about $200 \, \text{J} \, \text{h}^{-1} \, \text{kg}^{-0.85}$ after 3 h of oxygen lack, and this level remained constant for at least 8 subsequent hours of anoxia.

We must distinguish between the metabolic switch from lactate to ethanol on one hand and metabolic suppression on the other hand, although these two features might be triggered by the same mechanism. From Figure 4 we find that the rate of the initial decline of creatine phosphate in carp and goldfish muscle is of the same order of magnitude. In contrast, however, the increase of creatine phosphate during the recovery phase is three times slower in goldfish. This indicates that the metabolic rate of goldfish is high in the initial phase of the anoxia period and slows down later, which might coincide with the switch from lactate to ethanol. It is not possible to measure the heat flux in the initial phase with a calorimeter, but measurements of ATP turnover by saturation-transfer NMR are possible and may tell us in the near future whether or not both processes (i.e. metabolic depression and activation of ethanol production) are triggered by the same mechanism.

Van den Thillart & Kesbeke (1978) found that carbon dioxide production in goldfish is independent of oxygen availability. The substrate pyruvate is under normoxic conditions taken up by the mitochondria via an energy-driven transport mechanism. Upon anoxia exposure, the mitochondrial electrochemical potential is likely to be lowered, so pyruvate transport will become dependent mainly on a concentration gradient. This means that for the same pyruvate decarboxylation rate the pyruvate concentration has to increase together with lactate. Remarkably, though, it was found that the pyruvate level in the red and white myotomal muscles of goldfish is not raised after 12 h of anoxia (Van den Thillart *et al.*, 1982). The mitochondrial redox state was also unaltered (Van den Thillart *et al.*, 1982). Apparently, the energetic condition of the mitochondria has not changed very much, and accumulation of pyruvate cannot be the signal which switches on the ethanol pathway. Since the ΔG_0 value for conversion of pyruvate to ethanol$+CO_2$ is $-8.8\,\mathrm{kcal\,mol^{-1}}$ ($36.8\,\mathrm{kJ\,mol^{-1}}$), the equilibrium constant would suggest low pyruvate concentrations, which are indeed observed. The free energy decline of the lactate dehydrogenase reaction is $-6\,\mathrm{kcal\,mol^{-1}}$ ($25\,\mathrm{kJ\,mol^{-1}}$) and thus smaller than that of the route to ethanol. Taking into account that lactic acid accumulates whereas the levels of ethanol and CO_2 are kept low by passive efflux, it is obvious that the thermodynamic conditions for ethanol production are more favourable. The accumulation of lactate, which takes place especially in white muscle (Van den Thillart *et al.*, 1982, Figure 5), can therefore only be due to the fact that the glycolytic flux exceeds the capacity of the ethanol pathway.

In conclusion:

1. Ethanol and acetaldehyde are not produced under aerobic or hypoxic conditions, but only after the onset of anoxia (Figure 2). Therefore, a switch must operate which changes the course of metabolism from an Embden-Meyerhof glycolysis to alcoholic fermentation.

2. Since the mitochondrial NADH:NAD and pyruvate:acetyl CoA ratios are unaltered (see above), the major negative allosteric

modulators of the PDH complex (i.e. NADH and acetyl CoA) (Garland & Randle, 1964; Denton *et al.*, 1975) cannot be the agents which activate the switch.

3. Acidosis seems not to be involved in the trigger mechanism either, as J. Warnaar & G. Van den Thillart (unpublished) were able to reconstitute the lactate–ethanol shuttle in isolated mitochondria at a constant pH of 7.4.

4. Covalent modification of the PDH complex seems thus most likely. In mammals, PDH is subject to protein phosphorylation (Denton *et al.*, 1975). In the anoxic goldfish, such phosphorylation might lead to a decoupling of the decarboxylase part from the rest of the complex. This would initiate the formation of the anaerobic electron acceptor, acetaldehyde, and the production of ethanol by cytosolic alcohol dehydrogenase.

A. van Waarde was supported by a fellowship from the Royal Dutch Academy of Sciences.

References

Anderson, J. (1975). Anaerobic resistance of *Carassius auratus* (L.). Ph.D. thesis, Australian National University.

Blažka, P. (1958). The anaerobic metabolism of fish. *Physiological Zoology* **31**, 117–28.

Denton, R.M., Randle, P.J., Bridges, B.J., Cooper, R.H., Kerbey, A.L., Park, H.T., Severson, D.L., Stansbie, D. & Whitehouse, S. (1975). Regulation of mammalian pyruvate dehydrogenase. *Molecular and Cellular Biochemistry* **9**, 27–53.

De Zwaan, A. & van den Thillart, G. (1985). Low and high output modes of anaerobic metabolism of invertebrates and lower vertebrates. In *Proceedings of the First International Congress of Comparative Biochemistry and Physiology* (ed. R. Gilles), pp. 167–92. Springer-Verlag, Berlin.

Garland, P.B. & Randle, P.J. (1964). Control of pyruvate dehydrogenase in the perfused heart by the intracellular concentration of acetyl-coenzyme A. *Biochemical Journal* **91**, 6C–7C.

Hochachka, P.W. (1982). Metabolic arrest as a mechanism of protection against hypoxia. In *Protection of Tissues against Hypoxia* (ed. A. Wauquier, M. Borgers & W. K. Amery), pp. 1–14. Elsevier, Amsterdam.

Hochachka, P.W. (1985). Assessing metabolic strategies for surviving O_2-lack: role of metabolic arrest coupled with channel arrest. *Molecular Physiology* **8**, 331–50.

Hochachka, P.W. & Guppy, M. (1987). *Metabolic Arrest and the Control* of Biological Time. Harvard University Press, Cambridge, Massachusetts.

Hochachka, P.W. & Mommsen, T.P. (1983). Protons and anaerobiosis. *Science* **219**, 392–5.

Holopainen, I.J. & Hyvärinen, H. (1985). Ecology and physiology of crucian carp (*Carassius carassius* (L.)) in small Finnish ponds with anoxic conditions in winter. *Verhandlungen der Internationalen Verein für Limnologie* **22**, 2566–70.

Holopainen, I.J., Hyvärinen, H. & Piironen, J. (1986). Anaerobic wintering of crucian carp (*Carassius carassius* (L.)) II. Metabolic products. *Comparative Biochemistry and Physiology* **83A**, 239–42.

Johnston, I.A. & Bernard, L.M. (1983). Utilisation of the ethanol pathway in carp following exposure to anoxia. *Journal of Experimental Biology* **104**, 73–8.

Leaver, J. & Burt, J.R. (1981). Purification and properties of phosphofructokinase from cod (*Gadus morhua*) muscle. *Comparative Biochemistry and Physiology* **69B**, 127–32.

Linn, T.C., Pettit, F.H. & Reed, L.J. (1969). α-Keto acid dehydrogenase complexes X. Regulation of the activity of the pyruvate dehydrogenase complex from beef kidney mitochondria by phosphorylation and dephosphorylation. *Proceedings of the National Academy of Sciences of the USA* **62**, 234–41.

Mourik, J. (1982). Anaerobic metabolism in red skeletal muscle of goldfish, *Carassius auratus* (L.). Ph.D. thesis, University of Leiden.

Mourik, J., Raeven, P., Steur, K. & Addink, A.D.F. (1982). Anaerobic metabolism of red skeletal muscle of goldfish, *Carassius auratus* (L.): Mitochondrial produced acetaldehyde as anaerobic electron acceptor. *FEBS Letters* **137**, 111–14.

Nuccitelli, R. & Deamer, D.W. (eds) (1982). *Intracellular pH*. Alan R. Liss, New York.

Pörtner, H.O., Heisler, N. & Grieshaber, M.K. (1984). Anaerobiosis and acid–base status in marine invertebrates: a theoretical analysis of proton generation by anaerobic metabolism. *Journal of Comparative Physiology* **155**, 1–12.

Rahman, M.S. & Storey, K.B. (1988). Role of covalent modification in the control of glycolytic enzymes in response to environmental anoxia in goldfish. *Journal of Comparative Physiology* **157**, 813–20.

Seals, J.R. & Jarett, L. (1980). Activation of pyruvate dehydrogenase by direct addition of insulin to an isolated plasma membrane/mitochondria mixture: Evidence for generation of insulin's second messenger in a subcellular system. *Proceedings of the National Academy of Sciences of the USA* **77**, 77.

Shoubridge, E.A. (1980). The metabolic strategy of the anoxic goldfish. Ph.D. thesis, University of British Columbia.

Shoubridge, E.A. & Hochachka, P.W. (1981). The origin and significance of metabolic carbon dioxide production in the anoxic goldfish. *Molecular Physiology* **1**, 315–38.

Shoubridge, E.A. & Hochachka, P.W. (1983). The integration and control of metabolism in the anoxic goldfish. *Molecular Physiology* **4**, 165–95.

Storey, K.B. (1988). Tissue specific controls on carbohydrate metabolism during anoxia in goldfish. *Physiological Zoology* (in press).

Van den Thillart, G. (1977). Influence of oxygen availability on the energy metabolism of goldfish, *Carassius auratus* (L.). Ph.D. thesis, University of Leiden.

Van den Thillart, G. (1982). Adaptations of fish muscle energy metabolism to hypoxia and anoxia. *Molecular Physiology* **2**, 49–61.

Van den Thillart, G. (1986). Energy metabolism of swimming trout (*Salmo gairdneri*). Oxidation rates of palmitate, glucose, alanine, leucine and glutamate. *Journal of Comparative Physiology* **156**, 511–20.

Van den Thillart, G., De Wilde-van Berge Henegouwen, M. & Kesbeke, F. (1983). Anaerobic metabolism of goldfish, *Carassius auratus* (L.): Ethanol and CO_2-excretion rates and anoxia tolerances at 20, 10 and 5 °C. *Comparative Biochemistry and Physiology* **76A**, 295–300.

Van den Thillart, G. & Kesbeke, F. (1978). Anaerobic production of carbon dioxide and ammonia by goldfish, *Carassius auratus* L. *Comparative Biochemistry and Physiology* **59A**, 393–400.

Van den Thillart, G. & van Waarde, A. (1985). Teleosts in hypoxia: aspects of anaerobic metabolism. *Molecular Physiology* **8**, 393–409.

Van den Thillart, G., van Waarde, A., Dobbe, F. & Kesbeke, F. (1982). Anaerobic energy metabolism of goldfish, *Carassius auratus* (L.): Effects of anoxia on the measured and calculated NAD^+/NADH ratios in muscle and liver. *Journal of Comparative Physiology* **146**, 41–9.

Van den Thillart, G., van Waarde, A., Muller, H.J., Erkelens, C., Addink, A. & Lugtenburg, J. (1989). Fish muscle energy metabolism measured by *in vivo* ^{31}P-NMR during anoxia and recovery. *American Journal of Physiology* **256**, R922–9.

Van Waversveld, J., Addink, A., van den Thillart, G. & Smit, H. (1988). Direct calorimetry of free swimming goldfish at different oxygen levels. *Journal of Thermal Analysis* **33**, 1019–26.

Ward, G.R., Sutton, J.R., Jones, N.L. & Toews, C.J. (1982). Activation by exercise of human skeletal muscle pyruvate dehydrogenase in vivo. *Clinical Science* **63**, 87–92.

Wissing, J. & Zebe, E. (1988). The anaerobic metabolism of the bitterling *Rhodeus amarus* (Cyprinidae, Teleostei). *Comparative Biochemistry and Physiology* **89B**, 299–301.

ANDRÉ TOULMOND

Respiratory and metabolic adaptations of aquatic annelids to low environmental oxygen tensions

Abstract

Many marine and freshwater habitats are characterized by low or very low oxygen concentrations and partial pressures. Usually, the oxygen lack is correlated with high concentrations of carbon dioxide and of sulphides.

Many aquatic annelids are able to colonize and exploit these very special environments permanently, using complementary biochemical, anatomical, physiological and behavioural adaptations. Four respiratory and metabolic strategies can be suggested: (i) permanent aerobiosis with oxygen at normoxic partial pressure taken from an adjacent, well-oxygenated environment; (ii) permanent aerobiosis with oxygen at low partial pressures present in the inhabited medium; (iii) a periodic switch from aerobiosis to anaerobiosis; (iv) permanent anaerobiosis. These four solutions are analysed and discussed with reference to specific examples.

Introduction

Despite their relatively primitive, simple organization, annelids have successfully colonized all types of aquatic media: marine, brackish and fresh waters, as well as the terrestrial habitat. The plasticity of annelids made possible an extensive radiative evolution in which anatomical and physiological potentialities are exploited to the full (Clark, 1978). For example, a well-developed coelom, at least in polychaetes and oligochaetes, acting as a hydrostatic skeleton allows these forms to burrow deeply into sediments where they gain protection against predators and physiochemical variations in the water column (Mangum, 1976). Various types of respiratory pigment, with very different functional properties, certainly also constitute a favourable factor (Weber, 1980). Polychaetes are thus the most abundant invertebrates in soft marine substrates; their dominance is challenged by crustaceans only at the greatest oceanic depths (Sanders et al., 1965). Oligochaetes are normal dwellers of salt and fresh

191

waters and representative species of this class, the earthworms, are among the largest terrestrial invertebrates and constitute the most important animal biomass in continental ecosystems (Bouché, 1984). Leeches are less abundant, both as species and individuals, but these ectoparasitic worms can live in all aquatic media, and some have an aerial way of life in tropical rain forests (Fauvel, 1959).

In terrestrial environments, hypoxic conditions are rather rare, occurring only at high altitude and in very compact wet soils. Apparently earthworms, of which numerous species can burrow several meters down, encounter hypoxia very rarely and their resistance to anoxia is reduced (Cosgrove & Schwartz, 1965; Dales, 1969; Gruner & Zebe, 1978); in fact, they are considered as the main agents of soil aeration. Possible hypoxic conditions in their galleries may correspond to low oxygen partial pressures but, owing to the high capacitance of air for oxygen, the oxygen concentration may be rather high. Conversely, in aquatic media, the low capacitance of water for oxygen means that hypoxia always corresponds both to low partial pressures *and* low concentrations of oxygen (Dejours, 1981). Among hypoxic media of special interest here are the bottom waters, or even surface waters, of tropical marshes, small ponds, intertidal pools, stratified lakes and fjords (Carter & Beadle, 1930; Jones, 1961; Truchot & Duhamel-Jouve, 1980; Sverdrup *et al.*, 1949); interstitial water contained in sediments which is hypoxic or anoxic just below the oxidized surface layer (Jones, 1955; Brafield, 1964; Revsbeck *et al.*, 1980; Reimers *et al.*, 1986); anoxic waters of the hydrothermal vents along the mid-ocean ridges which, by mixing with the well-oxygenated deep seawater, create highly unstable microenvironments with respect to P_{O_2} (Johnson *et al.*, 1986).

Oxygen depletion is usually correlated with increased concentrations of carbon dioxide and low pH values. This is the case particularly at night in intertidal pools and vegetation-covered lakes where large plant and animal biomasses are only respiring (Truchot & Duhamel-Jouve, 1980; Heisler *et al.*, 1982). Hydrogen sulphide, a powerful inhibitor of the cytochrome *c* oxidase systems (see Powell & Somero, 1986), nitrites and ammonia often aggravate matters. When dead organic matter accumulates inside the substrate, anaerobic bacteria proliferate and anoxic, sulphide-rich black sediment is formed. Where dead organic matter is particularly abundant, as in eutrophic ponds or stratified lakes, in seagrass beds or at sewage outfalls, high sulphide concentrations extend beyond the sediment to the bottom water. Hydrogen sulphide concentrations may be even higher in large volumes of water around the deep-sea hydrothermal vents.

Despite the low oxygen concentration and the toxic compounds present, these aquatic environments may support abundant and complex animal communities. Annelids, usually present, often constitute the most important biomass. Most are partly or completely burrowed in the sediment but, at deep-sea hydrothermal vents, alvinellid polychaetes live in tubes directly fixed on the active black or white smokers (Fustec *et al.*, 1987).

How can these generally large, hypothetically aerobic metazoans (Dales, 1969; Weber, 1978), support such life conditions? Perhaps the first problem to be solved is that of protection against the deleterious effects of sulphides. The second problem is that of at least temporary access to oxygen.

Three principal respiratory and metabolic strategies are suggested: permanent aerobiosis with the oxygen present at normoxic partial pressure in the bottom water above the sediment, permanent aerobiosis with the oxygen at low partial pressures present in the medium, and periodic switching from aerobiosis to anaerobiosis. If none of these conditions is obtained, a fourth way of life, that of permanent anaerobiosis, must be considered.

Permanent aerobiosis with oxygen at normoxic partial pressure: *Chaetopterus variopedatus*

Subtidal annelids that live in a protective but anoxic or hypoxic sediment obtain enough oxygen for aerobiosis most simply by retaining access to the water column which, most usually, can be considered equilibrated with the atmosphere. *Chaetopterus variopedatus* (Renier) is a marine polychaete which lives in a parchment-like, U-shaped tube deeply buried in muddy sand or gravels (Fauvel, 1927). It is a typical suspension-feeder and, for both respiration and nutrition, is totally dependent on the current of water passing inside its tube and normally driven tailward by three piston-like segments of the middle body region (Brown, 1975, 1977). This current brings oxygen and nutritive particulate matter, which is collected inside the tube (McGinitie, 1939), and removes carbon dioxide and other wastes. *Chaetopterus* has neither respiratory pigment nor specialized respiratory surfaces, and its epidermis, devoid of the normal extracellular cuticle (Storch & Welsch, 1970), is most likely very permeable to gases.

Probably in relation with the type of nutrition, the tube is irrigated almost permanently. The specific ventilatory rate is high, about $110 \, \text{ml} \, \text{h}^{-1} \, \text{g}^{-1}$ at 15 °C, and the oxygen-extraction coefficient rather low, about 30%. Specific oxygen consumption rates are quite high, about $11 \, \mu\text{mol} \, \text{g}^{-1} \, \text{h}^{-1}$ in an actively ventilating animal of wet mass 4.3 g (Dales, 1969). Apparently, it is not known whether *Chaetopterus* can regulate its oxygen consumption when P_{O_2} decreases, nor whether it can resist anoxic periods. Concerning this last point, its capacities are almost surely not greater than that of planktonic polychaetes or of tube-dwelling species which are epibiotic on hard substrates and permanently live in the water column.

Chaetopterus's permanent U-shaped tube deserves special attention. It is cylindrical and each narrowed end opens above the sediment surface. Ultrastructural study reveals that the tube wall consists of plies of parallel protein fibres, tentatively analysed as collagen, imbedded in an unstructured polysaccharide matrix. Tube walls are 1- to 18-ply, the more the thicker the tube. Each ply has one to fifteen layers of parallel fibres, the orientation of which can vary greatly from one ply to another from virtually parallel to nearly perfectly orthogonal (Brown & McGee-Russel, 1971). The tube wall is mechanically and chemically very resistant and is considered to be impermeable to water and gases (Dales, 1969). In its tube, *Chaetopterus* is thus completely isolated from the anoxic and potentially toxic interstitial seawater in the sand but keeps permanent, direct contact with the water column from which it takes food and oxygen. The sand protects the soft-bodied worm from stresses originating in the open water and the tube likewise protects against stresses originating from the sand. It also increases the pumping efficiency since water cannot permeate from the tube into the sand.

Polychaete tubes often have simpler morphology and structure than that of *Chaetopterus*. They can be U-shaped but most are about vertical, with one end opening in the water column, the other buried sometimes one metre below the sand surface (Mangum, 1964; Myers, 1972). The tube walls' structure and composition vary greatly, sometimes within one polychaete family, and all intermediaries exist between the thick, tough tube of sabellids or eunicids and a burrow simply lined with a thin mucous sheet (Defretin, 1971). In the latter case, the worm is not isolated from the mud or sand and the burrow water remains distinct from the interstitial water only by virtue of its occupant's ventilatory activity. Obviously, in these conditions, water and dissolved substances, especially oxygen, are exchanged between the burrow and the interstitial environment and these exchanges are of special importance for the meiofauna (see Aller & Yingst, 1978; Meyers *et al.*, 1987).

Permanent aerobiosis with oxygen at low partial pressures: annelid members of the meiofauna and polychaetes of the family Alvinellidae

Meiofauna is composed of the highly specialized minute organisms, from all the invertebrate phyla, which permanently live in the water-filled pore-system of aquatic sediments (see Swedmark, 1964). Schematically, these sediments display a yellow upper oxidized layer of variable thickness, separated by the grey chemocline from a deeper anoxic, sulphide-rich black layer. There is a continuous gradient of decreasing oxygen concentration and increasing reducing conditions from the sediment surface down to the anoxic black layer (Fenchel & Riedl, 1970; Revsbeck *et al.*, 1980; Reimers *et al.*, 1986). If interstitial annelids are apparently absent from the grey and black layers (see Powell *et al.*, 1979), interstitial polychaetes and oligochaetes are fairly common in the yellow layer and, if present in the deeper reducing layers, live alongside the permeable burrows of the megafauna (Reise & Ax, 1979). In any of these situations, they live in an environment which, on a scale of millimetres to micrometres, must have highly fluctuating respiratory properties related to changes in water circulation, temperature, dead organic matter content and metabolic activity of both bacteria and megafauna.

Under declining oxygen tensions, four intertidal species, the polychaetes *Stygocapitella subterranea*, *Hesionides arenaria* and *Protodriloides symbioticus* and the oligochaete *Marionina achaeta*, are able to regulate their oxygen consumption down to a critical oxygen partial pressure (*Pc*) of about 10 Torr. Their specific O_2-consumption rates are very high, ranging from 4 (*Stygocapitella*) to 45 μmol O_2 g^{-1} h^{-1} (*Marionina*) but, owing to these animals' very small size, the oxygen needs per individual are very low, especially in *Stygocapitella* (Lasserre & Renaud-Mornant, 1973). Like most interstitial animals, these species have no specialized respiratory surfaces and oxygen diffuses through the entire body surface. Their very small size and cylindrical shape yield an extremely favourable surface-to-volume ratio: it has been calculated that in an adult gastrotrich 500 μm long, the oxygen flux through its body surface is more than 3 times its oxygen consumption when the oxygen pressure gradient between the water and the mitochondria is only 1 Torr (Colacino & Kraus, 1984).

Respiratory pigments are not, however, infrequent in interstitial animals. *Protodriloides* possesses an extracellular, vascular, high molecular mass haemoglobin (A. Toulmond, C. Jouin & J. de Frescheville, unpublished). A haemoglobin of very high oxygen affinity, P_{50} *ca.* 1 Torr, is present at high concentration in specialized cells of gastrotrichs (Colacino & Kraus, 1984). These authors consider that this haemoglobin functions as an oxygen store and conclude that a pigment role in oxygen transport or sulphide detoxification is unlikely. Among the species studied by Lasserre & Renaud-Mornant (1973), *Stygocapitella*, which lacks a respiratory pigment, is the most resistant to simultaneous oxygen deficiency and high sulphide concentration, but whether it detoxifies sulphides in its body wall as do some turbellarian species (Powell *et al.*, 1979) is unknown. The same low *Pc* value, low oxygen needs, efficient respiratory exchanger, respiratory pigment with high oxygen affinity and a capacity for regulation of oxygen uptake are characteristics of animals which, like the mysid crustacean *Gnathophausia ingens*, permanently live in the hypoxic oceanic oxygen minimum layer (Childress, 1975).

Some large polychaetes also colonize waters whose respiratory properties probably change widely in time and space. All six species so far described of the family Alvinellidae are uniquely associated with deep-sea hydrothermal vents (Desbruyères & Laubier, 1986). Two species, *Alvinella pompejana* and *Alvinella caudata*, secrete tortuous, parchment-like tubes loosely associated in an irregular honeycomb-like structure and embedded in a porous matrix of polymetallic sulphides (Desbruyères & Laubier, 1980). The tube wall is made of concentric fibrillar layers disposed as in plywood and contained in an unusually stable glycoprotein matrix having a high concentration of elemental sulphur (Gaill & Hunt, 1986). The inhabited tubes constitute the most exterior part of active white and black smoker walls. Hot, anoxic, acidic, sulphide-rich water from the vent continuously diffuses through the chimney wall and mixes with the deep, oxygenated, cold (2 °C) seawater. The worms are thus exposed to the steepest temperature gradient ever measured for an animal environment: the temperature inside the chimney may be 200 °C, about 50 to 40 °C at the base of the tubes and is only 20 °C at their periphery (Desbruyères *et al.*, 1982). A few centimetres above an alvinellid 'reef', the temperature may still be high and it oscillates at a high frequency (e.g. 1 Hz) between 2 and 16 °C, (D. Desbruyères, personal communication). Similar observations have been made in clumps of the vestimentiferan *Riftia*, where temperature varies from about 2 to 15 °C and fluctuates with no clear periodicity over time scales ranging from 1 second to 72 hours (Johnson *et al.*, 1988). These authors point out that temperature gives a good prediction of oxygen and sulphide concentrations: the higher the temperature, the higher the sulphide and the lower the oxygen which, at the site studied, was extrapolated by them to zero near 11 °C.

These results strongly suggest that alvinellids live in a hypoxic microenvironment, the degree of the hypoxia being unpredictable from one second to another. The anatomy and physiology of alvinellids is as yet incompletely studied. The morphology of their four pairs of gills is very special and rather uncommon, with a high mass-specific gill surface area (Table 1). The external segmentation is normal, but the corresponding internal septa are absent in the anterior half of the body. The resulting vast

Table 1. *Gills of the lugworm,* Arenicola marina, *and of* Alvinella
pompejana: *compared characteristics*

	Arenicola marina	*Alvinella pompejana*
Gill number	13 pairs	4 pairs
Type of gills	ramified, unciliated	stem with ciliated lamellae
	ca. 260 terminal	*ca.* 700 lamellae per gill
	branches per gill	
Epidermis thickness	6–14 μm	16 μm
Cuticle thickness	1–2 μm	1 μm
Diffusion distance	6–8 μm	1–3 μm
Mass-specific gill area	4 cm^{2}g^{-1} wet mass	12 cm^{2}g^{-1} wet mass

Data from Jouin & Toulmond (1989) and Jouin & Gaill (1990).

Table 2. *Comparison of the oxygen affinity and the Bohr effect of alvinellid
and lugworm extracellular haemoglobins*

Oxygen affinity is measured as P_{50} at pH 7.6 (oxygen partial pressure at which the
pigment is half saturated with oxygen). The Bohr effect ($\phi=\Delta\log P_{O_2}/\Delta pH$) is
measured when the pigment is half saturated ($\phi(P_{50})$), almost fully deoxygenated
($\phi(T)$) or almost fully oxygenated ($\phi(R)$). Experimental conditions: haem concentration *ca.* 60 μM in 0.05 M Bistris–HCl; pH between 6.6 and 7.6 at 20 °C;
$P_{CO_2}=0$; $[Cl^-]=0.47$ M; $[SO_4^{2-}]=0.03$ M; $[Mg^{2+}]=0.05$ M; $[Ca^{2+}]=0.01$ M; total
osmolarity *ca.* 1.05 Osm.

	P_{50} (Torr) (pH=7.6)	$\phi(P_{50})$	$\phi(T)$	$\phi(R)$
Alvinella pompejana	0.15	−1.18	−1.21	−0.24
Alvinella caudata	0.25	−0.90	−0.93	−0.30
Arenicola marina	2.05	−0.66	−0.34	−0.43

Data of A. Toulmond (unpublished).

coelomic cavity suggests that the alvinellids may ventilate their tubes in the
same way as, for example, terebellids. The well-developed, closed vascular
system of *A. pompejana* contains a high-molecular-mass, extracellular
haemoglobin with a quaternary structure similar to that of other annelids
(Terwilliger & Terwilliger, 1984). This haemoglobin's intrinsic oxygen
affinity, measured at one atmosphere and 20 °C, is quite high: P_{50} is
between 0.15 and 2 Torr at pH values between 7.6 and 6.6. Oxygen affinity
decreases and cooperativity increases with decreasing pH, and the normal
Bohr effect is maximum at low oxygen saturation values (Table 2 and

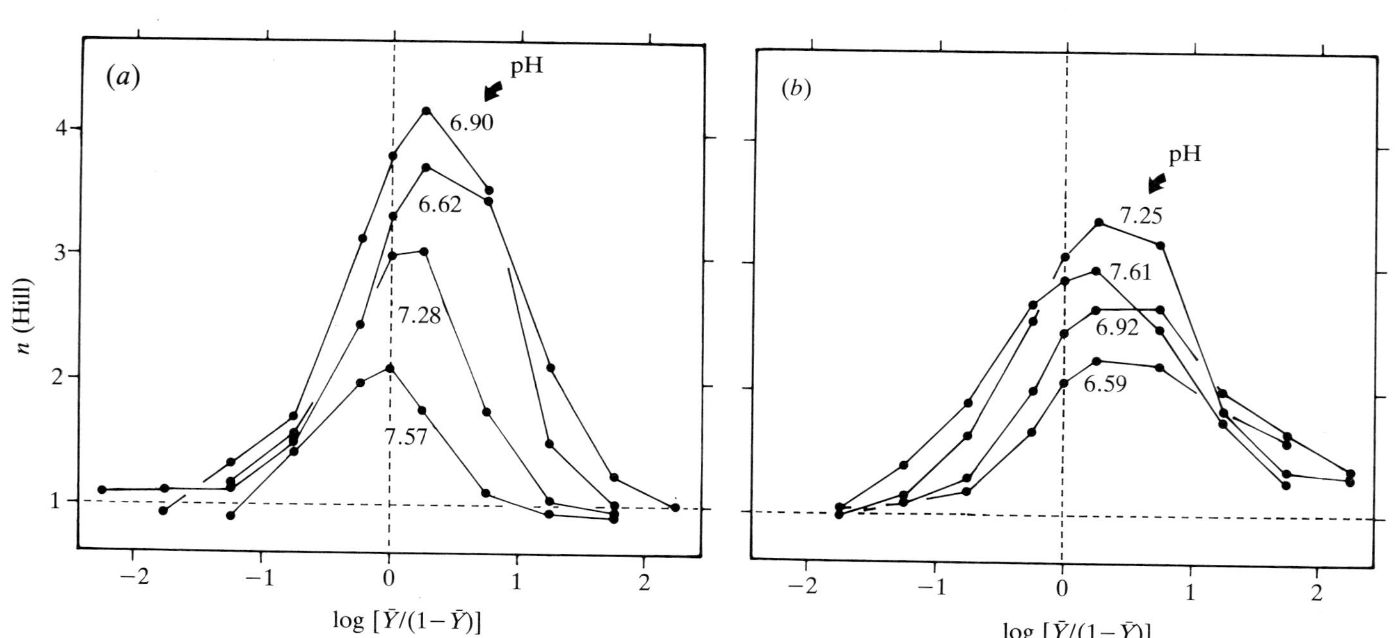

Figure 1. At four pH values, variations of the cooperativity, measured by Hill's coefficient n, as function of log [$\bar{Y}/(1-\bar{Y})$], where $\bar{Y}$ is the fractional oxygen saturation, of *Alvinella pompejana* (*a*) and *Arenicola marina* (*b*) extracellular haemoglobins. Experimental conditions as in Table 10.2 (A. Toulmond, unpublished).

Figure 1). The pigment, whose properties favor both loading and unloading of oxygen at very low P_{O_2}, seems thus well adapted to function in hypoxic conditions. But it normally functions at 250 atmospheres: what are the effects of hydrostatic pressure on the molecule? In sum, most aspects of the respiratory physiology of alvinellids are completely unknown, as are the mechanisms of protection against the action of sulphides. As for sulphide toxicity, the possibility of a detoxifying action of the bacteria living epibiontically on *A. pompejana* has been suggested (Gaill *et al.*, 1987) but as yet remains to be demonstrated.

Alternating aerobiosis and anaerobiosis: *Arenicola marina*

The lugworm, *A. marina*, is an intertidal polychaete which lives in a permanent L-shaped gallery simply consolidated by a thin mucous lining and with a single well-defined opening. The gallery is deeply dug into the anoxic black layer of fine to coarse sands, and is characterized by a long emersion time during low tide, and high organic matter and sulphide concentrations, but very low or null oxygen concentration in the interstitial water (Jones, 1955; Rullier, 1959; Brafield, 1964; Wells, 1966). The lugworm is a deposit-feeder, which swallows sand at the blind head-end of its burrow and consumes the organic matter it contains (Jacobsen, 1967).

The lugworm thus inhabits a permanent, stable hypoxic environment, but during high tide it has regular and easy access to the normoxic seawater which covers the sand. During this period the behavior of the lugworm is that of the subtidal *Chaetopterus* but with several differences: (i) it actively pumps seawater through its burrow to obtain oxygen and eliminate carbon dioxide and liquid wastes, but in a tail-to-head direction; (ii) water pumped by peristaltic, piston-like movements of the dorsal body wall musculature flows over 13 pairs of ramified gills; (iii) the blood contains an extracellular, high-molecular-mass haemoglobin, which can function as a short-term oxygen store; (iv) probably related to the fact that the lugworm does not feed from the water pumped, specific ventilatory rates at 15 °C are about ten times lower than in *Chaetopterus*, with an oxygen extraction coefficient higher than 80%; (v) specific oxygen consumption rates are very low, about 1.5 µmol O_2 g^{-1} h^{-1} for a worm of 17 g wet mass (Toulmond & Tchernigovtzeff, 1984).

The lugworm also differs strikingly from the interstitial annelids studied by Lasserre & Renaud-Mornant (1973). It regulates its specific metabolic rate from oxygen partial pressure of *ca.* 700 Torr down to a remarkably high *Pc* value of about 120 Torr, below which the oxygen consumption decreases linearly (Toulmond & Tchernigovtzeff, 1984). Complementary observations clearly show that a P_{O_2} below 40 Torr is for the lugworm a signal to stop ventilating. Ventilatory regulation seems to be rather complex, involving probably two types of epidermal O_2-sensitive chemoreceptors acting on a central nervous pacemaker (Wells, 1966; Toulmond *et al.*, 1984; Jouin *et al.*, 1985; Dejours & Toulmond, 1988).

In these conditions, what occurs during low tide, when the lugworm is confined to its burrow, filled with very hypoxic interstitial seawater? Field measurements show that the blood oxygen content drops nearly to zero within the first hour of tidal emersion. A blood acidosis then develops, which is the stronger the longer the emersion. This blood acidosis is both respiratory, by accumulation of carbon dioxide, and metabolic, by addition of the volatile fatty acids acetate and propionate, and it quickly disppears during the following high tide (Toulmond, 1973; Pionetti & Toulmond, 1980). The lugworm's respiratory exchanges are thus deeply modified during low tide and its energy metabolism periodically switches from aerobic to anaerobic pathways.

Clearly the lugworm is a typical facultative anoxybiont and as such has to deal with two main problems: conservation of fermentable substrates and avoidance of self-pollution by undesirable end-products, mainly protons. There are several possible, not incompatible solutions: more substrates can be stored; activity, metabolism, energy and ATP demands can decrease during the anaerobic period; anaerobic pathways can be used which produce more ATP per mole of fermentable substrate; more ATP can be turned over per mole of generated protons; proton production can be tolerated by improved buffering capacities of the tissues; end-product accumulation can be avoided by immediate recycling and/or excretion (Hochachka & Mommsen, 1983; Pörtner *et al.*, 1984; Kreutzer *et al.*, 1985; Hochachka, 1986). From an analysis of the literature, especially the detailed studies initiated by Zebe on the anaerobic metabolism of the lugworm (Zebe, 1975; Surholt, 1977; Pörtner *et al.*, 1979; Felbeck & Grieshaber, 1980; Schroff & Zebe, 1980; Schöttler *et al.*, 1984; Siegmund *et al.*, 1985; Holst & Zebe, 1986), it is possible to conclude that the lugworm uses all these possible solutions (for a more detailed analysis, see Toulmond, 1987).

The lugworm's metabolism thus seems perfectly adapted to life in intertidal anoxic sediments, particularly because of its efficient anaerobic metabolic pathways. The lugworm's extracellular and vascular haemoglobin also contributes essentially to its respiratory strategy. Besides particular buffering properties, which are of interest during low tide (Toulmond, 1977), this haemoglobin has a rather high intrinsic oxygen affinity which is, however, about ten times lower than that of *A. pompejana*'s respiratory pigment measured in the same conditions at 20 °C and 1 atmosphere (P_{50} between 2 and 10 Torr when pH varies between 7.6 and 6.6). The Bohr effect and cooperativity are also important but, unlike the situation in the alvinellid, the Bohr effect is maximum at high saturation values and cooperativity decreases with decreasing pH (Weber, 1981). Such properties and the presence of two body-wall myoglobins of very high oxygen affinity (P_{50} at 0.8 and 1.3 Torr at 20 °C) (Weber & Pauptit, 1972) must favour both oxygen loading and unloading during high tide (see Toulmond, 1985), but at higher internal oxygen partial pressures than in *Alvinella*, in accordance with the quite different properties of the medium respired.

The lugworm's haemoglobin may also protect the animal against sulphides. Patel & Spencer (1963) have reported on the presence in the lugworm's blood of a 'brown pigment' resulting from the autooxidation of the haemoglobin and catalysing the oxidation of sulphides *in vivo*. Wells & Pankhurst (1980) did not find such a sulphide detoxification system in the related species *Abarenicola affinis*, and their conclusions were not in favour of its existence in the lugworm. Toulmond *et al.* (1988) confirm that the haemoglobins of the lugworm and *A. pompejana* are especially resistant to oxidation. Sulphide detoxification systems in annelids remain as yet quite obscure.

Permanent anaerobiosis? Tubificid oligochaetes

Tubificid annelids form massive populations in marine and freshwater sediments heavily polluted with dead organic matter, where the concentration of oxygen is low and that of hydrogen sulphide high. Individuals of the best-known species, *Tubifex tubifex*, live head-down in vertical burrows dug in the bottom mud of ponds and slow-moving streams, where they feed on aerobic bacteria (Wavre & Brinkhurst, 1971). Respiratory exchanges take place through the unspecialized epidermis of the tail: in well-aerated water, the worm is completely retracted in its burrow, whereas in conditions of low oxygen concentration the tail waves freely in the water (Alsterberg, 1922). The oxygen consumption is independent of the external P_{O_2} down to a Pc of about 55 Torr in animals artificially grouped for experiments (Berg *et al.*, 1962; Seuss *et al.*, 1983) but only 11 Torr with Pc evaluated on single animals (Palmer, 1968). High oxygen concentrations have been reported to be toxic to *Tubifex* (Fox & Taylor, 1954; Walker, 1970) but one must consider that, at some high pressure, oxygen is toxic for all organisms, including mammals. The respiration of intact animals is only slightly affected by hydrogen sulphide and, curiously, *Tubifex* does not seem to have any sulphide detoxification system, despite the fact that isolated *Tubifex* mitochondria and cytochrome *c* oxydase are much more sensitive to hydrogen sulphide than the intact animal (Degn & Kristensen, 1981). As for *Pc* values, data concerning the extracellular, vascular, high-molecular-mass haemoglobin of *Tubifex* are rather contradictory. Oxygen affinity is probably high, but P_{50} values in the literature range from 0.6 to 23 Torr, and the existence of a Bohr effect remains controversial (see Weber, 1978).

Like interstitial polychaetes and oligochaetes, *Tubifex* thus appears to be well adapted to life in hypoxic waters, but, like *Arenicola*, it also seems well equipped to sustain long periods of complete anoxia. Its efficient anaerobic metabolism is very similar to that of the lugworm, leading to an increase in the concentrations of alanine, succinate and volatile fatty acids, two of which, acetate and propionate, are excreted in substantial amounts (Schöttler & Schroff, 1976). As in lugworms (Holst & Zebe, 1984), these acids can be reabsorbed by the integument and may possibly serve as an important nutritional supplement (Hipp *et al.*, 1986). Also as in lugworms, the succinate–fumarate metabolic pathway typical of long-term anaerobiosis begins to work at the rather high external P_{O_2} of about 50 Torr (Seuss *et al.*, 1983). However, unlike the lugworm, *Tubifex* seems able to withstand experimental anaerobiosis for months. Famme & Knudsen (1985*a*) have reported that adult *Tubifex* survived, grew and reproduced when experimentally incubated in completely anoxic conditions for more than 200 days. *Tubifex* can, if necessary, live as a strict anoxybiont.

Until recently, the only free-living (non-parasitic) metazoans considered to be permanent anoxybionts were the interstitial inhabitants of the anoxic, sulphide-rich black sediment layer, mainly turbellarians, gnathostomulids, nematodes and gastrotrichs (Fenchel & Riedl, 1970). They are members of the 'thiobios', a term coined by Boaden & Platt (1971) to designate the biome of bacteria, blue-green algae, protists and metazoans living in the black layer. Boaden (1975) later suggested that much early metazoan evolution took place in this habitat and postulated that the first metazoans were benthic anoxybionts, a thesis completely contradicting the generally agreed-on assumption that Metazoa were originally aerobic and free-swimming (see for example, Mangum, 1976). In vigorous contestation, Reise & Ax (1979), following a meticulous analysis of the microdistribution of the meiofauna alongside lugworms' burrows, concluded that the thiobios does not exist at all, that all members of the meiofauna are obligatory or facultative oxybionts living either in the normoxic superficial yellow layer or in discontinuous, eventually hypoxic sand pockets created in the deeper anoxic layers by the burrowing and respiratory activities of the megafauna.

Strict anaerobiosis among the free-living metazoans is in fact at the heart of the controversy. The observations of Famme & Knudsen (1985*a*) on *Tubifex* remain the only reported example of a species whose entire life cycle is spend totally anaerobically and the arguments developed by Boaden (1975) and others, to tentatively prove that metazoans of the thiobios are obligatory anoxybionts, are not really convincing. (i) The absence of mitochondria in an animal could be a good index of strict anaerobiosis, but Boaden (1975) himself agrees that the presence or absence of these organelles in thiobiotic animals is problematical. (ii) The presence of endosymbiotic bactcria is also not conclusive. The chemoautotrophic bacteria taken up by vestimentiferans, pogonophorans, bivalves or gutless oligochaetes are sulphide oxidizers (Felbeck *et al.*, 1983) and they use oxygen as the final electron acceptor. (iii) Vestimentiferans and bivalves of the hydrothermal vents show that life at high sulphide concentrations is possible, but in the presence of oxygen. (iv) Precise physiological and biochemical characteristics of a possibly anaerobic metabolism of thiobiotic animals are unknown. Even if new anaerobic pathways were discovered, they would not prove that the animal studied is a strict anoxybiont. (v) Negative aerotaxis as demonstrated in *Tubifex* (Famme & Knudsen, 1985*b*) may only show that the worm does prefer low oxygen tensions and is not a proof that it is an anoxybiont. Finally, in our opinion, free-living, strictly anaerobic metazoans, even those considered as primitive, have yet to be demonstrated. *Tubifex* must be considered as a facultative anoxybiont extremely well adapted to life in hypoxic and temporarily anoxic aquatic environments.

Conclusion

The idea of dividing annelids living in hypoxic environments into four categories corresponding to four different respiratory and metabolic strategies may appear to be artificial, and probably it is. Its greatest advantage is perhaps to point out that, despite the abundant literature of the last 25 years and despite real advances in the study of the metabolism of a small number of species, our understanding of the comparative physiology and biochemistry of annelid respiration and metabolism is still rather superficial. Even of the species discussed here, chosen because they are among the best studied, important data are lacking for all in major domains such as resistance to anoxia, regulation of oxygen uptake and ventilation, functional properties of the respiratory pigments or existence and characteristics of metabolic anaerobic pathways. Comparisons are thus difficult and often impossible to make. The absence of objective data on several critical points also implies that, even if the four categories correspond to reality, some of the specific examples may be quite wrong. Resistance to anoxic conditions is perhaps widely distributed among annelids. *Chaetopterus* is given as a type of strict oxybiont, but is perhaps a facultative anoxybiont, as are perhaps also those worms thought to use oxygen at low partial pressures present in their immediate environment. This last example also shows that the properties of the microenvironment, despite their potentially high explicative content, are often incompletely known. Progress in the field of the comparative respiratory physiology of annelids could be made by more thorough study of the few species for which an important corpus of data already exists, and on the micromilieu in which they live, a final comment which may apply to all invertebrates.

The author thanks Drs P. Dejours, S. Dejours and C. Jouin for valuable comments, discussion and editorial assistance. This work was supported by the Centre National de la Recherche Scientifique, LP 4601, Roscoff and Paris, the Université Pierre-et-Marie-Curie, Paris, and the Society for Experimental Biology, London.

References

Aller, R.C. & Yingst, Y. (1978). Biogeochemistry of tube-dwellings: a study of the sedentary polychaete *Amphitrite ornata* (Leidy). *Journal of Marine Research* **36**, 201–54.

Alsterberg, G. (1922). Die respiratorischen Mechanismen der Tubificiden. *Lunds Universtets Årsskrift* (N.F. Avd. 2) **18**, 1–176.

Berg, K., Jonasson, P.M. & Ockelmann, K.W. (1962). The respiration of some animals from the profundal zone of a lake. *Hydrobiologia* **19**, 1–39.

Boaden, P.J.S. (1975). Anaerobiosis, meiofauna and early metazoan evolution. *Zoologica Scripta* **4**, 21–4.

Boaden, P.J.S. & Platt, H.M. (1971). Daily migration patterns in an intertidal meiobenthic community. *Thalassia Jugoslavica* **7**, 1–12.

Bouché, M. (1984). Les vers de terre. *La Recherche* **15**, 796–04.

Brafield, A.E. (1964). The oxygen content of interstitial water in sandy shores. *Journal of Animal Ecology* **33**, 97–116.

Brown, S.C. (1975). Biomechanics of water–pumping by *Chaetopterus variopedatus* Renier. Skeletomusculature and kinematics. *Biological Bulletin* **149**, 136–50.

Brown, S.C. (1977). Biomechanics of water-pumping by *Chaetopterus variopedatus* Renier: kinetics and hydrodynamics. *Biological Bulletin* **153**, 121–32.

Brown, S.C. & McGee-Russel, S. (1971). *Chaetopterus* tubes: ultrastructural architecture. *Tissue and Cell* **3**, 65–70.

Carter, G.S. & Beadle, L.C. (1930). The fauna of the swamps of the Parguayan Chaco in relation to its environment. I. Physico-chemical nature of the environment. *Journal of the Linnean Society of London (Zoology)* **37**, 205–58.

Childress, J.J. (1975). The respiratory rates of midwater crustaceans as a function of depth of occurrence and relation to the oxygen minimum layer off Southern California. *Comparative Biochemistry and Physiology* **50A**, 787–99.

Clark, R.B. (1978). Composition and relationships. In *Physiology of Annelids* (ed. P.J. Mill), pp. 1–32. Academic Press, London and New York.

Colacino, J.M. & Kraus, D.W. (1984). Hemoglobin-containing cells of *Neodasys* (Gastrotricha, Chaetonotida): II. Respiratory significance. *Comparative Biochemistry and Physiology* **79A**, 363–9.

Cosgrove, W.B. & Schwartz, J.B. (1965). The properties and function of the blood pigment of the earthworm, *Lumbricus terrestris*. *Physiological Zoology* **38**, 206–12.

Dales, R.P. (1969). Respiration and energy metabolism in annelids. In *Chemical Zoology*, vol. 4 (ed. M. Florkin & B.T. Scheer), pp. 93–109. Academic Press, New York and London.

Defretin, R. (1971). The tubes of polychaete annelids. In *Comprehensive Biochemistry*, vol. 26C (ed. M. Florkin & E.H. Stotz), pp. 713–47. Elsevier, Amsterdam.

Dejours, P. (1981). *Principles of Comparative Respiratory Physiology*. Elsevier/North-Holland, Amsterdam.

Dejours, P. & Toulmond, A. (1988). Ventilatory reactions of the lugworm *Arenicola marina* (L.) to ambient water oxygenation changes: a possible mechanism. *Physiological Zoology* **61**, 407–14.

Degn, H. & Kristensen, B. (1981). Low sensitivity of *Tubifex* sp. respiration to hydrogen sulfide and other inhibitors. *Comparative Biochemistry and Physiology* **69B**, 809–17.

Desbruyères, D. & Laubier, L. (1980). *Alvinella pompejana* gen. sp. nov., Ampharetidae aberrant des sources hydrothermales de la ride Est-Pacifique. *Oceanologica Acta* **3**, 267–74.

Desbruyères, D. & Laubier, L. (1986). Les Alvineliidae, une famille nouvelle d'annélides polychètes inféodées aux sources hydrothermales sous-marines: systématique, biologie et écologie. *Canadian Journal of Zoology* **64**, 2227–45.

Desbruyères, D., Crassous, P., Grassle, J., Khripounoff, A., Reyss, D., Rio, M. & Van Praet, M. (1982). Données écologiques sur un nouveau site d'hydrothermalisme actif de la ride du Pacifique oriental. *Comptes Rendus des Séances Hebdomadaires de l'Académie des Sciences, Paris, Ser. III* **295**, 489–94.

Famme, P. & Knudsen, J. (1985a). Anoxic survival, growth and reproduction by the freshwater annelid, *Tubifex* sp., demonstrated using a new simple anoxic chemostat. *Comparative Biochemistry and Physiology* **81A**, 251–3.

Famme, P. & Knudsen, J. (1985*b*). Aerotaxis by the freshwater oligochaete *Tubifex* sp. *Oecologia* **65**, 599–601.

Fauvel, P. (1927). Polychètes sédentaires. In *Faune de France*, vol. 16 (ed. Office Central de Faunistique), pp. 1–494. Lechevalier, Paris.

Fauvel, P. (1959). Polychètes. In *Traité de Zoologie*, vol. 5, part 1 (ed. P.P. Grassé), pp. 13–196. Masson, Paris.

Felbeck, H. & Grieshaber, M.K. (1980). Investigations on some enzymes involved in the anaerobic metabolism of amino acids of *Arenicola marina*. *Comparative Biochemistry and Physiology* **66B**, 205–13.

Felbeck, H., Liebezeit, G., Dawson, R. & Giere, O. (1983). CO_2 fixation in tissues of marine oligochaetes (*Phallodrilus leukodermatus* and *P. planus*) containing symbiotic, chemoautotrophic bacteria. *Marine Biology* **75**, 187–91.

Fenchel, T.M. & Riedl, J. (1970). The sulfide system: a new biotic community underneath the oxidised layer of marine sand bottom. *Marine Biology* **7**, 255–68.

Fox, H.M. & Taylor, A.E. R. (1954). The tolerance of oxygen by aquatic invertebrates. *Proceedings of the Royal Society of London* **B143**, 214–25.

Fustec, A., Desbruyères, D. & Juniper, S.K. (1987). Deep-sea hydrothermal vent communities at 13°N on the East Pacific rise: microdistributions and temporal variations. *Biological Oceanography* **4**, 121–64.

Gaill, F., Desbruyères, D. & Prieur, D. (1987). Bacterial communities associated with 'Pompei worms' from the East Pacific rise hydrothermal vents: SEM, TEM observations. *Microbial Ecology* **13**, 129–39.

Gaill, F. & Hunt, S. (1986). Tubes of deep sea hydrothermal vent worms *Riftia pachyptila* (Vestimentifera) and *Alvinella pompejana* (Annelida). *Marine Ecology Progress Series* **34**, 267–74.

Gruner, B. & Zebe, E. (1978). Studies on the anaerobic metabolism of earthworms. *Comparative Biochemistry and Physiology* **60B**, 441–5.

Heisler, N., Forcht, G.R., Ultsch, G.R. & Anderson, J.F. (1982). Acid-base regulation in response to environmental hypercapnia in two aquatic salamanders, *Siren lacertina* and *Amphiuma means*. *Respiration Physiology* **49**, 141–58.

Hipp, E., Bickel, U., Mustafa, T. & Hoffmann, K.H. (1986). Integumentary intake of acetate and propionate (VFA) by *Tubifex* sp., a freshwater oligochaete. II. Role of VFA as nutritional resources and effects of anaerobiosis. *Journal of Experimental Zoology* **240**, 299–308.

Hochachka, P.W. (1986). Defense strategies against hypoxia and hypothermia. *Science* **231**, 234–41.

Hochachka, P.W. & Mommsen, T.P. (1983). Protons and anacrobiosis. *Science* **219**, 1391–7.

Holst, H. & Zebe, E. (1984). Absorption of volatile fatty acids from ambient water by the lugworm *Arenicola marina*. *Marine Biology* **80**, 125–34.

Holst, H. & Zebe, E. (1986). Volatile fatty excretion during anaerobiosis in the lugworm *Arenicola marina*. *Comparative Biochemistry and Physiology* **83A**, 189–96.

Jacobsen, V.H. (1967). The feeding of the lugworm, (*Arenicola marina*) (L.). Quantitative studies. *Ophelia* **4**, 91–109.

Johnson, K.S., Beehler, C.L., Sakamoto-Arnold, C.M. & Childress, J.J. (1986). In situ measurements of chemical distributions in a deep-sea hydrothermal vent field. *Science* **231**, 1139–41.

Johnson, K.S., Childress, J.J. & Beehler, C.L. (1988). Short term temperature variability in the Rose Garden hydrothermal vent field: an unstable deep-sea environment. *Deep Sea Research* **35**, 1711–21.

Jones, J.D. (1955). Observations on the respiratory physiology and on the haemoglobin of the polychaete genus *Nephthys* with special reference to *N. hombergii* (Aud. et M. Edw.). *Journal of Experimental Biology* **32**, 110–25.

Jones, J.D. (1961). Aspects of respiration in *Planorbis corneus* L. and *Lymnaea stagnalis* L. (Gastropoda: Pulmonata). *Comparative Biochemistry and Physiology* **4**, 1–29.

Jouin, C. & Gaill, F. (1990). Gills of hydrothermal vent annelids: functional anatomy and ultrastructure in two alvinellid species. *Progress in Oceanography* (in press).

Jouin, C., Tchernigovtzeff, C., Baucher, M.F. & Toulmond, A. (1985). Fine structure of probable mechano- and chemoreceptors in the caudal epidermis of the lugworm *Arenicola marina* (Annelida, Polychaeta). *Zoomorphology* **105**, 76–82.

Jouin, C. & Toulmond, A. (1989). The ultrastructure of the gill of the lugworm *Arenicola marina*. *Acta Zoologica* **70**, 121–9.

Kreutzer, U., Siegmund, B. & Grieshaber, M.K. (1985). Role of coupled substrates and alternative end products during hypoxia tolerance in marine invertebrates. *Molecular Physiology* **8**, 371–92.

Lasserre, P. & Renaud-Mornant, J. (1973). Resistance and respiratory physiology of intertidal meiofauna to oxygen-deficiency. *Netherlands Journal of Sea Research* **7**, 290–302.

McGinitie, G.E. (1939). The method of feeding in *Chaetopterus*. *Biological Bulletin* **77**, 115–18.

Mangum, C.P. (1964). Activity patterns in metabolism and ecology of polychaetes. *Comparative Biochemistry and Physiology* **11**, 239–50.

Mangum, C.P. (1976). Primitive respiratory adaptations. In *Adaptations to environment. Essays on the Physiology of Marine Animals* (ed. R.C. Newell), pp. 191–278. Butterworth, London.

Meyers, M.B., Fossing, H. & Powell, E.N. (1987). Microdistribution of interstitial meiofauna, oxygen and sulfide gradients, and the tubes of macro-infauna. *Marine Ecology Progress Series* **35**, 223–41.

Myers, A.C. (1972). Tube worm-sediment relationships of *Diopatra cuprea* (Polychaeta: Onuphidae). *Marine Biology* **17**, 350–6.

Palmer, M.F. (1968). Aspects of the respiratory physiology of *Tubifex tubifex* in relation to its ecology. *Journal of Zoology (London)* **154**, 463–73.

Patel, S. & Spencer, C.P. (1963). The oxidation of sulphide by the haem compounds from the blood of *Arenicola marina*. *Journal of the Marine Biological Association of the United Kingdom* **43**, 167–75.

Pionetti, J.M. & Toulmond, A. (1980). Tide-related changes of volatile fatty acids in the blood of the lugworm, *Arenicola marina* (L.). *Canadian Journal of Zoology* **58**, 1723–7.

Pörtner, H.O., Heisler, N. & Grieshaber, M.K. (1984). Anaerobiosis and acid-base status in marine invertebrates: a theoretical analysis of proton generation by anaerobic metabolism. *Journal of Comparative Physiology* **155**, 1–12.

Pörtner, H.O., Surholt, B. & Grieshaber, M.K. (1979). Recovery from anaerobiosis of the lugworm, *Arenicola marina* L.: changes of metabolic concentrations in the body-wall musculature. *Journal of Comparative Physiology* **133**, 227–31.

Powell, E.N., Crenshaw, M.A. & Rieger, R.M. (1979). Adaptations to sulfide in the meiofauna of the sulfide system. I. ^{35}S-sulfide accumulation and the presence of a sulfide detoxification system. *Journal of Experimental Marine Biology and Ecology* **37**, 57–76.

Powell, M.A. & Somero, G.N. (1986). Adaptations to sulfide by hydrothermal vent animals: sites and mechanisms of detoxification and metabolism. *Biological Bulletin* **171**, 274–90.

Reimers, C.E., Fischer, K.M., Merewether, R., Smith, K.L. & Jahnke, R.A. (1986). Oxygen microprofiles measured in situ in deep ocean sediments. *Nature* **320**, 741–44.

Reise, K. & Ax, P. (1979). A meiofaunal 'thiobios' limited to the anaerobic sulfide system of marine sand does not exist. *Marine Biology* **54**, 225–37.

Revsbeck, N.P., Sørensen, J., Blakburn, T.H. & Lomholt, J.P. (1980). Distribution of oxygen in marine sediments measured with microelectrodes. *Limnology and Oceanography* **25**, 403–11.

Rullier, F. (1959). Etude bionomique de l'Aber de Roscoff. *Travaux de la Station Biologique de Roscoff* **10**, 1–350.

Sanders, H.L., Hessler, R.R. & Hampson, G.R. (1965). An introduction to the study of deep-sea benthic faunal assemblages along the Gay Head-Bermuda transect. *Deep-Sea Research* **12**, 845–67.

Schöttler, U. & Schroff, G. (1976). Untersuchungen zum anaeroben Glycogenabbau bei *Tubifex tubifex* M. *Journal of Comparative Physiology* **108**, 243–54.

Schöttler, U., Wienhausen, G. & Westermann, J. (1984). Anaerobic metabolism in the lugworm *Arenicola marina* L.: the transition from aerobic to anaerobic metabolism. *Comparative Biochemistry and Physiology* **79B**, 93–103.

Schroff, G. & Zebe, E. (1980). The anaerobic formation of propionic acid in the mitchrondria of the lugworm *Arenicola marina. Journal of Comparative Physiology* **138**, 35–41.

Seuss, J., Hipp, E. & Hoffmann, K.H. (1983). Oxygen consumption, glycogen content and the accumulation of metabolites in *Tubifex* during aerobic-anaerobic shift and under progressing hypoxia. *Comparative Biochemistry and Physiology* **75A**, 557–62.

Siegmund, B., Grieshaber, M.K., Reitze, M. & Zebe, E. (1985). Alanopine and strombine are end products of anaerobic glycolysis in the lugworm, *Arenicola marina. Comparative Biochemistry and Physiology* **82B**, 337–45.

Storch, V. & Welsch, U. (1970). Über die Feinstruktur der Polychaeten-Epidermis (Annelida). *Zeitschrift für Morphologie der Tiere* **66**, 310–22.

Surholt, B. (1977). Production of volatile fatty acids in the anaerobic carbohydrate catabolism of *Arenicola marina. Comparative Biochemistry and Physiology* **58B**, 147–50.

Sverdrup, H.U., Johnson, M.W. & Fleming, R.H. (1949). *The Oceans. Their Physics, Chemistry and General Biology.* Prentice-Hall, New York.

Swedmark, B. (1964). The interstitial fauna of marine sand. *Biological Review* **39**, 1–42.

Terwilliger, N.B. & Terwilliger, R.C. (1984). Hemoglobin from the 'Pompeii worm', *Alvinella pompejana*, an annelid from a deep sea hot hydrothermal vent environment. *Marine Biology Letters* **5**, 191–201.

Toulmond, A. (1973). Tide-related changes of blood respiratory variables in the lugworm *Arenicola marina* (L.). *Respiration Physiology* **19**, 130–44.

Toulmond, A. (1977). Temperature-induced variations of blood acid-base status in the lugworm. *Arenicola marina* (L.): In vitro study. *Respiration Physiology* **31**, 139–49.

Toulmond, A. (1985). Circulating respiratory pigments in marine animals. In *Physiological Adaptations of Marine Animals* (ed. M.S. Laverack). *Symposia of the Society for Experimental Biology* **39**, 163–206. Cambridge University Press.

Toulmond, A. (1987). Adaptations to extreme environmental hypoxia in water breathers. In *Comparative Physiology of Environmental Adaptations*, vol. 2 (ed. P. Dejours), *8th ESCP Conference, Strasbourg, 1986* pp. 123–36. Karger, Basel.

Toulmond, A., De Frescheville, J., Frisch, M.H. & Jouin, C. (1988). Les pigments respiratoires de la faune inféodée à l'hydrothermalisme profond. *Oceanologica Acta* (Vol. Sp.) **8**, 195–201.

Toulmond, A. & Tchernigovtzeff, C. (1984). Ventilation and respiratory gas exchanges of the lugworm *Arenicola marina* (L.) as functions of ambient P_{O_2} (20-700 Torr). *Respiration Physiology* **57**, 349–63.

Toulmond, A., Tchernigovtzeff, C., Greber, P. & Jouin, C. (1984). Epidermal sensitivity to hypoxia in the lugworm. *Experientia* **40**, 541–3.

Truchot, J.P. & Duhamel-Jouve, A. (1980). Oxygen and carbon dioxide in the marine intertidal environment: diurnal and tidal changes in rock-pools. *Respiration Physiology* **39**, 241–54.

Walker, J.G. (1970). Oxygen poisoning in the annelid *Tubifex tubifex*. I. Response to oxygen exposure. *Biological Bulletin* **138**, 235–44.

Wavre, M. & Brinkhurst, R.O. (1971). Interactions between some tubificid oligochaetes and bacteria found in the sediments of Toronto Harbour, Ontario. *Journal of the Fisheries Research Board of Canada* **28**, 335–41.

Weber, R.E. (1978). Respiration. In *Physiology of Annelids* (ed. P.J. Mill), pp. 369–92. Academic Press, London and New York.

Weber, R.E. (1980). Functions of invertebrate hemoglobins with special reference to adaptations to environmental hypoxia. *American Zoologist* **20**, 79–101.

Weber, R.E. (1981). Cationic control of O_2 affinity in lugworm erythrocruorin. *Nature* **292**, 386–7.

Weber, R.E. & Pauptit, E. (1972). Molecular and functional heterogeneity in myoglobin from the polychaete *Arenicola marina* (L.). *Archives of Biochemistry and Biophysics* **148**, 322–4.

Wells, G.P. (1966). The lugworm (*Arenicola*). A study of adaptation. *Netherlands Journal of Sea Research* **3**, 294–313.

Wells, R.M.G. & Pankhurst, N.W. (1980). An investigation into the formation of sulphide and oxidation compounds from the haemoglobins of the lugworm *Abarenicola affinis* (Ashworth). *Comparative Biochemistry and Physiology* **66C**, 255–9.

Zebe, E. (1975). In vivo-Untersuchungen über den Glucose-Abbau bei *Arenicola marina* (Annelida, Polychaeta). *Journal of Comparative Physiology* **101**, 133–45.

A.C. TAYLOR & R.J.A. ATKINSON

Respiratory adaptations of aquatic decapod crustaceans and fish to a burrowing mode of life

Introduction

A burrowing mode of life has been adopted by many species of decapod Crustacea and fish, but knowledge of their physiological adaptations is limited to relatively few species. Among the decapods, perhaps the best known are the semi-terrestrial species for which there have been numerous studies on burrowing behaviour and on burrow structure, but comparatively few studies of a physiological nature. The present paper is concerned mainly with aquatic species, since the special problems associated with a semi-terrestrial mode of life have been reviewed elsewhere (Little, 1983; Powers & Bliss, 1983; Atkinson & Taylor, 1988; Burggren & McMahon, 1988). Our use of the literature is selective owing to the limitations of space.

We consider it to be important to make a distinction between species which construct burrows and those which bury themselves for concealment, though the term 'burrowing' is sometimes used in relation to both in the literature. The distinction is important since their behavioural strategies differ and different adaptations may be involved (Nayar, 1951; Nye, 1974). In a recent review, Atkinson & Taylor (1988) discussed these concepts in relation to burrowing decapods and their conclusions are equally applicable to burrowing fish. Although the term 'burrowing' may legitimately be used for species that form intrastratal trails rather than burrows (see Frey, 1973), in the present paper, the term 'burrowing' is restricted to those species that construct semi-permanent burrows within the sediment. This contrasts with the term 'burying', which we use to describe animals that excavate for concealment, do not migrate through the sediment and do not construct burrows.

Among the natant decapods, burrow construction is characteristic of the family Alpheidae but burying behaviour is common among other natantians associated with soft sediments. Both behavioural strategies are perhaps more widespread among the reptant decapods; burrow construction is common in members of the infraorders Astacidea, Thalassinidea and Brachyura and burying behaviour occurs in many Brachyura as well as some anomurans (Abele & Felgenhaur, 1982). Unfortunately, our knowledge of the behaviour and physiological ecology of burrowing decapods is restricted primarily to semi-terrestrial marine and shallow water species whereas much less is known about deep-water marine species or freshwater species.

There are comparatively few species of fish that have adopted a burrow-dwelling mode of life. Such fish are mostly known from observations in shallow inshore waters, but there are some deep water examples, e.g. tilefish (Branchiostegidae) (Able *et al.*, 1982, 1987; Grimes *et al.* 1986). Species that have adopted a burrowing mode of life for at least part of their lives occur among the Agnatha and among a number of orders of bony fish including the Lepidosereniformes, Anguilliformes, Siluriformes, Synbranchiformes and Perciformes. This last group contains the best known burrow-dwelling species, which occur in the families Branchiostegidae, Cepolidae, Lumpenidae, Opistognathidae, and Gobiidae (including Periophthalminae). Notable among the Anguilliformes are the garden eels (Heterocongrinae), whose burrows are mucus-lined tubes.

More numerous than species which construct burrows are those which simply bury themselves for concealment in particulate sediments. This behaviour is common in skates and rays (Rajiformes) and in flatfish (Pleuronectiformes), but also occurs in many other orders and numerous families. Particularly good examples occur among the freshwater loaches (Cobitidae), the sandfishes and their relatives (Trichodontidae and related families) and the sandeels (Ammodytidae).

Burrow construction

Decapod Crustacea excavate their burrows by variously using their first three pairs of pereiopods and often their third maxillipeds. There may be considerable variation, however, between species in the use of these appendages and the pleopods and telson may also be employed (Atkinson & Taylor, 1988).

In the case of fish, burrow construction is usually achieved by mouth, either by expelling mouthfuls of sediment as pellets (Atkinson *et al.*, 1977), or as a sediment plume (Rao, 1939; Nayar, 1951; Hirano *et al.*, 1984). Head or tail probing and body undulations may also be used (Clark, 1971; Blaber & Whitfield, 1977; Atkinson *et al.*, 1987).

The respiratory apertures of decapod crustaceans frequently become obscured by sediment during burrowing and, in fish, excavation may also inhibit respiration, particularly in the cases where mouth-excavated sediment is expelled through the operculae (see Nayar, 1951; Hirano *et al.*, 1984). Unfortunately, there is no information on these relationships.

Burrow structure

The burrows of decapod Crustacea and those of fish vary greatly in complexity and in some cases their burrows are associated (Atkinson *et al.*, 1977; Colin & Arneson, 1978). Burrows may be simple shafts with a single opening to the surface, for example, many littoral and supralittoral crab burrows (Vannini, 1980), lungfish nesting burrows (Schultz & Stern, 1948) and the vertical burrows of garden eels (Clark, 1971) and the red band-fish, *Cepola rubescens* (Atkinson *et al.*, 1977). Often, burrows are J- or U-shaped, e.g. those of many crabs (Vannini, 1980) and those of larval lampreys and adult hagfish (Hardisty, 1979). U-shaped burrows may be elaborated with the addition of one or more branch tunnels, e.g. burrows of nephropid lobsters, crabs, gobies and the snake blenny, *Lumpenus lampretaeformis* (Atkinson, 1986; Atkinson *et al.*, 1987). Y-shaped burrows are also common, for example, those of some thalassinid mud-shrimps (Dworschak, 1983) and mud-skippers (Periophthalminae) (Clayton & Vaughan, 1986). Complex branched systems with multiple openings also occur, e.g. the burrows of many thalassinids (Dworschak, 1983; Nash *et al.*, 1984), those of some freshwater crayfish (Horwitz *et al.*, 1985) and some eel goby (Gobioididae) burrows (Rao, 1939). Examples from the range of burrow forms among decapod Crustacea and fish are given in Figure 1.

The sometimes complex burrows of alpheid shrimps are often co-occupied by gobiid fish (Karplus *et al.*, 1972, 1974; Yanagisawa, 1984), which do not themselves actively contribute to the process of excavation. Also, various gobies are frequently found in the burrows constructed by callianassid mud-shrimps (MacGinitie, 1934; MacGinitie & MacGinitie, 1968; Grossman, 1979) but the relationships are not as close as those with alpheid shrimps (Hoffman, 1981). The burrows of the red band-fish, *C. rubescens*, may intersect with those of crabs and thalassinids (Atkinson *et al.*, 1977) and the burrows of opistognathid fish and periophthalmins may also interconnect with those of crabs (Colin & Arneson, 1978; Clayton & Vaughan, 1986). The benefits of such associations may well include those of efficient burrow irrigation.

Differences in the complexity of burrows (in particular the number and configuration of openings) have important behavioural and physiological implications for the occupant (see below).

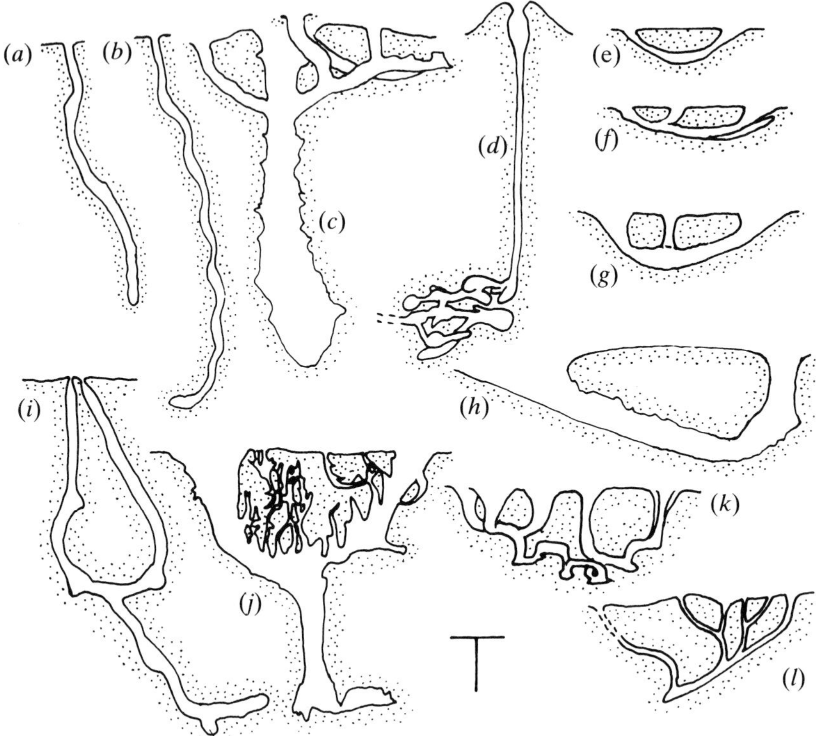

Figure 1. Side elevations of the burrows of decapod crustaceans (*a, d, g, h, i, k*) and fish (*b, c, e, f, j, l*). (*a*) *Uca inversa inversa* (original); (*b*) *Gorgasia sillneri* (derived from Clark, 1971); (*c*) *Cepola rubescens* (derived from Atkinson *et al.*, 1977); (*d*) *Callianassa subterranea* (derived from Atkinson, 1986); (*e*) *Lesueurigobius friesii* (dervied from Rice & Johnstone, 1972); (*f*) *Lumpenus lampretaeformis* (derived from Atkinson *et al.*, 1987); (*g*) *Goneplax rhomboides* (derived from Atkinson, 1974); (*h*) *Nephrops norvegicus* (derived from Rice & Chapman, 1971); (*i*) *Upogebia pusilla* (derived from Dworschak, 1983); (*j*) *Boleophthalmus boddarti* (derived from Clayton & Vaughan, 1986); (*k*) *Calocaris macandreae* (derived from Nash *et al.*, 1984); (*l*) *Taenioides* sp. (derived from Rao, 1939). Crab burrows branch from fish burrows in (*c*) and (*j*). Horizontal and vertical scale bar lengths, 10 cm.

Burrow function

The most important burrow function is concealment from predators. For some species, however, the burrow provides protection from adverse environmental conditions and, by providing a more stable microenvironment, enables them to survive periods when conditions outside the burrow become extreme, as in semi-terrestrial crabs (Little, 1983;

Powers & Bliss, 1983; Eshky *et al.*, 1988), aestivating lungfish (Schultz & Stern, 1948) and intertidal gobies which enter thalassinid burrows at low tide (MacGinitie & MacGinitie, 1968; Grossman, 1979; Hoffman, 1981). Burrows may also act as a locus for reproductive activity, egg incubation and co-occupancy by juveniles, in decapod crustaceans (Forbes, 1973; Chapman, 1980; Horwitz *et al.*, 1985) and in fish (Brillet, 1969, 1984; Kobayashi *et al.*, 1971; Hudson, 1977; Gibson & Ezzi, 1978; Hoffman & Robertson, 1983; Gordon *et al.*, 1985).

The burrow environment

In contrast to the burrows of semi-terrestrial decapods, which offer considerable protection from extreme environmental conditions (Atkinson & Taylor, 1988), those inhabited by aquatic burrowers may expose their occupants to hypoxic and hypercapnic conditions (Bridges, 1987; Atkinson & Taylor, 1988). Many aquatic burrowing decapods use their pleopods to irrigate their burrows, for example, thalassinids such as *Callianassa* spp. (MacGinitie, 1934; Farley & Case, 1968), *Upogebia* spp. (MacGinitie & MacGinitie, 1968; Hill, 1981; Mukai & Koike, 1984) and *Calocaris macandreae* (S.J. Anderson, A.C. Taylor & R.J.A. Atkinson, unpublished observations). Similar behaviour is seen in the Norway lobster, *Nephrops norvegicus* (R.J.A. Atkinson & A.C. Taylor, unpublished observations). It is possible that the ventilatory currents produced by the beating of the scaphognathites may also help to circulate water within the burrow. This may be very important in burrowing crabs. In *Goneplax rhomboides*, the respiratory current is deflected by the flagella of the maxillipeds so that it flows along the burrow (Rice & Chapman, 1971).

Suspension-feeding species such as *Upogebia* spp. maintain a flow of water through the burrow whenever possible (MacGinitie & MacGinitie, 1968; Jorgensen, 1966), but most species irrigate their burrows only intermittently under normoxic conditions. In *Callianassa californiensis* and in *C. jamaicense* pleopod activity increases as the P_{O_2} of the water is reduced. (Farley & Case, 1968; Torres *et al.*, 1977; Felder, 1979). Recent studies on the thalassinid, *Calocaris macandreae*, have shown that this species does not increase burrow irrigation until the P_{O_2} of the water declines to approximately 70 Torr (Figure 2). Further reduction of the P_{O_2} of the burrow water results in a progressive increase in pleopod activity until, at very low oxygen tensions, burrow irrigation ceases (S.J. Anderson *et al.*, unpubl. obs.). Since the energy costs of burrow irrigation are likely to be high, intermittent irrigation may represent a behavioural adaptation to minimize these costs.

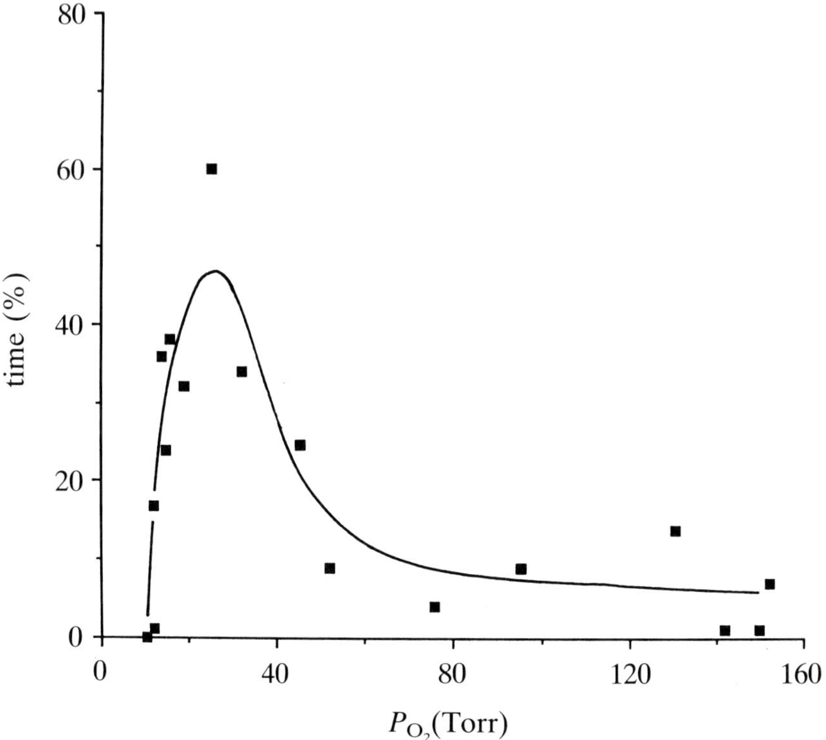

Figure 2. The relationship between oxygen tension and the percentage time spent irrigating the burrow in the decapod, *Calocaris macandreae*. The line through the data points has been fitted by eye.

Information on the physicochemical conditions within fish burrows is scarce. Burrowing gobies such as *Clevelandia ios*, *Lepidogobius lepidus* and *Typhlogobius californiensis*, which occupy the burrows of intertidal thalassinids may experience severe hypoxia at low tide: oxygen tensions of less than 20 Torr have been recorded in these burrows (Farley & Case, 1968; Thompson & Pritchard, 1969; Congleton, 1974; Torres *et al.*, 1977). Similarly, Gordon *et al.* (1978) recorded on oxygen concentrations of 0.2–0.7 ml l^{-1} (P_{O_2} approximately 5–20 Torr) at depths of 5–15 cm into the burrows of *Periophthalmus cantonensis*; conditions may be totally anoxic at the bottom of their deep (up to 1 m) burrows (Gordon *et al.*, 1985). El-Ziady *et al.* (1979) have recorded oxygen concentrations of 1 ppm (approximately 25–30 Torr) in the burrow of *P. koelreuteri*.

Conditions within the burrows of other species may not be quite so severe. Figure 3 presents laboratory recordings of the P_{O_2} and P_{CO_2} within the burrows of two species of sublittoral fish, the red band-fish, *Cepola*

(*a*)

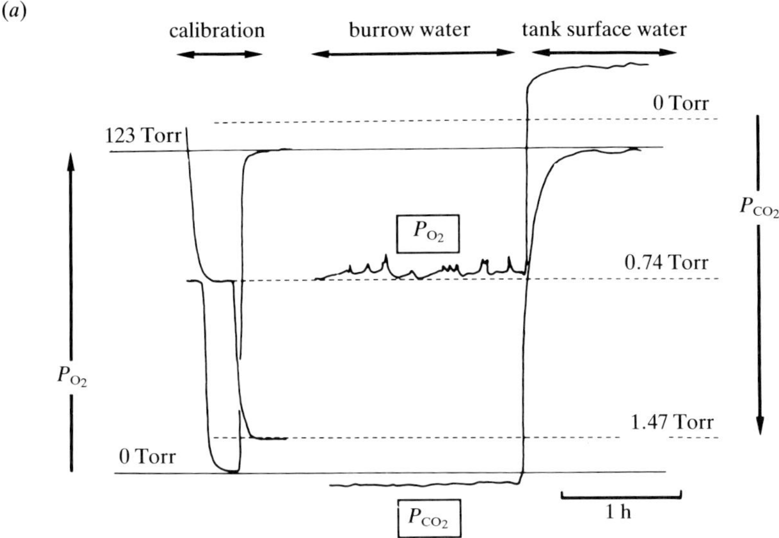

(*b*)

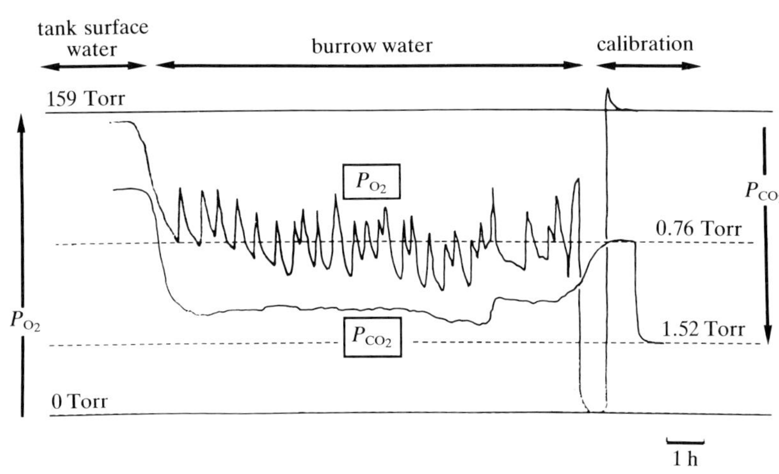

Figure 3. The P_{O_2} and P_{CO_2} of the water within, at the entrance to and above the burrows of (*a*) *Cepola rubescens* and (*b*) *Lumpenus lampretaeformis*. The recordings were obtained by inserting narrow bore catheters into the burrows and siphoning water past oxygen and carbon dioxide electrodes held in thermostatted cells. Parts of the traces show the calibration of the electrodes using known partial pressures of the gases. The figure is redrawn from Atkinson *et al.* (1987).

rubescens (Cepolidae) and the snake blenny *Lumpenus lampretaeformis* (Lumpenidae) (Pullin *et al.*, 1980; Atkinson *et al.*, 1987). The recordings shown in Figure 3 were obtained by inserting narrow catheters into the burrows and indicate that the P_{O_2} within the burrow of *C. rubescens* may be much lower (P_{O_2}=60 Torr) than the P_{O_2} of the water at the surface of the mud (P_{O_2}=120 Torr). Conditions within the burrow were also hypercapnic; the P_{CO_2} of the burrow water was over 1.5 Torr, compared with a P_{CO_2} of 0.3 Torr at the surface of the mud (Atkinson *et al.*, 1987). The oxygen tensions within the burrows constructed by fish maintained in the laboratory were similar to those recorded in the field (P_{O_2}=80 Torr) (Pullin *et al.*, 1980).

The oxygen tension of the water within the burrows of *L. lampretaeformis* fluctuated between 60 and 129 Torr compared with a P_{O_2} of approximately 150 Torr in the water above the burrow. Conditions within the burrows of *L. lampretaeformis* were also hypercapnic (P_{CO_2}>1.5 Torr). Increases in the P_{O_2} of the burrow water could be correlated with short periods of burrow irrigation (mean duration 21 s, periodicity *ca.* 39 h^{-1}) produced by undulating body movements (Atkinson *et al.*, 1987). Burrow irrigation is also effected by body undulations in *C. rubescens*. The irrigation periods are longer (over 30 s) and more frequent than in *L. lampretaeformis* and are dependent on P_{O_2} (Pullin *et al.*, 1980). Maximum flow rates were approximately 40 ml min^{-1} in the narrow (2 cm diameter) burrows of *L. lampretaeformis* (Atkinson *et al.*, 1987) compared with only 10 ml min^{-1} in the wider (5 cm diameter) burrows of *C. rubescens* (Pullin *et al.*, 1980) (see Figure 1). This irrigation behaviour is clearly of importance in preventing the conditions within the burrow from becoming too extreme. Such behaviour is probably common to many other sublittoral burrowing species but, unfortunately, information on other species is almost totally lacking.

In areas where there is sufficient water flow across the burrow entrances, the arrangement and shape of the entrances may set up pressure gradients which draw water through the burrows (Vogel, 1977). Grimes *et al.* (1986) have speculated that passive ventilation may occur in the burrows of the tilefish, *Lopholatilus chamaeleonticeps*, although active irrigation by the fish is likely to be of greater importance. Movement of the burrow occupant is also likely to contribute to burrow irrigation, but its importance is difficult to assess.

Physiological adaptations to life within burrows

Aquatic decapod Crustacea which burrow into soft sediments are, in general, highly tolerant of hypoxia. Many decapods are able to maintain their rates of oxygen consumption independent of ambient oxygen tension and this ability is particularly well-developed in burrowing species; the 'critical' P_{O_2} (Pc) at which respiratory independence can no longer be maintained is often low in burrowing species. For example, Pc values are approximately 10–20 Torr in *Callianassa californiensis* (Thompson & Pritchard, 1969; Miller *et al.*, 1976; Torres *et al.*, 1977) and *C. jamaicense* (Felder, 1979), 45–50 Torr in *Upogebia pugettensis* (Thompson & Pritchard, 1969), 20–40 Torr in *Goneplax rhomboides* (A.C. Taylor, unpublished observations), 40 Torr in *Nephrops norvegicus* (Hagerman & Uglow, 1985), 10–20 Torr in *Calocaris macandreae* (S.J. Anderson *et al.*, unpublished observations) and 30 Torr in *Parastacoides tasmanicus* (Swain *et al.*, 1987)

The information currently available suggests that burrowing species do not possess any unique respiratory adaptations; the respiratory responses to hypoxia appear to be almost identical in burrowing and non-burrowing species. There is some evidence, however, that burrowers often possess respiratory pigments which have a greater oxygen affinity and exhibit moderately high Bohr values (Miller *et al.*, 1976; Miller & Van Holde, 1981; Taylor, *et al.*, 1985; Bridges, 1986; Pelster *et al.*, 1988*a*; S.J. Anderson *et al.*, unpublished observations), properties that will facilitate oxygen uptake during hypoxia.

Some burrowing decapods regularly experience very severe hypoxia or even total anoxia. For example, intertidal thalassinid decapods may experience very low burrow water oxygen tensions at low tide when burrow irrigation is impossible. Some of these species are tolerant of severe hypoxia and can even survive total anoxia for many hours or sometimes days. *Callianassa californiensis* and *Upogebia pugettensis* achieve this by resorting to anaerobic metabolism (Thompson & Pritchard, 1969; Hill, 1981; Mukai & Koike, 1984; reviewed by Atkinson & Taylor, 1988). Hill (1981) has shown, however, that some *Upogebia* spp. from South Africa moved up to the air–water interface at low tide so that the cephalothorax was above the water level. Water pumped through the gill chambers trickled down the animal's ventral surface where some reoxygenation probably occurred. The utilization of air as a source of oxygen has also been observed in the burrow-dwelling freshwater crayfish *Cambarus diogenes diogenes* and *Parastacoides tasmanicus*, which partly emerge at the air–water interface within their burrows when the P_{O_2} of the burrow water becomes excessively hypoxic (Grow & Merchant, 1980; Swain *et al.*, 1987).

Burrowing fish must also be adapted to the hypoxic and hypercapnic conditions within the burrow. On the somewhat limited information currently available, it would appear that burrowing species exhibit a high tolerance of hypoxia but little tolerance of anoxia (Gordon *et al.*, 1978). Thus *Cepola rubescens* and *Lumpenus lampretaeformis* are both able to maintain their rate of aerobic respiration approximately constant down to critical oxygen tensions (Pc) of 50–70 Torr (Pullin *et al.*, 1980; Pelster *et al.*, 1988*b*). In the intertidal goby, *Typhlogobius californiensis*, however, the Pc is as low as 9–16 Torr (Congleton, 1974). In these three species, maintenance of the rates of aerobic respiration under hypoxic conditions is perhaps helped by the metabolic rates being lower than in some non-burrowing species. This may also be true for burrowing decapods (see Thompson & Pritchard, 1969; Felder, 1979; Hill, 1981), but, for both groups, accurate comparative studies are required to confirm this. Some species of fish avoid severe hypoxic or anoxic conditions within the burrow by utilizing air as a source of oxygen, for example, *Synbranchus marmoratus* (Synbranchidae) (Graham & Baird, 1984).

As in burrowing decapods, the respiratory responses of burrowing fish to hypoxia appear to be almost identical to those of non-burrowing species. Thus *Lumpenus lampretaeformis* and *Cepola rubescens* both show an increase in opercular ventilation rate when exposed to hypoxia. In *L. lampretaeformis* the ventilation rate increased from 57 ± 5 beats min^{-1} at 150 Torr to 88 ± 5 beats min^{-1} at 40 Torr (Pelster, 1985). Similarly, the ventilation rate of *C. rubescens* increased from 59 ± 2.5 beats min^{-1} at 153 Torr to 77 ± 1.8 beats min^{-1} at 60 torr (Bridges, 1987). Unfortunately, almost no information is available on the cardiovascular responses of burrowing fish to hypoxia. In *C. rubescens*, the heart rate remains approximately constant at 61 ± 0.5 beats min^{-1} during hypoxia and declines only when the P_{O_2} is less than 80 Torr (heart rate$=49\pm3$ beats min^{-1}) (Bridges, 1987).

The oxygen and carbon dioxide transporting properties of the blood of *C. rubescens* and *L. lampretaeformis* have been the subject of recent studies (Pullin *et al.*, 1980; Bridges *et al.*, 1982; Pelster *et al.*, 1988*a*). Both species have similar haematocrit and haemoglobin concentrations and these are comparable to those of other sedentary fish. ATP is present in the blood of both species and the ATP:Hb ratio is similar to that measured in many other fish (Weber, 1982). Small amounts of GTP are also present in the blood of *C. rubescens* but not in *L. lampretaeformis* (Pelster *et al.*, 1988*b*). ATP has a marked effect on haemoglobin oxygen affinity in *C. rubescens*; the P_{50} of stripped blood is 4.9 Torr compared with 13.4 Torr in whole blood (Bridges *et al.*, 1982). Pelster *et al.* (1988*b*) have shown that the ATP concentrations in the blood of *L. lampretaeformis* are unaffected, however, by moderate hypoxia, which indicates that in this species, at least, haemoglobin oxygen affinity is not modulated by variations in blood ATP concentrations caused by the reduced P_{O_2} of the burrow water.

The oxygen affinity of the haemoglobin of *Cepola rubescens* (P_{50}=13.4 Torr, pH=7.8) is higher than in *Lumpenus lampretaeformis* (P_{50}=20 Torr, pH=7.8) (Bridges *et al.*, 1984; Pelster *et al.*, 1988*a*). The blood of both species was also found to exhibit a high Bohr value which may be a considerable advantage to species which regularly experience hypoxia (Bridges *et al.*, 1982; Pelster *et al.*, 1988*a*). Modulation of haemoglobin oxygen affinity by changes in blood pH is dependent on the buffer value of the blood. It may be significant, therefore, that blood buffer values are relatively low (3.9–5.4 mmol pH^{-1}) compared with much higher values in white muscle (43–48 mmol pH^{-1}) (Pelster *et al.*, 1988*a*).

The oxygen affinity of the haemoglobin of the blood of these two species is not as high, however, as that of some rockpool fish (Powers, 1980; Bridges *et al.*, 1984) or some freshwater species (Powers *et al.*, 1979; Jensen & Weber, 1982) which regularly experience hypoxia and suggests that the blood of burrowing species is not necessarily characterized by having a high oxygen affinity. There is considerable variation between species in the P_{O_2} of the burrow water, resulting both from differences in burrow structure and irrigation behaviour. The evidence currently available suggests that, in both fish and decapod Crustacea, blood oxygen affinity is more closely correlated with the degree of hypoxia normally experienced.

Burying behaviour

Burying by decapod Crustacea is achieved by limb movements, often assisted by expulsion of water from the posterior branchial apertures (Garstang, 1896; McLay & Osborne, 1985). Crabs dig backwards into the sediment (Warner, 1977), whereas shrimps excavate the sediment directly beneath them (Dyer & Uglow, 1978).

Burying by fish can be achieved by undulatory movements of the body or by fin movements, sometimes assisted by downwards ejection of water from the opercular or buccal cavities (Trueman & Ansell, 1969; Lewis, 1976). Penetration of the sediment may therefore be head-first (e.g. Ammodytidae), tail-first (e.g. Ophidiinae) or horizontally (e.g. Rajidae and Pleuronectiformes) (Greenfield, 1968; Trueman & Ansell, 1969; Lewis, 1976; Hobson, 1986).

Adaptations associated with burying

Although many species of fish and decapod Crustacea bury into soft sediments, rather few studies of their respiratory physiology have been carried out. The information currently available would appear to suggest, however, that they show few physiological adaptations that can be considered to be specific adaptations to this mode of life. Instead, many species show behavioural or morphological adaptations which enable them to maintain a flow of water across the gills when buried. The magnitude of this problem will depend, of course, on the extent to which the animal is buried in the substratum.

Garstang (1896, 1897*a–c*,) was the first to study the ventilatory behaviour of crabs buried in sediment. He noted that, in many buried crabs, the orientation of each flexed cheliped provided a channel between it and the body, termed an 'exostegal channel'. This provided a route for inhalant water and kept the Milne–Edwards openings at the base of the chelipeds free from sand. Periodically, the direction of the scaphognathite-driven ventilation current is reversed and one result of this is to keep the exostegal channels free from sediment (Garstang, 1897*b*; Cumberlidge & Uglow, 1978; McDonald *et al.*, 1977; Taylor, 1984). The portunid crabs, *Carcinus maenas* and *Liocarcinus depurator*, also demonstrate this behaviour (Cumberlidge & Uglow, 1978).

More complex adaptations are seen in the crab *Ebalia tuberosa* (Leucosiidae) in which inhalant respiratory water flows down channels formed by the mouthparts and buccal wall to enter the branchial chambers at the bases of the third maxillipeds (Schembri, 1981). In the case of the crabs *Corystes cassivelaunus* (Corystidae) and *Atelecyclus rotundatus* (Atelecyclidae) which bury more deeply in the sediment, inhalant water flows down a tube formed by the apposed second antennae. When buried, these two species mainly ventilate their branchial chambers in a reverse direction, i.e. inhalant water enters the prebranchial apertures adjacent to the mouth and exhalant water is expelled through the apertures between the limbs (Garstang 1896, 1897*a*; Bohn, 1902; Arudpragasam & Naylor, 1966; Taylor, 1984). This expulsion of water aids the crab during digging.

Many decapod crustaceans exhibit an intermittent pattern of branchial ventilation when buried in the substratum (Bridges, 1979; Taylor, 1984; A.C. Taylor, unpublished observations), though this behaviour is also known to occur among non-burying species (McMahon & Wilkens, 1983). Periods of active ventilation alternate with periods of apnoea which may be

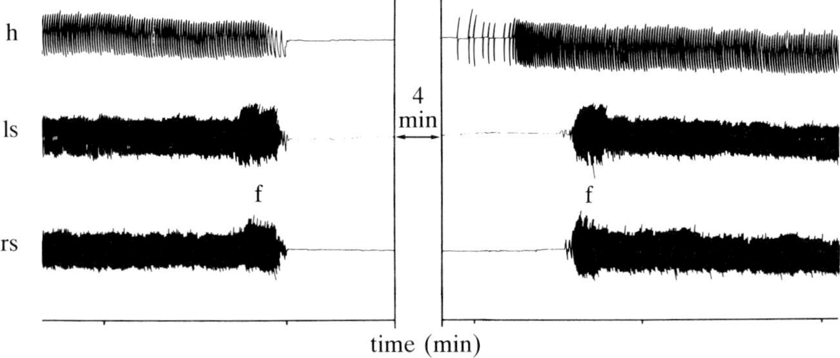

Figure 4. Part of a continuous recording of the activity of the heart (h) and of both left and right scaphognathites (ls and rs) in an individual *Atelecyclus rotundatus* buried in sand. Ventilation of the branchial chambers was in a reversed direction except for brief periods of forward pumping (f) at the beginning and at the end of a period of apnoea.

unilateral, involving inactivity of only one scaphognathite, or bilateral, during which there is complete cessation of ventilatory activity. These periods of apnoea are often accompanied by a reduction in heart rate. During brief ventilatory pauses, the reduction in heart rate may be only slight but during longer pauses the animals frequently undergo periods of complete cardiac arrest.

Intermittent ventilation in decapods appears to occur only in quiescent animals. The most satisfactory explanation for the occurrence of a rhythm of alternating periods of ventilation and apnoea is that it enables the animals to make a saving in metabolic energy during periods of inactivity by reducing the energy costs of ventilation. It is perhaps significant, therefore, that the duration of these ventilatory pauses is greatest among burying species. A good example of this is seen when the crab *A. rotundatus* is buried in sand (Figure 4). The duration of the ventilatory pauses varies considerably among individuals. In some crabs the pauses last for only 2–3 min whereas in others they last for approximately 30 min (Taylor, 1984). During ventilatory pauses, the P_{O_2} of the water in the branchial chambers declines. Figure 5 shows a recording of the P_{O_2} from one of the branchial chambers of a buried *A. rotundatus*. The troughs represent periods of apnoea during which there is a progressive reduction in branchial P_{O_2} and the peaks represent periods of ventilation during which the P_{O_2} is dramatically increased. A similar ventilatory pattern is seen when *C. cassivelaunus* is buried in sand (Bridges, 1979). *Ebalia tuberosa*, when buried in muddy

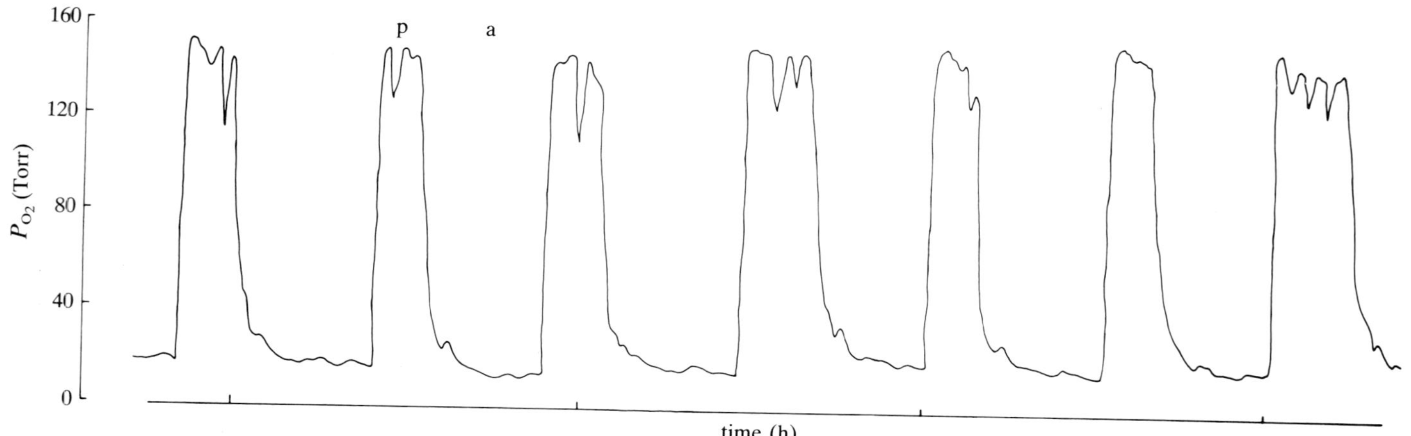

Figure 5. Recording of the P_{O_2} of the water in one of the branchial chambers of *Atelecyclus rotundatus* when buried in sand: p, periods of active ventilation; a, periods of apnoea.

gravel, also exhibits a regular rhythm of alternating periods of ventilation and apnoea, but in this species water flow is primarily in a forward direction (Schembri, 1979).

Burying behaviour is not restricted to reptant decapods. The crangonid shrimp, *Crangon crangon*, also spends much time buried in sediment. It creates small inhalant and exhalant openings at the bases of the antennal scales and, using its scaphognathites in combination with the flagella of the maxillipeds, draws respiratory water beneath the animal and into the branchial chambers at the ventral edges of its branchiostegites (Dyer & Uglow, 1978).

Since burying species ventilate their gills with water drawn directly from above the plane of the sediment surface, they are not normally subjected to the same degree of hypoxia that may be faced by species which live within burrows. Nevertheless, some have haemocyanin with a high oxygen affinity, e.g. *Corystes cassivelaunus* and *Atelecyclus rotundatus* (Taylor *et al.*, 1985; Bridges, 1986) and this would clearly be advantageous should hypoxic conditions arise.

Many species of burying fish utilize some form of sieve or filter to prevent particles of sediment from entering the mouth with the respiratory current. This is well illustrated in families within the Trachinoidei (Bond, 1979). For example, members of the genus *Trichodon* possess a fringe of very fine teeth, which can be used for this purpose (Schultz & Stern, 1948). Other species such as the weever fish, *Echiichthys* (=*Trachinus*) *vipera* possess numerous papillae on the margins of the lips which can interlock to prevent sediment particles from entering the mouth (Lewis, 1976). In *E. vipera*, the mouth is also upwardly-directed which enables this species to maintain an inhalant flow of water through the mouth even when almost completely buried in the substratum. The weever fish shows another modification to maintain a respiratory current when completely buried in the substratum. When the flow of inhalant water is reduced by sand over the mouth, the supracleithrum is raised and water passes between it and the side of the fish. Sand particles are prevented from entering by the presence of further papillae along the margin of the supracleithrum (Lewis, 1976).

Respiratory adaptations to a burying mode of life are also seen in the flathead, *Platycephalus fuscus* (Platycephalidae). This fish maintains a very narrow oral aperture when buried and this, together with the liberal secretion of mucus into the sediments adjacent to the mouth, prevents sediment particles entering the mouth with the respiratory current. To prevent the opercular cavities from filling with sediment and to prevent sediment disturbance during exhalation (which might jeopardize concealment), the exhalant current leaves dorsally via two specialized opercular valves (Douglas & Lanzing, 1981). An analogous modification of exhalant apertures appears to occur in dragonets (Callionymidae) in which the opening from each opercular cavity is single and dorsally-directed (Hughes & Umezawa, 1968). Another adaptation is seen in the spotted stargazer, *Genyagnus monopterygius* (Trachinoidei: Uranoscopidae) in which the exhalant water passes along temporary, mucus-lined channels in the sand which open at a distance of up to 20 cm from the fish; this may help prevent detection by predators or prey (Forster & Starling, 1982).

More specialized morphological adaptations to maintain a respiratory current when buried are seen in the rays and skates. Since the mouth is situated ventrally the inhalant respiratory current now enters via the spiracles (Daniel, 1922). In these species, the exhalant current leaves via the gill slits on the ventral surface. In certain flatfish (Pleuronectidae and Soleidae), however, although water is pumped over the gills on both sides, the exhalant current normally leaves only via the upper opercular opening. This is possible because the opercular cavities communicate ventrally. It has been suggested that this evolved because water leaving via the vental opercular opening would have to percolate through the sand which might cause a considerable resistance to the ventilatory current (Yazdani & Alexander, 1967).

Although there is some information on ventilatory adaptations in buried fish, very little is known about other aspects of their respiratory physiology. From studies of agnathans and flatfish, it would appear that their rates of oxygen consumption are low, as might be expected from their state of inactivity, and that they may be tolerant of hypoxia (Kersten *et al.*, 1979; Steffensen *et al..* 1984).

Conclusions

Species of decapods and fish that bury in particulate substrata show behavioural and morphological adaptations to ensure that branchial ventilation is unimpaired. Burrow-dwelling species may regularly experience hypoxic and hypercapnic conditions and for them, burrow irrigation strategies are important in limiting the magnitude of these problems. Both burying and burrow-dwelling species may show considerable tolerance of hypoxic conditions, or even anoxia in the case of some decapods. The physiological adaptations which enable them to survive these conditions are, however, not unique to these groups since they may be shared by species which regularly experience oxygen deficiency in other environments.

Financial assistance from the Royal Society and from the Society for Experimental Biology is gratefully acknowledged.

References

Abele, L.G. & Felgenhaur, B.E. (1982). Decapoda. In *Synopsis and Classification of Living Organisms* vol. 2 (Ed. S.B. Parker), pp. 296–326. McGraw-Hill Book Co., New York.

Able, K.W., Grimes, C.B., Cooper, R.A. & Uzmann, J.R. (1982). Burrow construction and behaviour of tilefish, *Lopholatilus chamaeleonticeps*, in Hudson Submarine Canyon. *Environmental Biology of Fishes* **7**, 199–205.

Able, K.W., Twichell, D.C., Grimes, C.B. & Jones, R.S. (1987). Tilefishes of the genus *Caulolatilus* construct burrows in the sea floor. *Bulletin of Marine Science* **40**, 1–10.

Arudpragasam, K.D. & Naylor, E. (1966). Patterns of gill ventilation in some decapod Crustacea. *Journal of Zoology* **150**, 401–11.

Atkinson, R.J.A. (1974). Behavioural ecology of the mud-burrowing crab *Goneplax rhomboides*. *Marine Biology* **25**, 239–52.

Atkinson, R.J.A. (1986). Mud-burrowing megafauna of the Clyde Sea Area. *Proceedings of the Royal Society of Edinburgh* **90B**, 351–61.

Atkinson, R.J.A., Pelster, B., Bridges, C.R., Taylor, A.C. & Morris, S. (1987). Behavioural and physiological adaptations to a burrowing lifestyle in the snake blenny, *Lumpenus lampretaeformis* and the red band fish, *Cepola rubescens*. *Journal of Fish Biology* **31**, 639–59.

Atkinson, R.J.A., Pullin, R.S.V. & Dipper, F.A. (1977). Studies on the red band fish, *Cepola rubescens*. *Journal of Zoology* **182**, 369–84.

Atkinson, R.J.A. & Taylor, A.C. (1988). Physiological ecology of burrowing decapods. In: *Aspects of the Biology of Decapod Crustacea* (ed. A.A. Fincham & P.S. Rainbow), pp. 201–26. (Symposium of the Zoological Society of London, no. 59). Oxford University Press.

Blaber, S.J.M. & Whitfield, A.K. (1977). The biology of the burrowing goby *Croilia mossambica* Smith (Teleostei, Gobiidae). *Environmental Biology of Fishes* **1**, 197–204.

Bohn, G. (1902). Des méchanismes respiratoires chez les Crustacés Décapodes. *Bulletin Scientifique de la France et de la Belgique* **36**, 178–551.

Bond, C.E. (1979). *Biology of fishes*. W.B. Saunders & Co., Philadelphia.

Bridges, C.R. (1979). Adaptations of *Corystes cassivelaunus* to an arenicolous mode of life. In *Cyclic phenomena in marine plants and animals. Proceedings of the 13th European Marine Biology Symposium* (ed. E. Naylor & R.G., Hartnoll), pp. 317–324. Pergamon Press, Oxford.

Bridges, C.R. (1986). A comparative study of the respiratory properties and physiological function of haemocyanin in two burrowing and two non-burrowing crustaceans. *Comparative Biochemistry and Physiology* **83A**, 261–70.

Bridges, C.R. (1987). Environmental extremes – The respiratory physiology of intertidal rockpool fish and sublittoral burrowing fish. *Zoologische Beiträge* **30**, 65–84.

Bridges, C.R., Taylor, A.C. & Atkinson, R.J.A. (1982). Respiratory properties of the blood of the burrowing red band fish *Cepola rubescens*. *Journal of Experimental Marine Biology and Ecology* **59**, 51–60.

Bridges, C.R., Taylor, A.C., Morris, S.J. & Grieshaber, M.K. (1984). Ecophysiological adaptations in *Blennius pholis* (L.) blood to intertidal rockpool environments. *Journal of Experimental Marine Biology and Ecology* **77**, 151–67.

Brillet, C. (1969). Etude du comportement constructeur des poissons amphibies Periophthalmidae. *Revue d'Ecologie la Terre et la Vie* **116**, 496–520.

Brillet, C. (1984). Habitat et comportement du poisson amphibie *Periophthalmus koelreuteri africanus* Eggert a Tulear. Comparison avec le *P. sobrinus* Eggert sympatrique. *Revue d'Ecologie la Terre et la Vie* **39**, 337–45.

Burggren, W.W. & McMahon, B.R. (1988). *Biology of the land crabs.* Cambridge University Press.

Chapman, C.J. (1980). Ecology of juvenile and adult *Nephrops*. In *The Biology and Management of Lobsters*, vol. 2, (ed. J.S. Cobb & B.F. Phillips), pp. 143–78. Academic Press, London.

Clark, E. (1971). The Red Sea garden eel. *Underwater Naturalist* **7**, 4–10.

Clayton, D.A. & Vaughan, T.C. (1986). Territorial acquisition in the mudskipper *Boleophthalmus boddarti* (Teleostei, Gobiidae) on the mudflats of Kuwait. *Journal of Zoology* **209**, 501–19.

Colin, P.L. & Arneson, D.W. (1978). Aspects of the natural history of the swordtail jawfish, *Lonchopisthus micrognathus* (Poey) (Pisces: Opistognathidae), in south-western Puerto Rico. *Journal of Natural History* **13**, 689–97.

Congleton, J.L. (1974). The respiratory response to asphyxia of *Typhlogobius californiensis* (Teleostei: Gobiidae) and some related gobies. *Biological Bulletin of the Marine Biological Laboratory, Woods Hole* **146**, 186–205.

Cumberlidge, N. & Uglow, R.F. (1978). Heart and scaphognathite activity during the digging behaviour of the shore crab *Carcinus maenas*. In *Physiology and Behaviour of Marine Organisms, Proceedings of the 12th European Marine Biology Symposium* (ed. D.S. McLusky & A.J. Berry), pp. 23–30. Pergamon Press, Oxford.

Daniel, J.F. (1922). *The elasmobranch fishes.* University of California Press.

Douglas, W.A. & Lanzing, W.J.R. (1981). The respiratory mechanisms of the dusky flathead, *Platycephalus fuscus* (Platycephalidae, Scorpaeniformes). *Journal of Fish Biology* **18**, 545–52.

Dworschak, P.C. (1983). The biology of *Upogebia pusilla* (Petagna) (Decapoda, Thalassinidea). 1. The burrows. *Pubblicazioni della Stazione zoologica di Napoli* **4**, 19–43.

Dyer, M.F. & Uglow, R. (1978). Gill chamber ventilation and

scaphognathite movements in *Crangon crangon*. *Journal of Experimental Marine Biology and Ecology* **31**, 95–207.

El-Ziady, S., Eissa, S.M. & Al-Naqi, Z. (1979). Ecological studies on the mud-skipper fish, *Periophthalmus chrysospilos* (Bleeker) in Kuwait. *Bulletin of the Faculty of Science, Cairo University* **48**, 113–33.

Eshky, A.A., Atkinson, R.J.A. & Taylor, A.C. (1988). Effects of temperature on oxygen consumption and heart rate in the semi-terrestrial crab, *Ocypode saratan* (Forskål). *Marine Behaviour and Physiology* **13**, 341–58.

Farley, R.D. & Case, J.F. (1968). Perception of external oxygen by the burrowing shrimp, *Callianassa californiensis* Dana and *C. affinis* Dana. *Biological Bulletin of the Marine Biological Laboratory, Woods Hole* **134**, 261–5.

Felder, D.L. (1979). Respiratory adaptations of the estuarine mud shrimp, *Callianassa jamaicense* (Schmitt, 1935) (Crustacea, Decapoda, Thalassinidea). *Biological Bulletin of the Marine Biological Laboratory, Woods Hole* **157**, 125–38.

Forbes, A.T. (1973). An unusual abbreviated larval life in the estuarine burrowing prawn *Callianassa kraussi* (Crustacea: Decapoda: Thalassinidea). *Marine Biology* **22**, 361–5.

Forster, M.E. & Starling, L. (1982). Ventilation and oxygen consumption in the spotted stargazer *Genyagnus monopterygius*. *New Zealand Journal of Marine & Freshwater Research* **16**, 135–40.

Frey, R.W. (1973). Concepts in the study of biogenic sedimentary structures. *Journal of Sedimentary Petrology* **43**, 6–19.

Garstang, W. (1896). Contributions to marine bionomics. 1. The habits and respiratory mechanisms of *Corystes cassivelaunus*. *Journal of the Marine Biological Association of the U.K.* **4**, 233–32.

Garstang, W. (1897*a*). Contributions to marine bionomics. 2. The function of the antero-lateral denticulations of the carapace in sand-burrowing crabs. *Journal of the Marine Biological Association of the U.K.* **4**, 396–401.

Garstang, W. (1897*b*). Contributions to marine bionomics. 3. The systematic features, habits, and respiratory phenomena of *Portumnus nasutus* (Latreille). *Journal of the Marine Biological Association of the U.K.* **4**, 402–7.

Garstang, W. (1897*c*). On some modifications of structure subservient to respiration in decapod Crustacea which burrow in sand; with some remarks on the utility of specific characters in the genus *Calappa* and the description of a new species of *Albunea*. *Quarterly Journal of Microscopical Science* **40**, 211–32.

Gibson, R.N. & Ezzi, I.A. (1978). The biology of a Scottish population of Fries' goby, *Lesueurigobius friesii*. *Journal of Fish Biology* **12**, 371–89.

Gordon, M.S., Gabaldon, D.J. & Yip, A.Y.-w. (1985). Exploratory observations on microhabitat selection within the intertidal zone by the Chinese mudskipper fish *Periophthalmus cantonensis*. *Marine Biology* **85**, 209–15.

Gordon, M.S., Ng, W. & Yip, A. (1978). Aspects of the physiology of terrestrial life in amphibious fishes. III. The Chinese mudskipper *Periophthalmus cantonensis. Journal of Experimental Biology* **72**, 57–75.

Graham, J.B. & Baird, T.A. (1984). The transition to air breathing in fishes III. Effects of body size and aquatic hypoxia on aerial gas exchange of the swamp eel *Synbranchus marmoratus. Journal of Experimental Biology* **108**, 357–75.

Greenfield, D.W. (1968). Observations on the behaviour of the basket-weave cusk-eel *Otophidium scrippsi*. Hubbs. *California Fish and Game* **54**, 108–14.

Grimes, C.B., Able, K.W. & Jones, R.S. (1986). Tilefish, *Lopholatilus chamaeleonticeps*, habitat, behaviour and community structure in Mid-Atlantic and southern New England waters. *Environmental Biology of Fishes* **15**, 273–92.

Grossman, G.D. (1979). Symbiotic burrow-occupying behaviour in the bay goby, *Lepidogobius lepidus. California Fish and Game* **65**, 122–4.

Grow, L. & Merchant, H. (1980). The burrow habit of the crayfish, *Cambarus diogenes diogenes* (Girard). *American Midland Naturalist* **103**, 231–7.

Hagerman, L. & Uglow, R.F. (1985). Effects of hypoxia on the respiratory and circulatory regulation of *Nephrops norvegicus. Marine Biology* **87**, 273–8.

Hardisty, M.W. (1979). *Biology of Cyclostomes*. Chapman & Hall, London.

Hill, B. (1981). Respiratory adaptations of three species of *Upogebia* (Thalassinidea, Crustacea) with special reference to low tide periods. *Biological Bulletin of the Marine Biological Laboratory, Woods Hole* **160**, 272–9.

Hirano, O., Yamamoto, K. & Konno, H. (1984). Borrowing behaviour and preference of the bottom conditions in the *Pseudogobio esocinus. Journal of Shimonoseki University Fisheries* **32**, 75–81.

Hobson, E.S. (1986). Predation on the Pacific sand-lance *Ammodytes hexapterus* Pisces Ammodytidae during the transition between day and night in South eastern Alaska U.S.A. *Copeia* **1**, 223–6.

Hoffman, C.J. (1981). Associations between the arrow goby *Clevelandia ios* (Jordon and Gilbert) and the ghost shrimp *Callianassa californiensis* Dana in natural and artificial burrows. *Pacific Science* **35**, 211–16.

Hoffman, S.G. & Robertson, D.R. (1983). Foraging and reproduction of two Caribbean reef toadfishes (Batrachoididae). *Bulletin of Marine Science* **33**, 919–27.

Horwitz, P.H.J., Richardson, A.M.M. & Boulton, A. (1985). The burrowing habit of two sympatric species of land crayfish, *Engaeus urostrictus* and *E. tuberculatus* (Decapoda: Parastacidae). *Victorian Naturalist* **102**, 188–97.

Hudson, R.C.L. (1977). Preliminary observations on the behaviour of the gobiid fish *Signigobius biocellatus* Hocse and Allan, with particular reference to its burrowing behaviour. *Zeitschrift für Tierpsychologie* **43**, 214–20.

Hughes, G.M. & Umezawa, S. (1968). On respiration in the dragonet *Callionymus lyra* L. *Journal of Experimental Biology* **49**, 565–82.

Jensen, F.B. & Weber, R.E. (1982). Respiratory properties of tench blood and hemoglobin adaptation to hypoxic-, hypercapnic water. *Molecular Physiology* **2**, 235–50.

Jorgensen, C.B. (1966). *Biology of suspension feeding*. Pergamon Press, London.

Karplus, I., Szlep, R. & Tsurnamal, M. (1972). Associative behaviour of the fish *Cryptocentrus cryptocentrus* (Gobiidae) and the pistol shrimp *Alpheus djiboutensis* (Alpheidae) in artificial burrows. *Marine Biology* **15**, 95–104.

Karplus, I., Szlep, R. & Tsurnamal, M. (1974). The burrows of alpheid shrimp associated with gobiid fish in the northern Red Sea. *Marine Biology* **24**, 259–68.

Kersten, A., Lomholt, J.P. & Johansen, K. (1979). The ventilation, extraction and uptake of oxygen in undisturbed flounders, *Platichthys flesus*: responses to hypoxia acclimation. *Journal of Experimental Biology* **83**, 169–79.

Kobayashi, T., Dotsu, Y. & Takata, T. (1971). Nest and nesting behaviour of the mudskipper *Periophthalmus cantonensis*, in Ariake Sound. *Bulletin of the Faculty of Fisheries, Nagasaki University* **32**, 27–39.

Lewis, D.B. (1976). Studies of the biology of the lesser weever fish *Trachinus vipera* Cuvier. *Journal of Fish Biology* **8**, 127–38.

Little, C. (1983). *The Colonisation of Land. Origins and Adaptations of Terrestrial Animals*. Cambridge University Press.

MacGinitie, G.E. (1934). The natural history of *Callianassa californiensis* Dana. *American Midland Naturalist* **16**, 166–77.

MacGinitie, G.E. & MacGinitie, N. (1968). *Natural History of Marine Animals*, 2nd edn. McGraw-Hill Book Co, New York.

McDonald, D.G., McMahon, B.R. & Wood, C.M. (1977). Patterns of heart and scaphognathite activity in the crab *Cancer magister*. *Journal of Experimental Zoology* **202**, 33–44.

McLay, C.L. & Osborne, T.A. (1985). Burrowing behaviour of the paddle crab *Ovalipes catharus* (White, 1843) (Brachyura: Portunidae). *New Zealand Journal of Marine & Freshwater Research* **19**, 125–30.

McMahon, B.R. & Wilkens, J.L. (1983). Ventilation, perfusion and oxygen uptake. In *The Biology of Crustacea* (ed. D.E. Bliss), vol. 5: *Internal Anatomy and Physiological Regulation* (ed. L.H. Mantel), pp. 289–372. Academic Press, New York.

Miller, K.I., Pritchard, A.W. & Rutledge, P.S. (1976). Respiratory regulation and the role of the blood in the burrowing shrimp *Callianassa californiensis* (Decapoda: Thalassinidea). *Marine Biology* **36**, 233–42.

Miller, K.I. & Van Holde, K.E. (1981). The effect of environmental variables on the structure and function of hemocyanin from *Callianassa californiensis*. 1. Oxygen binding. *Journal of Comparative Physiology* **143**, 253–60.

Mukai, H. & Koike, I. (1984). Behaviour and respiration of the burrowing shrimps *Upogebia major* (de Haan) and *Callianassa japonica* (de Haan). *Journal of Crustacean Biology* **4**, 191–200.

Nash, R.D.M., Chapman, C.J., Atkinson, R.J.A. & Morgan, P.J. (1984). Observations on the burrows and burrowing behaviour of *Calocaris macandreae* (Crustacea: Decapoda: Thalassinoidea). *Journal of Zoology* **202**, 425–39.

Nayar, K.K. (1951). Some burrowing fishes from Travancore. *Proceedings of the Indian Academy of Science* **34B**, 311–28.

Nye, P.A. (1974). Burrowing and burying by the crab *Macrophthalmus hirtipes*. *New Zealand Journal of Marine and Freshwater Research* **8**, 243–54.

Pelster, B. (1985). Mechanismen der Anpassung an das Leben in extremen Biotopen: Vergleichende Studien zur Atmungsphysiologie bei *Lumpenus lampretaeformis* und *Blennius pholis*. Ph.D. Thesis, University of Düsseldorf.

Pelster, B., Bridges, C.R., Taylor, A.C., Morris, S. & Atkinson, R.J.A. (1988*a*). Respiratory adaptations of the burrowing marine teleost *Lumpenus lampretaeformis* (Walbaum). I. O_2 and CO_2 transport, acid-base balance: a comparison with *Cepola rubescens* L. *Journal of Experimental Marine Biology and Ecology* **124**, 31–42.

Pelster, B., Bridges, C.R. & Grieshaber, M.K. (1988*b*). Respiratory adaptations of the burrowing marine teleost *Lumpenus lampretaeformis*. (Walbaum). II. Metabolic adaptations. *Journal of Experimental Marine Biology and Ecology* **124**, 43–55.

Powers, D.A. (1980). Molecular ecology of Teleost fish hemoglobin: Strategies for adapting to changing environments. *American Zoologist* **20**, 139–62.

Powers, D.A., Fyhn, H.J., Fyhn, U.E.H., Martin, J.P., Garlick, R.L. & Wood, S.C. (1979). A comparative study of the oxygen equilibria of blood from 40 genera of Amazonian fishes. *Comparative Biochemistry and Physiology* **62A**, 67–85.

Powers, L.W. & Bliss, D.E. (1983). Terrestrial adaptations. In *The Biology of Crustacea* (ed. D.E. Bliss), vol. 8 (ed. F.J. Vernberg & W.B. Vernberg), pp. 271–333. Academic Press, New York.

Pullin, R.S.V., Morris, D.J., Bridges, C.R. & Atkinson, R.J.A. (1980). Aspects of the respiratory physiology of the burrowing fish *Cepola rubescens* L. *Comparative Biochemistry and Physiology* **66A**, 35–42.

Rao, H.S. (1939). On the burrowing habits of a Gobioid fish of the genus *Taenoides* in the Andamans. *Proceedings of the National Institute of Sciences of India* **5**, 275–9.

Rice, A.L. & Chapman, C.J. (1971). Observations on the burrows and burrowing behaviour of two mud-dwelling decapod crustaceans, *Nephrops norvegicus* and *Goneplax rhomboides*. *Marine Biology* **10**, 330–42.

Rice, A.L. & Johnstone, A.D.F. (1972). The burrowing behaviour of the gobiid fish *Lesueurigobius friesii* (Collet). *Zeitschrift für Tierpyschologie* **30**, 431–8.

Schembri, P.J. (1979). An unusual respiratory rhythm in the crab *Ebalia tuberosa* (Pennant) (Crustacea: Decapoda: Leucosiidae). In *Cyclic phenomena in marine plants and animals. Proceedings of the 13th European Marine Biology Symposium* (ed. E. Naylor & R.G. Hartnoll), pp. 327–335. Pergamon Press, Oxford.

Schembri, P.J. (1981). The functional morphology of the branchial chambers and associated structures of *Ebalia turberosa* (Crustacea: Decapoda: Leucosiidae), with special reference to ventilation of the egg-mass. *Journal of Zoology* **195**, 423–36.

Schultz, L.P. & Stern, E.M. (1948). *The Ways of Fishes*. Van Nostrand Co, Toronto.

Steffensen, J.F., Johansen, K., Sindberg, C.D., Sorensen, J.H. & Møller, J.L. (1984). Ventilation and oxygen consumption in the hagfish, *Myxine glutinosa* L. *Journal of Experimental Marine Biology and Ecology* **84**, 173–8.

Swain, R., Marker, P.F. & Richardson, A.M.M. (1987). Respiratory responses to hypoxia in stream-dwelling (*Astacopsis franklinii*) and burrowing (*Parastacoides tasmanicus*) parastacid crayfish. *Comparative Biochemistry and Physiology* **87A**, 813–17.

Taylor, A.C. (1984). Branchial ventilation in the burrowing crab, *Atelecyclus rotundatus*. *Journal of the Marine Biological Association of the U.K.* **64**, 7–20.

Taylor, A.C., Morris, S. & Bridges, C.R. (1985). Oxygen and carbon dioxide transporting properties of the blood of three sublittoral species of burrowing crab. *Journal of Comparative Physiology* **155**, 733–42.

Thompson, R.K. & Pritchard, A.W. (1969). Respiratory adaptations of two burrowing crustaceans, *Callianassa californiensis* and *Upogebia pugettensis* (Decapoda, Thalassinidea). *Biological Bulletin of the Marine Biological Laboratory, Woods Hole* **136**, 274–87.

Torres, J.J., Gluck, D.L. & Childress, J.J. (1977). Activity and physiological significance of the pleopods in the respiration of *Callianassa californiensis* (Dana) (Crustacea: Thalassinidea). *Biological Bulletin of the Marine Biological Laboratory, Woods Hole* **152**, 134–46.

Trueman, E.R. & Ansell, A.D. (1969). The mechanisms of burrowing into soft substrata by marine animals. *Oceanography & Marine Biology Annual Review* **7**, 315–66.

Vannini, M. (1980). Researches on the coast of Somalia. The shore and the dune of Sar Uanle. 27. Burrows and digging behaviour in *Ocypode* and other crabs (Crustacea, Brachyura). *Monitore zoologico italiano* **13**, 11–44.

Vogel, S. (1977). Flows in organisms induced by movements of the external medium. In *Symposium on Scale Effects in Animal Locomotion* (ed. T.J. Pedley), pp. 285–97. Academic Press, London.

Warner, G.F. (1977). *The Biology of Crabs*. Elek Science, London.

Weber, R.E. (1982). Intraspecific adaptation of hemoglobin function in fish to oxygen availability. In *Exogenous and Endogenous Influences on Metabolic and Neural Control* (ed. A.D.F. Addink & N. Spronk), pp. 87–102. Pergamon Press, Oxford.

Yanagisawa, Y. (1984). Studies on the interspecific relationship between gobiid fish and snapping shrimp II. Life history and pair formation of snapping shrimp *Alpheus bellulus. Publications of the Seto Marine Biological Laboratory* **29**, 93–116.

Yazdani, G.M. & Alexander, R. McN. (1967). Respiratory currents in flatfish. *Nature* **213**, 96–7.

P. J. BUTLER

Respiratory adaptations to limited oxygen supply during diving in birds and mammals

Introduction

Although aquatic birds and mammals are unable to exchange gases with water, they do spend substantial proportions of their time submerged under water. However, for all species studied the majority of dives are well short of the maximum and are thought to be aerobic in nature (Butler, 1989). Despite these generalities, there are large differences between species with respect to the duration of dives and the rate of energy expenditure associated with underwater activities. These two factors are of obvious importance in terms of the amount of oxygen that is consumed. For diving animals, the other important factors are the amount of oxygen that can be stored in the body and the rate at which gases can be exchanged (i.e. the rate at which the oxygen stores can be replenished and the rate at which CO_2 can be eliminated) when the animal is at the surface.

This review will deal mainly with the last two factors as they encompass adaptations to the respiratory system, although the cardiovascular and muscular systems are inextricably involved also. However, to put these adaptations into context, there follows a brief section on dive duration and rate of oxygen consumption during submersion in a range of birds and mammals.

Aerobic metabolism and dive duration

In ducks and small mammals (e.g. mink), diving for food can be a very expensive business, partly because they are buoyant and/or relatively inefficient at underwater locomotion. When diving for a mean duration of 14.4 s, mean rate of oxygen uptake of tufted ducks, *Aythya fuligula*, is not significantly different from that when the animals are swimming at the surface at maximum sustainable velocity (Woakes & Butler, 1983). These levels are 3.5–4 times the resting value. However, as dive duration increases, rate of oxygen consumption may decline (Butler, 1989). These animals can remain submerged during voluntary diving for 46 s, but the

235

preferred duration appears to be approximately 20 s (Stephenson *et al.*, 1986).

Muskrats, *Ondatra zibethica*, also remain submerged for around 20 s (MacArthur, 1984) and, like tufted ducks, feed on sedentary material. On the other hand, mink and otters, *Mustela vison* and *Lutra lutra* respectively, hunt under water. Mink travel at 0.5 m s^{-1} and successful feeding dives last for an average of 10 s, whereas if no prey is caught the animal remains submerged for an average of only 5 s. Thus, if they do not locate prey within a few seconds, they surface, presumably because there would be insufficient time for pursuit even if a fish were spotted (Dunstone, 1978). When feeding in fresh water, otters dive, on average, for 13 s, irrespective of whether prey is caught or not, and hunting lasts for an average of 10 min (Conroy & Jenkins, 1986). However, when hunting in the sea, successful dives are 13–16 s duration whereas unsuccessful ones are 23–25 s and hunting lasts for 24 min (Kruuk & Hewson, 1977; Conroy & Jenkins, 1986). The reasons for these differences in dive behaviour are not clear, although the latter authors suggest that one possibility is the different type of prey in the two environments, with the slow-swimming eel being more common in fresh water. Certainly an otter in pursuit of prey generally remains submerged for approximately 30 s, surfaces for a quick breath and then continues the chase (Erlinge, 1968). This is clearly a different strategy from that adopted by mink. In muskrats and mink, oxygen uptake during diving is equivalent to swimming at the surface at approximately 0.4 m s^{-1} and 0.5 m s^{-1} respectively and is 2.4 times the resting values (Williams, 1983; Butler, 1988).

A number of birds such as grebes, divers, cormorants and a range of auks are, to a greater or lesser extent, active predators of fish and remain submerged for an average of 20–70 s, with the imperial cormorant (blue-eyed shag), *Phalacrocorax atriceps*, having a reported maximum dive duration of 95 s and the guillemot, *Uria aalge*, of 202 s (Butler & Jones, 1982; Cooper, 1985; Wanless *et al.*, 1988). From birds trapped in nets, it appears that common murres (guillemots) can dive to 180 m, razorbills, *Alca torda*, to 120 m, the great northern diver, *Gavia immer*, to 60 m and the imperial cormorant to 25 m (Schorger, 1947; Conroy & Twelves, 1972; Piatt & Nettleship, 1985). These times and depths compare favourably with those of the smaller penguins. When inshore, the little penguin, *Eudyptula minor*, remains submerged for very short periods of 10–15 s, and the greatest recorded depth is 60 m (Mill & Baldwin, 1983), whereas the larger chinstrap, *Pygoscelis antarctica*, and gentoo, *P. papua*, penguins have mean dive durations of 91 to 128 s and maxima of 130 and 189 s respectively (Trivelpiece *et al.*, 1986). The greatest recorded depths for these birds are 70 and 100 m respectively (Lishman & Croxall, 1983; Conroy & Twelves, 1972).

It is the larger penguins, the kings and the emperors, *Aptenodytes patagonica* and *A. forsteri*, which are the most impressive avian divers. When diving naturally and presumably feeding, the emperors remain submerged for 2.5–9 min with a maximum recorded depth of 265 m (Kooyman *et al.*, 1971*a*). King penguins can dive to between 240 and 290 m, although 50% of the dives are probably to less than 50 m (Kooyman *et al.*, 1982).

From direct measurements of oxygen uptake in Humboldt penguins, *Spheniscus humboldti*, diving for food in a pool and in little penguins swimming at $0.7\,\mathrm{m\,s^{-1}}$ in a water channel, it appears that energy expenditure during underwater swimming in penguins is approximately 30% greater than, but not significantly different from, that in resting birds (Butler & Woakes, 1984; Baudinette & Gill, 1985). Clearly these experimental conditions were somewhat different from those that exist at the feeding grounds. None the less, Jones & Furilla (1987) point out that the average daily metabolic rate (ADMR) measured with labelled water in a number of species of penguins when foraging is between 40 and 60% greater than the fasting rate. Unfortunately, in some of the field studies it was not possible to estimate metabolic rate when the birds were actually diving, so the value for ADMR probably includes periods of rest (Kooyman *et al.*, 1982; Davis *et al.* 1983). By incorporating information on time budgets, Nagy *et al.* (1984) estimated that, when swimming under water at sea at an average velocity of $1.75\,\mathrm{m\,s^{-1}}$, energy expenditure in jackass penguins, *Spheniscus demersus*, is 6 times the resting value. This can be compared with a metabolic rate of 4.3 times the lowest resting value in Humboldt penguins swimming at $1.25\,\mathrm{m\,s^{-1}}$ (Hui, 1988). It would appear, therefore, that metabolism during diving in penguins can be as elevated above the resting value as it is in ducks. The difference is that penguins can swim faster.

Pinnipeds and cetaceans also exhibit a range of diving abilities. Fur seals, *Callorhinus ursinus* and *Arctocephalus* spp., have an average dive duration of approximately 2 min and a maximum duration of approximately 6 min. Average and maximum depths are approximately 50 m and 150 m respectively (Gentry & Kooyman, 1986). Scholander (1940) reported that the normal dive duration of the common porpoise, *Phocaena*, is 4–5 min, although dives of up to 12 min may occur. During long periods of diving in the open sea, the mean depth of dolphins, *Delphinus*, is 64 m (Evans, 1971), although the bottlenose dolphin, *Tursiops truncatus*, has been trained to dive down to 300 m (Ridgway *et al.*, 1969). Adult Weddell seals, *Leptonychotes weddelli*, can remain submerged for 70 min (Kooyman *et al.*, 1971*b*), but approximately 85% of their dives last for 20 min or less and there is no increase in blood lactate after surfacing from such dives (Kooyman *et al.*, 1980). Dive depth is usually between 200 and 400 m but a depth of 600 m has been reported (Kooyman, 1966).

The post-dive lactate data from Weddell seals indicate that feeding dives in these animals are largely, if not entirely, aerobic in nature. However, there is evidence from direct measurements of oxygen uptake by Kooyman *et al.* (1973*a*) and M.A. Castellini (personal communication) and from calculations by Guppy *et al.* (1986) that aerobic metabolism during diving in these animals is similar to or even less than the resting value. If this is so, it is clearly an important factor in determining the diving performance of these animals.

Like most other divers, the Weddell seal performs a series of dives in fairly quick succession (a bout) and then rests for a period. In this particular species a bout may last for 8–12 h (Kooyman *et al.*, 1980), but female northern elephant seals, *Mirounga angustirostris*, dive more or less continuously during the first 14–27 days at sea following lactation (Le Boeuf *et al.*, 1988). Mean dive duration is 19.2 min and maximum duration 48 min. Mean depth is 400 m, but maximum (estimated) depth is a record (for pinnipeds) 894 m. Even so this palls into insignificance against the depths of 2000 and 3000 m claimed for sperm whales, *Physeter catodon* (Clarke, 1976, 1979). Lockyer (1977) substantiated that these animals can indeed dive to over 1000 m depth and remain submerged for 60 min although most dives are less than 10 min in duration and to less than 400 m.

The size and replenishment of the oxygen stores

If aerobic metabolism is the major, or indeed the sole, source of ATP production during normal feeding dives of aquatic birds and mammals, then the amount of oxygen stored in the body before submersion and the rate at which it can be replenished upon surfacing are of crucial importance to these animals.

Muscles

As shown in Table 1*a* the concentration of myoglobin (Mb) in locomotory muscles is certainly higher in some aquatic birds and mammals compared with their non-diving relatives, even those that are very active such as hummingbirds and thoroughbred horses. As the affinity of Mb for oxygen is higher than that of haemoglobin (Hb), oxygen bound to the Mb in a particular muscle will only be available to that muscle. For aquatic mammals, penguins, auks and other birds that use their wings for under-water locomotion, this means that most of this oxygen is where it is needed. For those birds that use their legs under water, any oxygen combined with the Mb of the pectoral muscles would be unavailable to the locomotory muscles.

As Snyder (1983) points out, the amount of oxygen dissolved in tissue is approximately 0.8 ml kg^{-1} and is sufficient to support aerobic metabolism for only a few seconds. However, the amount of oxygen that can be stored in myoglobin increases with increased diving ability. In terrestrial mammals it ranges from 4 to 15 ml (kg tissue)$^{-1}$; in shallow diving rodents, the manatee and otariid seals it ranges from 16 to 47 ml (kg tissue)$^{-1}$, whereas in deep-diving phocid seals and large cetaceans the values are 55–110 ml (kg tissue)$^{-1}$. The last value (from the ribbon seal) is over half of the carrying capacity of the blood in man. Because mass-specific oxygen consumption decreases with increased body mass, the oxygen stored in the muscle of the larger cetaceans may itself be sufficient to allow aerobic metabolism to continue for the duration of normal dives (Snyder, 1983). For example, the oxygen stored in the muscles of a 450 kg Weddell seal would maintain aerobic metabolism (of 4.2 ml O$_2$ kg^{-1} min^{-1}) for approximately 15 min during normal diving activity (Kooyman, 1985). Assuming that metabolic rate in marine mammals scales to body mass raised to the power of approximately 0.75 (Lavigne *et al.*, 1986), aerobic metabolism during diving in a 20 000 kg sperm whale (Lockyer, 1976) would, all other things being equal, be approximately 1.6 ml O$_2$ kg^{-1}. This is some 38% of that of the Weddell seal. However, the Mb concentration is 1.3 times greater in the sperm whale. Thus, the oxygen stored in the muscle of sperm whales should maintain aerobic metabolism during diving for approximately 50 min. Also, in those muscles where the Mb concentration is above 1.7% it could have a significant effect on the diffusion of oxygen from the capillaries to the mitochrondria (see Weber *et al.*, 1981 for discussion), and this could be of immense benefit as the partial pressure of oxygen in arterial blood (Pa_{O_2}) declines during a dive. As well as possessing increased stores of oxygen in their muscles, diving birds and mammals may also have unusually high levels of phosphocreatine in their muscles (T. Williams, personal communication).

Respiratory system

The use of the respiratory system as an oxygen store presents two problems. Firstly, the greater the volume of the respiratory system, the greater the buoyancy of the animal, and therefore the greater the energy cost of diving. Secondly, air is compressible. Any reduction in volume of

240   *P. J. Butler*

Table 1. *Sites of oxygen storage in diving and non-diving homeotherms*

(a) Muscles

	Myoglobin [g (100 g tissue)$^{-1}$]
Gull (pectoralis)[a]	0.55
Hummingbird (pectoralis)[b]	0.47
Tufted duck (gastrocnemius)[c]	0.98
Tufted duck (pectoralis)[c]	0.73
Gentoo penguin (pectoralis)[d]	4.4
Thoroughbred horse (psoas)[e]	0.88
Beaver[f]	1.2
Bottlenose dolphin[g]	3.2
Porpoise (psoas)[g]	4.1
Weddell seal[h]	4.5
Sperm whale[i]	5.7
Bottlenose whale[i]	6.3
Ribbon seal[j]	8.1

(b) Respiratory system

	Volume (ml BTPS kg^{-1})
Mallard duck[k]	112
Tufted duck[k]	180
Adélie penguin[l]	165
Dog[m]	61
Horse[m]	74
Man[m]	74
Beaver[f]	60
Harbour porpoise[i]	59
Weddell seal[n]	48
Bottlenose whale[i]	25

(c) Circulatory system

	Volume [ml (100 g body mass)$^{-1}$]	Hb [g (100 ml blood)$^{-1}$]	Oxygen capacity [ml O$_2$ (100 ml blood)$^{-1}$]
Pigeon[o,p]	9.2	15.3	21.2
Mallard[k]	9.1	17.1	22.9
Tufted duck[k]	11.4	18.4	24.6
Chinstrap penguin[q]	—	19.6	26.3
Emperor penguin[r]	—	16.8	23
Man[s,t]	7.7	14.8	20.2
Beaver[f,u]	6.5	11.9	16.1
Weddell seal[v]	14.8	23.7	31.6
Bladdernose seal[w]	—	26.4	36.0
Bottlenose porpoise[x,y,z]	7.1	14.4	19.7
Dall porpoise[y,z]	14.3	20.3	26.5
Sperm whale[x]	—	14.7	18.5

Table 1 (cont.)

Sources of data:

[a]Pages & Planas, 1983	[n]Kooyman *et al.*, 1971*b*
[b]Johansen *et al.*, 1987	[o]Baumann & Baumann, 1977
[c]Butler & Turner, 1988	[p]Bond & Gilbert, 1958
[d]Weber *et al.*, 1974	[q]Milsom *et al.*, 1973
[e]Lawrie, 1950	[r]Lenfant *et al.*, 1969*a*
[f]McKean & Carlton, 1977	[s]Farhi & Rahn, 1955
[g]Snyder, 1983	[t]Butler & Jones, 1982
[h]Castellini & Somero, 1981	[u]Clausen & Ersland, 1968
[i]Scholander, 1940	[v]Lenfant *et al.*, 1969*b*
[j]Lenfant *et al.*, 1970*b*	[w]Clausen & Ersland, 1969
[k]Keijer & Butler, 1982	[x]Lenfant, 1969
[l]Kooyman *et al.*, 1973*b*	[y]Ridgway & Johnston, 1966
[m]Gehr *et al.*, 1981	[z]Horvath *et al.*, 1968

this air as a result of thoracic compression during descent would cause an increase in partial pressure of the component gases and thus enhance gas exchange across the alveoli. The increase in the partial pressure of oxygen in arterial blood, Pa_{O_2}, may be of some benefit during submersion, but upon ascent thoracic volume will increase and Pa_{O_2} will decline. This could give rise to severe hypoxia (Lanphier & Rahn, 1963). As nitrogen is inert, it will accumulate in the blood and tissues and may cause narcosis. If ascent is too rapid to allow this nitrogen to leave the blood via the lung, it will form bubbles in the tissues (decompression sickness or caisson disease) and may lead to death. Repetitive diving to shallow depths with insufficient time at the surface to recover completely can also lead to decompression sickness (Paulev, 1965). Also, all the body fluids are at ambient pressure, so if the volume of the gas is not reduced sufficiently during descent (e.g. if the chest wall is not compliant enough) and if it is in contact with soft body tissues, there will be a pressure difference across the tissue–gas interface. This could cause pulmonary oedema and blood vessels to rupture (commonly known as 'the squeeze'). Despite these potential problems, birds that dive have respiratory systems with larger (approximately 50%) volumes than those of their non-diving relatives (Table 1*b*).

Penguins and diving ducks hyperventilate before submersion; whatever its primary function, this undoubtedly increases the oxygen stores (Kooyman *et al.*, 1971*a*; Butler & Woakes, 1979; Woakes & Butler, 1983). However, the partial expiration before a dive by tufted ducks may not reduce the effective volume of the respiratory system much below 180 ml kg^{-1}. Penguins, however, dive on inspiration and do not appear to

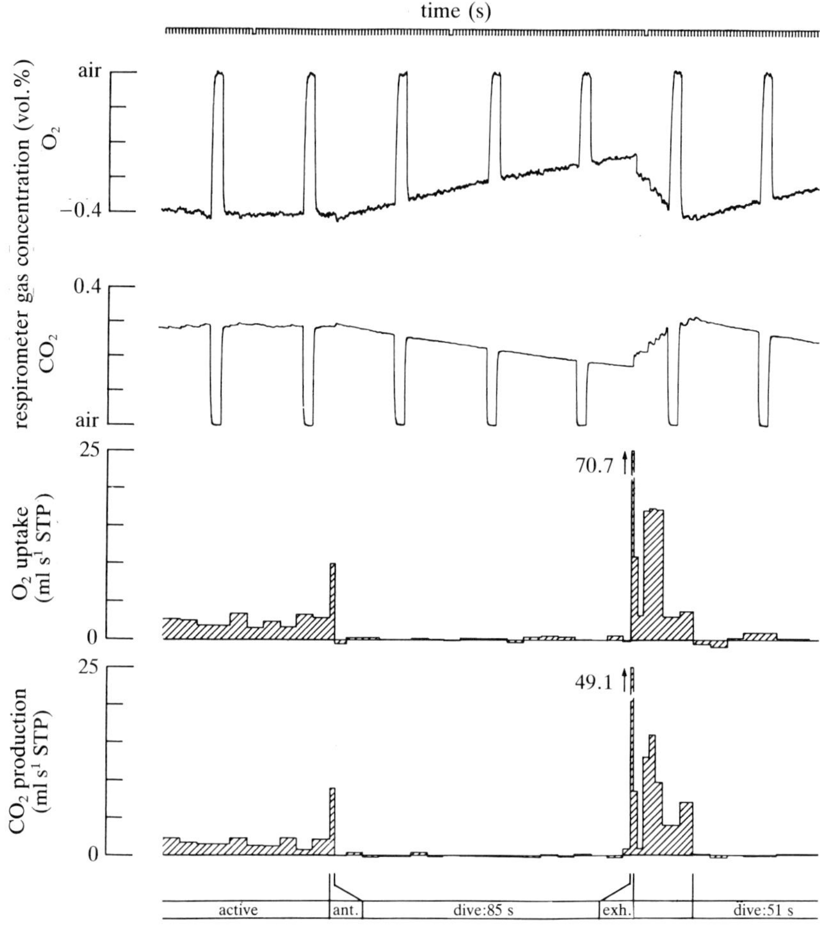

Figure 1. Oxygen uptake and carbon dioxide production of a Humboldt penguin before and after a voluntary dive of 85 s duration. Note the brief increase in gas exchange immediately preceding the dive (ant) and also the large changes in gas concentrations associated with the first exhalation (exh) upon surfacing (actual values are given next to the histogram). (From Butler & Woakes, 1984.)

exhale under water (Kooyman *et al.*, 1971*a*) so the volume of their respiratory system will probably exceed 160 ml kg^{-1}. In both of these animals, therefore, there is a substantial proportion of the available oxygen present in the respiratory system (approximately 50% in the tufted duck) (Keijer & Butler, 1982), and there is clear evidence from the Humboldt penguin that at least some of this oxygen is used during shallow voluntary dives (Figure 1). Whether or not the movements of the limbs cause movement of air between the anterior and posterior groups of air sacs (and, therefore, through the lungs) (Eliassen, 1960) remains to be seen. Although of some apparent significance as an oxygen store in diving birds, the respiratory system of tufted ducks exhibits a lower volume after several months of long duration diving (Stephenson *et al.*, 1988). This would effectively reduce the energetic cost of diving.

When birds dive deeply, as some auks and penguins do, there is evidence from work on Adélie and gentoo penguins that gas exchange across the lung is impaired at a depth of approximately 40 m (Kooyman *et al.*, 1973*b*). This will not only prevent problems associated with nitrogen narcosis, decompression sickness and the squeeze but will also render any oxygen in the respiratory system unusable. Precisely how this impairment of gas exchange in the lungs of birds is achieved is unclear (Jones & Furilla, 1987).

There seems to be some disagreement in the interpretation of the data on the relative size of the lungs in aquatic and terrestrial mammals. Kooyman (1973) states that lung volume of marine mammals is greater than that in terrestrial species, especially if the comparison is made on a lean weight basis. Synder (1983) is of the view that as a general rule there is not an increased lung volume in marine mammals. It is interesting to note, however, that resting volume of the lung is not necessarily the important feature.

The Weddell seal has an inspiratory lung volume that is similar to that of terrestrial mammals yet it exhales upon submersion (Kooyman *et al.*, 1971*b*). On average, diving lung volume is 60% of inspiratory volume. Both phocid seals and large cetaceans dive to great depths and it may be more important for these to expel air from the alveoli to prevention nitrogen narcosis, decompression sickness or the squeeze than to use the lung as an oxygen store. Data from freely diving Weddell seals indicate that their alveoli are fully collapsed at depths of 30–35 m (Falke *et al.*, 1985). Small cetaceans, dolphins and porpoises are relatively shallow divers and appear to dive upon inspiration (Scholander, 1940). Dolphins, *Tursiops truncatus*, trained to dive to 100 m, showed no signs of lung collapse to depths of 70 m (Ridgway & Howard, 1979). So these animals may use the lung as an oxygen store.

Penguins and marine mammals not only swim under water when feeding, they also do so when moving from one place to another. They still have to surface periodically to ventilate their lungs and if, for any reason, they need to swim very rapidly then it may be energetically more effective if they leap clear of the water when ventilating and swim at least 3 body

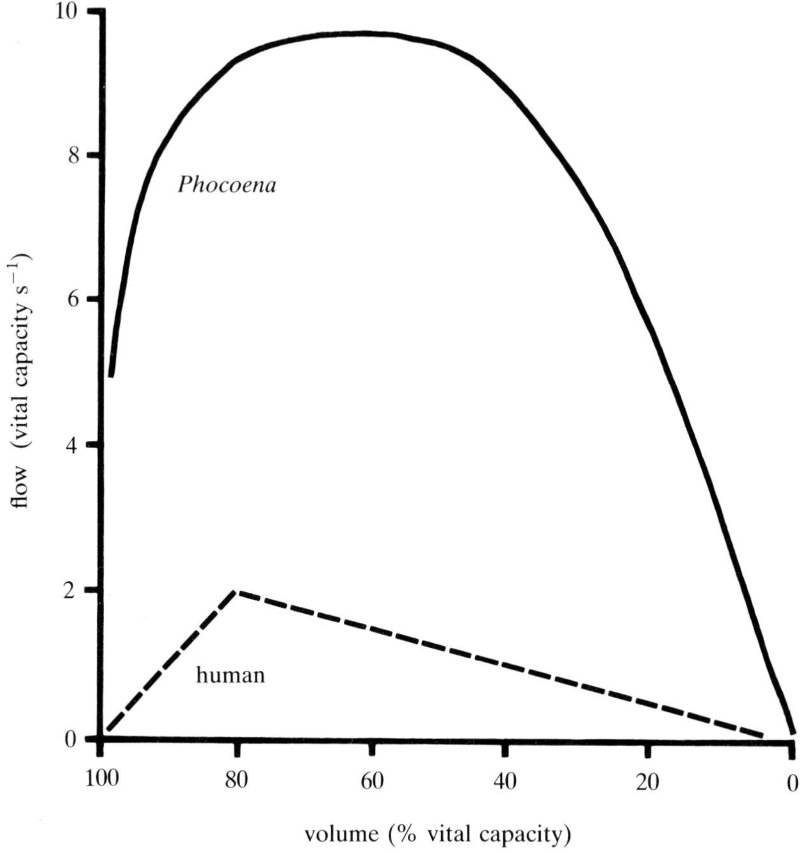

Figure 2. A comparison of the expiratory flow-volume curves of the lungs of man (broken lines) and of the harbour porpoise, *Phocoena phocoena* (solid lines). Note that, in man, after vital capacity has fallen below about 80%, the flow rate steadily declines. It does not do so in the porpoise. (Modified from Kooyman & Cornell, 1981.)

diameters beneath the surface in between times (Au & Weihs, 1980). Certainly a number of penguins and cetaceans do leap clear of the water ('porpoise') when swimming fast (Blake, 1983; Kooyman & Davis, 1987). This means that the movement of gases in and out of the respiratory system must occur very rapidly, and this is, no doubt, a problem associated with surfacing in some aquatic homeotherms even in the absence of porpoising.

The few studies that have been performed on marine mammals indicate that, in some small cetaceans at least, lungs are capable of emptying to 10–15% of total lung capacity (20–40% in man), and emptying time may be as

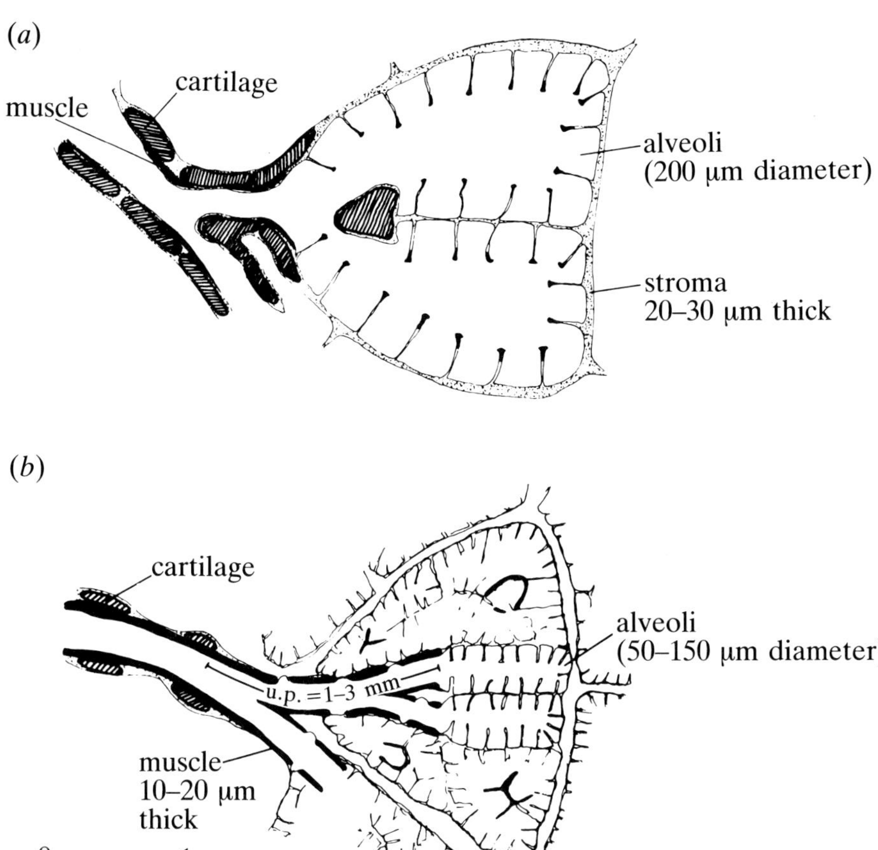

Figure 3. The peripheral structure of the lungs of pinnipeds. (*a*) In otariids (sea lions and fur seals) cartilage extends to the mouths of the alveolar sacs. (*b*) In phocids (true seals) the final airways are muscular tubes and the length of airway from the last piece of cartilage (i.e. the unprotected path length, u.p.) is 1–3 mm. (Denison & Kooyman, 1973.)

short as 0.3 s (1.5–2.0 s in man) (Olsen *et al.*, 1969; Kooyman & Sinnett, 1979; Kooyman & Cornell, 1981). During expiration (and inspiration), flow rates rapidly reach a peak and remain near maximum for a large proportion of the time, then rapidly decline to zero. This is not the case for man, where flow rate during expiration declines steadily after the peak has been reached (Figure 2). The reason for the latter is probably that the airways narrow, and therefore their resistance to flow increases, as pleural pressure exceeds airway pressure. This is prevented in small cetaceans and otariid seals by cartilage reinforcement of the peripheral airways (Figure

3*a*) (Denison & Kooyman, 1973; Kooyman & Sinnett, 1979). This reinforcement may also be important in preventing collapse of the lower airways when the animals are diving deeply, thus allowing the alveoli to empty and obviating the problems of nitrogen narcosis, decompression sickness and the squeeze. Strangely, though, in the lungs of phocid seals, some of which dive to extreme depths, the lower airways are muscular tubes with no cartilage (Figure 3*b*). Denison & Kooyman (1973) suggested that this may be sufficient reinforcement to prevent collapse at depth but insufficient to prevent collapse during rapid expiration. The former part of this suggestion has been substantiated by a study on pulmonary shunts in harbour seals and sea lions during simulated deep dives (Kooyman & Sinnett, 1982).

Circulatory system

The maximum oxygen storage capacity of the blood is, of course, related to the product of Hb concentration [Hb] and blood volume. Among birds, the chinstrap penguin with a haemoglobin concentration of 19.6 g% has a calculated oxygen carrying capacity of 26.3 ml $(100 \text{ ml})^{-1}$ (Table 1*c*). This is substantially greater than that recorded for domestic ducks (Meyer *et al.*, 1978; Black & Tenney, 1980) and somewhat greater than that in the pigeon and mallard duck (Table 1*c*). Total blood volume appears to be highest in the red-throated diver *G. stellata* at 13.2 ml $(100 \text{ g})^{-1}$ (Bond & Gilbert, 1958). Important though the oxygen in the blood undoubtedly is in birds, the relatively large volume of the respiratory system in these animals means that it has a usable oxygen store that is similar to or 1.5 times as large as that of the blood (Kooyman, 1975; Keijer & Butler, 1982).

As with Mb concentration, the phocid seals have the highest blood oxygen storage capacity and this relates to two features. One is the enlarged hepatic sinus in the venous system, which accommodates a large blood volume (Butler & Jones, 1982), the other is the large spleen (Qvist *et al.*, 1986). There is a 60–70% increase in Hb concentration in the blood of Weddell seals during voluntary submersion (Figure 4*b*). The conclusion of Qvist *et al.* (1986) is that oxygenated red blood cells (RBCs) are released from the spleen and that these substantially counteract the dramatic fall in partial pressure of oxygen in arterial blood (Pa_{O_2}) that occurs (figure 4*a,c*). It is assumed that the RBCs are fully oxygenated upon recovery and stored in the spleen until required again. Such storage appears not to occur between dives within a bout (Castellini *et al.*, 1988) as haematocrit and therefore Hb concentration are elevated throughout an entire bout. Uptake of excess RBCs into the spleen most likely occurs at the end of a bout.

Thus seals, which dive deeply and for long duration, have a greatly enhanced oxygen storage capacity of the blood compared with terrestrial mammals (Table 1*c*), particularly if calculated on a lean-mass basis. As blubber may account for up to 30% of body mass in seals, on a lean-mass basis blood volume can be above 20 ml $(100 \text{ g})^{-1}$. Most of the oxygen is stored in the blood and tissues of phocid seals, but there is proportionately more in the lungs, in seal lions and walruses (Lenfant *et al.*, 1970*b*). On a mass basis, the harbour seal, *Phoca vitulina*, can store 2.7 times the amount of oxygen that man can (Figure 5*a*). Its blood has a mass-specific oxygen capacity 2.3 times that of man, whereas the longer and deeper diving Weddell seal has a mass-specific oxygen capacity 1.5 times that of the harbour seal and 3.6 times that of man (figure 5*b*).

There are some interesting anomalies among the cetaceans. The larger species, such as the sperm whale, which are known to have diving abilities which exceed those of the Weddell seal, have a blood oxygen storage capacity similar to that of terrestrial mammals. On the other hand, the Dall porpoise, *Phocoenoides dalli*, which is a relatively shallow diver, has a high blood oxygen storage capacity (Ridgway & Johnston, 1966; Horvath *et al.*, 1968). As mentioned earlier though, the very large cetaceans, because of their relatively low mass-specific oxygen requirements, may have sufficient oxygen in their myoglobin to maintain aerobic metabolism during most of their feeding dives. The blood may, therefore, only be required to supply non-muscular parts of the body during the majority of dives and the muscles during exceptionally long periods of submersion.

In some aquatic homeotherms, the oxygen storage capacity of the blood varies with diving activity. When performing dives to a mean distance of 6 m and for a duration of 25 s, tufted ducks have a blood volume of 141 ml kg^{-1}, whereas when diving to a mean distance of 0.6 m for a mean duration of 11 s, blood volume is 107 ml kg (Stephenson *et al.*, 1988). The muskrat has a [Hb] of 20 g $(100 \text{ ml})^{-1}$ during the winter months when it dives beneath ice-covered ponds to feed, but in the summer when it dives for much shorter periods, [Hb] is approximately 14 g $(100 \text{ ml})^{-1}$ (Aleksiuk & Frohlinger, 1971; Snyder, 1983). On the other hand, oxygen transport and storage in the harbour seal are inherent and not the result of diving experience (Kodama *et al.*, 1977).

Thus, in some diving mammals at least, increased [Hb] and/or blood

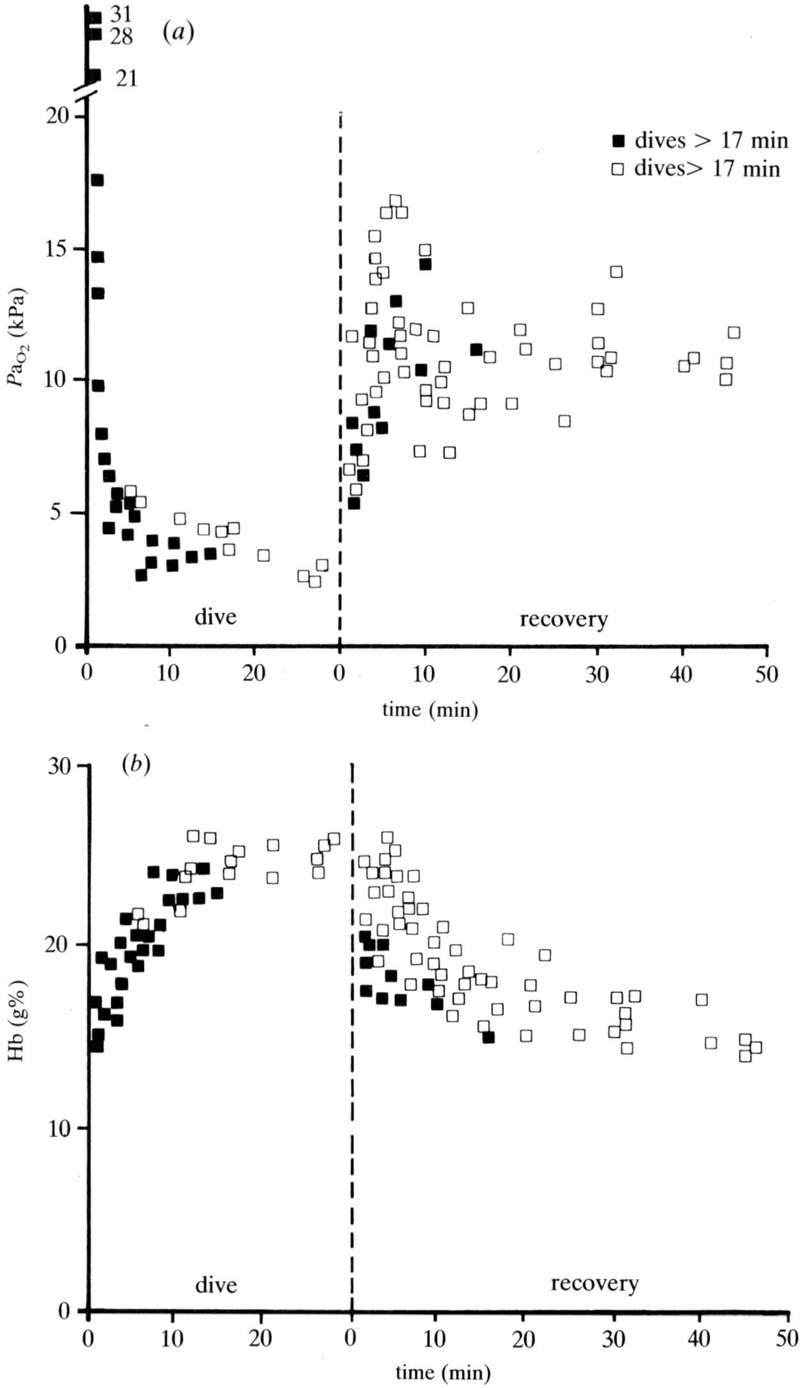
(a)
31
28
21
Pa_{O_2} (kPa)
■ dives > 17 min
□ dives> 17 min
dive
recovery
time (min)
(b)
Hb (g%)
dive
recovery
time (min)

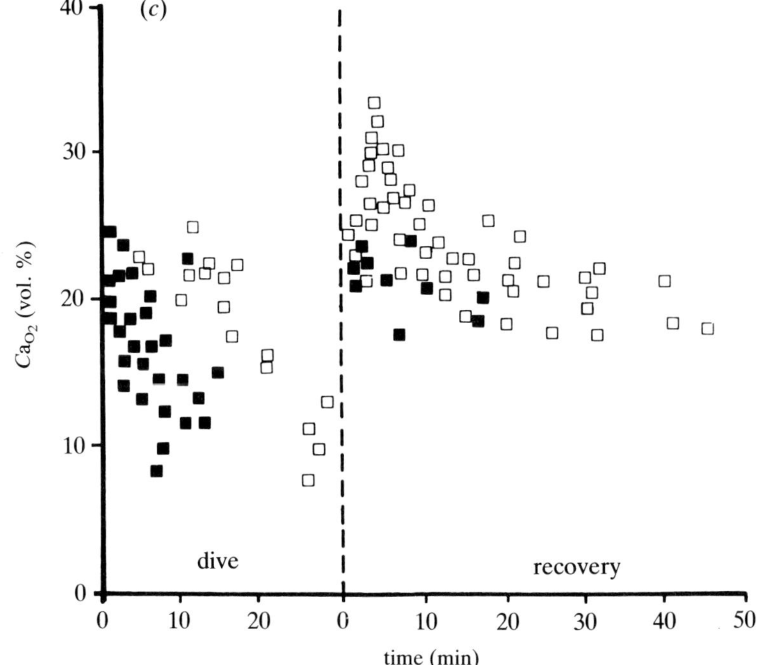

Figure 4. Changes in (*a*) partial pressure of oxygen, (*b*) haemoglobin concentration and (*c*) concentration of oxygen in arterial blood (Pa_{O_2}, Hb, Ca_{O_2}) of four Weddell seals during diving and after resurfacing. Dives were divided into those of short (less than 17 min) and long (more than 17 min) durations. In (*a*) high values of Pa_{O_2} obtained within the first minute of descent are the result of compression of the gases in the lung and those upon surfacing may be the result of the low body temperatures of the seals (i.e. Pa_{O_2} would be measured at a higher temperature than that of the blood). In (*c*) it is clear that, during long dives, Ca_{O_2} remained above the resting values of 18–19 vol.% for 15–17 min. From then on the rate of decrease was not significantly different from that measured during short dives. (Redrawn from Qvist *et al.*, 1986.)

volume enhance the oxygen storage ability of the blood . As far as the metabolizing cells are concerned, however, there is another important factor, which is the flow of blood to the tissues. As blood flow is exponentially related to blood viscosity, and blood viscosity is itself exponentially related to haematocrit, changes in [Hb] may have a significant effect on perfusion. Studies on the blood of elephant seals indicate that it has a haematocrit of 65% whereas in rabbits the value is 35% (Hedrick *et al.*, 1986). This, together with significantly greater plasma protein concentration, causes a three times greater viscosity in the seal (Figure 6). These

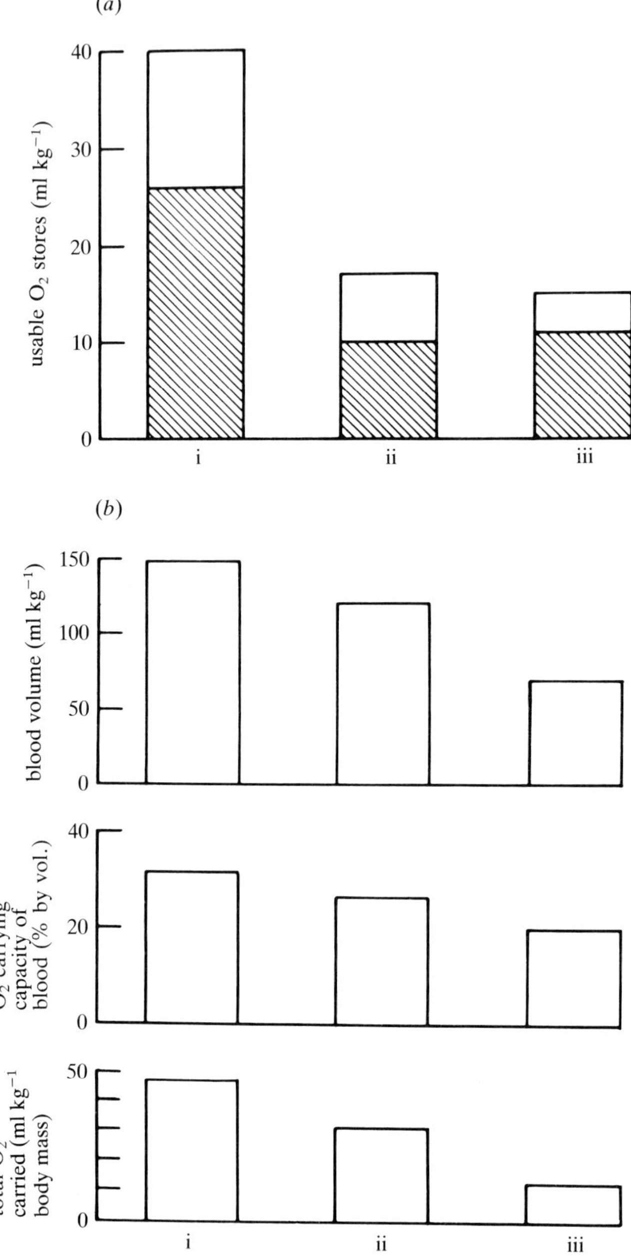

Figure 5. Oxygen stores in aquatic and terrestrial mammals. (*a*) Usable
oxygen stores in (i) harbour seal, (ii) dog and (iii) man. Cross-hatched
areas refer to nonpulmonary stores. (*b*) Total blood volume per unit body

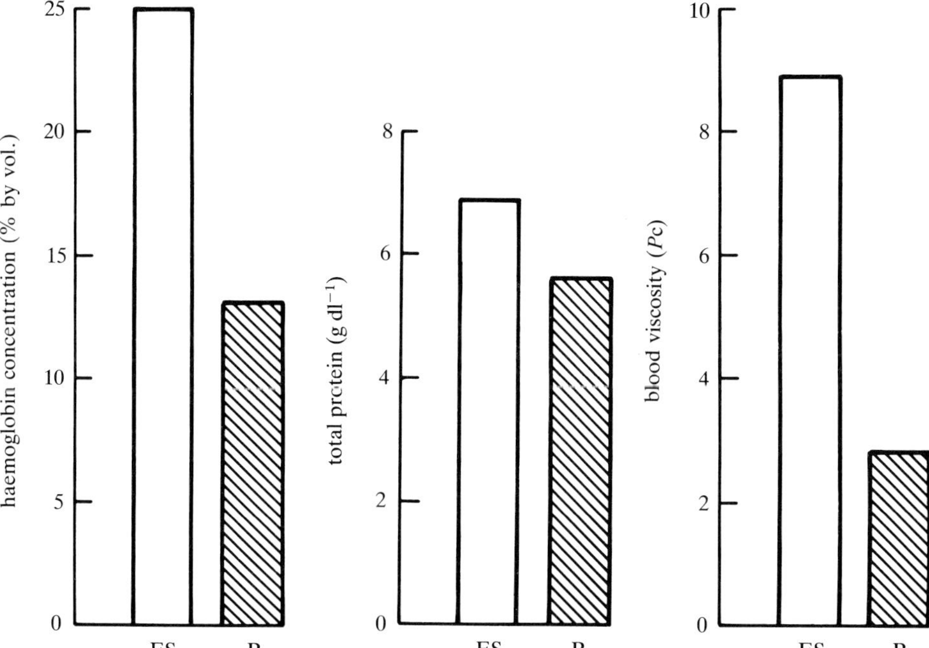

Figure 6. Haematological and rheological values for the blood of elephant seals (ES) and rabbits (R). (Data from Hedrick *et al.*, 1986.)

authors suggest that by increasing [Hb], animals like the elephant and Weddell seals may increase storage capacity at the expense of aerobic scope (because of viscosity-related limitations to perfusion). However, because of the large amount of O_2 stored in Mb, perfusion of the active locomotory muscles need not be that high anyway during normal feeding dives. By reducing [Hb] at the end of a bout, the Weddell seal also reduces blood viscosity and hence the energetic cost of perfusion.

Thus, the respiratory system may act as an important store of usable oxygen in shallow-diving birds and mammals. However, because of the problems associated with its compressibility, in deeper-diving species its main function is probably to reload the stores associated with myoglobin and haemoglobin. This is facilitated, not only by high respiratory flow rates and the ability to exchange most of the lung gases, but also by increased

Caption for fig. 5. (cont.)
mass, oxygen carrying capacity of a given volume of blood, and total amount of oxygen in blood per unit body mass of (i) Weddell seal, (ii) harbour seal and (iii) man. (Butler & Jones, 1982.)

circulation as a result of increased cardiac output (indicated by a large tachycardia upon surfacing) and, presumably, by peripheral vasodilatation (Butler & Woakes, 1979; Hill *et al.*, 1987; Fedak *et al.*, 1988). By optimizing all of these, some birds and mammals are not only able to perform remarkable feats on the basis of individual dives, but also to spend remarkable proportions of their daily activities under water (Butler, 1989).

References

Aleksiuk, M. & Frohlinger, A. (1971). Seasonal metabolic organization in the muskrat (*Ondatra zibethica*). I. Changes in growth, thyroid activity, brown adipose tissue, and organ weights in nature. *Canadian Journal of Zoology* **49**, 1143–54.

Au, D. & Weihs, D. (1980). At high speeds dolphins save energy by leaping. *Nature* **284**, 548–50.

Baudinette, R.V. & Gill, P. (1985). The energetics of 'flying' and 'paddling' in water: locomotion in penguins and ducks. *Journal of Comparative Physiology* **155**, 373–80.

Baumann, F.H. & Baumann, R. (1977). A comparative study of the respiratory properties of bird blood. *Respiration Physiology* **31**, 333–43.

Black, C.P. & Tenney, S.M. (1980). Oxygen transport during progressive hypoxia in high altitude and sea-level water fowl. *Respiratory Physiology* **39**, 217–39.

Blake, R.W. (1983). Energetics of leaping in dolphins and other aquatic animals. *Journal of the Marine Biological Association, U.K.* **63**, 61–70.

Bond, C.F. & Gilbert, P.W. (1958). Comparative study of blood volume in representative aquatic and nonaquatic birds. *American Journal of Physiology* **194**, 519–21.

Butler, P.J. (1988). The exercise response and the 'classical' diving response during natural submersion in birds and mammals. *Canadian Journal of Zoology* **66,** 29–39.

Butler, P.J. (1989). Metabolic adjustments to breath-holding in higher vertebrates. *Canadian Journal of Zoology* **63**, 3024–31.

Butler, P.J. & Jones, D.R. (1982). The comparative physiology of diving in vertebrates. *Advances in Comparative Physiology and Biochemistry* **8**, 179–364.

Butler, P.J. & Turner, D.L. (1988). Effect of training on maximal oxygen uptake and aerobic capacity of locomotory muscles in tufted ducks, *Aythya fuligula*. *Journal of Physiology* **401**, 347–59.

Butler, P.J. & Woakes, A.J. (1979). Changes in heart rate and respiratory frequency during natural behaviour of ducks, with particular reference to diving. *Journal of Experimental Biology* **79**, 283–300.

Butler, P.J. & Woakes, A.J. (1984). Heart rate and aerobic metabolism in Humboldt penguins, *Spheniscus humboldti*, during voluntary dives. *Journal of Experimental Biology* **108**, 419–28.

Castellini, M.A., Davis, R.W. & Kooyman, G.L. (1988). Blood chemistry regulation during repetitive diving in Weddell seals. *Physiological Zoology* **61**, 379–86.

Castellini, M.A. & Somero, G.N. (1981). Buffering capacity of vertebrate muscle: correlations with potentials for anaerobic function. *Journal of Comparative Physiology* **143**, 191–8.

Clarke, M.R. (1976). Observation on sperm whale diving. *Journal of the Marine Biological Association, U.K.* **56**, 809–10.

Clarke, M.R. (1979). The head of the sperm whale. *Scientific American* **240**, 128–41.

Clausen, G. & Ersland, A. (1968). The respiratory properties of the blood of two diving rodents, the beaver and the water vole. *Respiration Physiology*, **5**, 350–9.

Clausen, G. & Ersland, A. (1969). The respiratory properties of the blood of the bladdernose seal (*Cystophora cristata*). *Respiration Physiology* **7**, 1–6.

Conroy, J.W.H. & Jenkins, D. (1986). Ecology of otters in northern Scotland. VI. Diving times and hunting success of otters (*Lutra lutra*) at Dinnet Lochs, Aberdeenshire and in Yell Sound, Shetland. *Journal of Zoology, London* **209**, 341–6.

Conroy, J.W.H. & Twelves, E.L. (1972). Diving depths of the gentoo penguin (*Pygoscelis papua*) and blue-eyed shag (*Phalacrocorax atriceps*) from the South Orkney Islands. *British Antarctic Survey Bulletin* **30**, 106–8.

Cooper, J. (1985). Diving patterns of cormorants Phalacrocoracidae. *Ibis* **128**, 562–70.

Davis, R.W. Kooyman, G.L. & Croxall, J.P. (1983). Water flux and estimated metabolism of free-ranging gentoo and macaroni penguins at South Georgia. *Polar Biology* **2**, 41–6.

Denison, D.M. & Kooyman, G.L. (1973). The structure and function of the small airways in pinniped and sea otter lungs. *Respiration Physiology* **17**, 1–10.

Dunstone, N. (1978). The fishing strategy of the mink (*Mustela vison*); time budgeting of hunting effort? *Behaviour* **67**, 3–4.

Eliassen, E. (1960). Cardiovascular responses to submersion asphyxia in avian dives. *Årbok for Universitetet I Bergen Mat* (Naturv Serie) **2**, 1–100.

Erlinge, S. (1968). Food studies on captive otters *Lutra lutra*. *Oikos* **19**, 259–70.

Evans, W.E. (1971). Orientation behavior of delphinids: radio telemetric studies. *Annals of the New York Academy of Science* **188**, 142–60.

Falke, K J , Hill, R.D., Schneider, R.C., Guppy, M., Liggins, G.C., Hochachka, P.W., Elliott, R.E. & Zapol, W.M. (1985). Seal lungs collapse during free diving: evidence from arterial nitrogen tensions. *Science* **229**, 556–7.

Farhi, L.E. & Rahn, H. (1955). Gas stores of the body and the unsteady state. *Journal of Applied Physiology* **7**, 472–84.

Fedak, M.A., Pullen, M.R. & Kanwisher, J. (1988). Circulatory responses of seals to periodic breathing: heart rate and breathing during exercise and diving in the laboratory and open sea. *Canadian Journal of Zoology* **66**, 53–60.

Gehr, P., Mwangi, D.K., Ammann, A., Maloiy, G.M.O., Taylor, C.R. & Weibel, E.R. (1981). Design of the mammalian respiratory system. V. Scaling morphometric pulmonary diffusing capacity to body mass: wild and domestic mammals. *Respiration Physiology* **44**, 61–86.

Gentry, R.L. & Kooyman, G.L. (1986). *Fur Seals. Maternal Strategies on Land and at Sea.* Princeton University Press.

Guppy, M., Hill, R.D., Schneider, R.C., Qvist, J., Liggins, G.C., Zapol, W.M. & Hochachka, P.W. (1986). Microcomputer-assisted metabolic studies of voluntary diving of Weddell seals. *American Journal of Physiology* **250**, R175–87.

Hedrick, M.S. & Duffield, D.A. (1986). Blood viscosity and optimal hematocrit in a deep-diving mammal, the northern elephant seal (*Mirounga angustirostris*). *Canadian Journal of Zoology* **64**, 2081–5.

Hill, R.D., Schneider, R.C., Liggins, G.C., Schuette, A.H., Eliott, R.L., Guppy, M., Hochachka, P.W., Qvist, J., Falke, K.J. & Zapol, W.M. (1987). Heart rate and body temperature during free diving of Weddell seals. *American Journal of Physiology* **253**, R344–51.

Horvath, S.M., Chiodi, H., Ridgway, S.H. & Azar, S. (1968). Respiratory and electrophoretic characteristics of hemoglobin of porpoises and sea lions. *Comparative Biochemistry and Physiology* **24**, 1027–33.

Hui, C.A. (1988). Penguins swimming. II Energetics and behaviour. *Physiological Zoology* **61**, 344–50.

Johansen, K., Berger, M., Bicudo, J.E.P.W., Ruschi, A. & de Almeida, P.J. (1987). Respiratory properties of blood and myoglobin in hummingbirds. *Physiological Zoology* **60**, 269–78.

Jones, D.R. & Furilla, R.A. (1987). The anatomical, physiological, behavioral, and metabolic consequences of voluntary and forced diving. In *Bird Respiration*, vol. 2 (Ed. T.J. Seller), pp. 75–125. Boca Raton, Florida: CRC Press.

Keijer, E. & Butler, P.J. (1982). Volumes of the respiratory and circulatory systems in tufted and mallard ducks. *Journal of Experimental Biology* **101**, 213–20.

Kodama, A.M., Elsner, R. & Pace, N. (1987). Effects of growth, diving history and high altitude on blood oxygen capacity in harbor seals. *Journal of Applied Physiology* **42**(6), 852–8.

Kooyman, G.L. (1966). Maximum diving capacities of the Weddell seal, *Leptonychotes weddelli*. *Science* **151**, 1553–4.

Kooyman, G.L. (1973). Respiratory adaptations in marine mammals. *American Zoology* **13**, 457–68.

Kooyman, G.L. (1975). Behaviour and physiology of diving. In *Biology of Penguins* (ed. B. Stonehouse), pp. 115–37. Macmillan, New York.

Kooyman, G.L. (1985). Physiology without restraint in diving mammals. *Marine Mammal Science* **1**, 166–78.

Kooyman, G.L. & Cornell, L.H. (1981). Flow properties of expiration and inspiration in a trained bottlenosed porpoise. *Physiological Zoology* **54**, 55–61.

Kooyman, G.L. & Davis, R.W. (1987). Diving behavior and performance, with special reference to penguins. In *Seabirds: Feeding Biology and Role in Marine Ecosystems* (ed. J.P. Croxall), pp. 63–75. Cambridge University press.

Kooyman, G.L., Davis, R.W., Croxall, J.P. & Costa, D.P. (1982). Diving depths and energy requirements of king penguins. *Science* **217**, 726–7.

Kooyman, G.L., Drabek, C.M., Elsner, R. & Campbell, W.B. (1971*a*). Diving behavior of the emperor penguin, *Aptenodytes forsteri. The Auk* **88**, 775–95.

Kooyman, G.L., Kerem, D.H., Campbell, W.B. & Wright, J.J. (1971*b*). Pulmonary function in freely diving Weddell seals, *Leptonychotes weddelli. Respiration Physiology* **12**, 271–82.

Kooyman, G.L., Kerem, D.H., Campbell, W.B. & Wright, J.J. (1973*a*). Pulmonary gas exchange in freely diving Weddell seals, *Leptonychotes weddelli. Respiration Physiology* **17**, 283–90.

Kooyman, G.L., Schroeder, J.P., Greene, D.G. & Smith, V.A. (1973*b*). Gas exchange in penguins during simulated dives to 30 and 68 m. *American Journal of Physiology* **225**, 1467–71.

Kooyman, G.L. & Sinnett, E.E. (1979). Mechanical properties of the harbor porpoise lung, *Phocoena phocoena. Respiration Physiology* **36**, 287–300.

Kooyman, G.L. & Sinnett, E.E. (1982). Pulmonary shunts in harbor seals and sea lions during simulated dives to depth. *Physiological Zoology* **55**, 105–11.

Kooyman, G.L., Wahrenbrock, E.A., Castellini, M.A., Davis, R.W. & Sinnett, E.E. (1980). Aerobic and anaerobic metabolism during voluntary diving in Weddell seals: evidence of preferred pathways from blood chemistry and behavior. *Journal of Comparative Physiology* **138**, 335–46.

Kruuk, H. & Hewson, R. (1977). Spacing and foraging of otters (*Lutra lutra*) in a marine habitat. *Journal of Zoology, London* **185**, 205–12.

Lanphier, E.H. & Rahn, H. (1963). Alveolar gas exchange during breath-hold diving. *Journal of Applied Physiology* **18**, 471–7.

Lavigne, D.M., Innes, S., Worthy, G.A.J. & Kovacs, K.M. (1986). Metabolic rate – body size relations in marine mammals. *Journal of Theoretical Biology* **122**, 123–4.

Lawrie, R.A. (1950). Some observations on factors affecting myoglobin concentrations in muscle. *Journal of Agricultural Science* **40**, 356–66.

Le Boeuf, B.J., Costa, D.P., Huntley, A.C. & Feldcamp, S.D. (1988).

256

Continuous, deep diving in female northern elephant seals, *Mirounga angustirostris*. *Canadian Journal of Zoology* **66**, 446–58.

Lenfant, C. (1969). Physiological properties of the blood of marine mammals. In *The Biology of Marine Mammals* (ed. H.T. Andersen), pp. 95–116. New York: Academic Press.

Lenfant, C., Elsner, R., Kooyman, G.L. & Drabek, C.M. (1969*a*). Respiratory function of the blood of the Adélie penguin *Pygoscelis adeliae*. *American Journal of Physiology* **216**, 1598–600.

Lenfant, C., Elsner, R., Kooyman, G.L. & Drabek, C.M. (1969*b*). Respiratory function of blood of the adult and fetus Weddell seal *Leptonychotes weddelli*. *American Journal of Physiology* **216**, 1595–7.

Lenfant, C., Elsner, R., Kooyman, G.L. & Drabek, C.M. (1970*a*). Tolerance to sustained hypoxia in the Weddell seal, *Leptonychotes weddelli*. In *Antarctic Ecology*, vol. 1 (ed. M.W. Holdgate), pp. 471–6. London: Academic Press.

Lenfant, C., Johansen, K. & Torrance, J.D. (1970*b*). Gas transport and oxygen storage capacity in some pinnipeds and the sea otter. *Respiration Physiology* **9**, 277–86.

Lishman, G.S. & Croxall, J.P. (1983). Diving depths of the chinstrap penguin *Pygoscelis antarctica*. *British Antarctic Survey Bulletin* **61**, 21–5.

Lockyer, C. (1976). Body weights of some species of large whales. *Journal du Conseil International pour l'Exploration de la Mer* **36**, 71–81.

Lockyer, C. (1977). Observations on diving behaviour of the sperm whale *Physeter catodon*. In *A Voyage of Discovery* (ed. M. Angel), pp. 591–609. Oxford: Pergamon Press.

MacArthur, R.A. (1984). Aquatic thermoregulation in the muskrat (*Ondatra zibethicus*): energy demands of swimming and diving. *Canadian Journal of Zoology* **62**, 241–8.

McKean, T. & Carlton, C. (1977). Oxygen storage in beavers. *Journal of Applied Physiology* **42**, 545–7.

Meyer, M., Holle, J.P. & Scheid, P. (1978). Bohr effect induced by CO_2 and fixed acid at various levels of O_2 saturation in duck blood. *Pflügers Archiv* **376**, 237–40.

Mill, G.K. & Baldwin, J. (1983). Biochemical correlates of swimming and diving behavior in the little penguin *Eudyptula minor*. *Physiological Zoology* **56**, 242–54.

Milsom, W.K., Johansen, K. & Millard, R.W. (1973). Blood respiratory properties in some Antarctic birds. *Condor* **75**, 472–4.

Nagy, K.A., Siegfried, W.R. & Wilson, R.P. (1984). Energy utilization by free-ranging Jackass penguins, *Spheniscus demersus*. *Ecology* **65**, 1648–55.

Olsen, C.R., Hale, F.C. & Elsner, R. (1969). Mechanics of ventilation in the pilot whale. *Respiration Physiology* **7**, 137–49.

Packer, B.S., Altman, M., Cross, C.E., Murdaugh, H.V., Linta, J.M. & Robin, E.D. (1969). Adaptations to diving in the harbor seal: oxygen stores and supply. *American Journal of Physiology* **217**, 903–6.

Pages, T. & Planas, J. (1983). Muscle myoglobin and flying habits in birds. *Comparative Biochemistry and Physiology* **74A**, 289–94.

Paulev, P. (1965). Decompression sickness following repeated breath-hold dives. *Journal of Applied Physiology* **20**, 1028–31.

Piatt, J.F. & Nettleship, D.N. (1985). Diving depths of four alcids. *The Auk* **102**, 293–7.

Qvist, J., Hill, R.D., Schneider, R.C., Falke, K.J., Liggins, G.C., Guppy, M., Elliot, R.L., Hochachka, P.W. & Zapol, W.M. (1986). Hemoglobin concentrations and blood gas tensions of free-diving Weddell seals. *Journal of Applied Physiology* **61**, 1560–9.

Ridgway, S.H. & Howard, R. (1979). Dolphin lung collapse and intramuscular circulation during free diving: evidence from nitrogen washout. *Science* **206**, 1182–3.

Ridgway, S.H. & Johnston, D.G. (1966). Blood oxygen and ecology of porpoises of three genera. *Science* **151**, 456–7.

Ridgway, S.H., Scronce, B.L. & Kanwisher, J. (1969). Respiration and deep diving in the bottlenose porpoise. *Science* **166**, 1651–4.

Scholander, P.F. (1940). Experimental investigations on the respiratory function in diving mammals and birds. *Hvalradets Skrifter* **22**, 1–131.

Schorger, A.W. (1947). The deep diving of the loon and old-squaw and its mechanism. *The Wilson Bulletin* **59**, 151–9.

Snyder, G.K. (1983). Respiratory adaptations in diving mammals. *Respiration Physiology* **54**, 269–94.

Stephenson, R., Butler, P.J. & Woakes, A.J. (1986). Diving behaviour and heart rate in tufted ducks (*Aythya fuligula*). *Journal of Experimental Biology* **126**, 341–59.

Stephenson, R., Turner, D.L. & Butler, P.J. (1988). The relationship between diving activity and oxygen storage capacity in the tufted duck, *Aythya fuligula*. *Journal of Experimental Biology* **141**, 265–75.

Trivelpiece, W.Z., Bengtson, J.L., Trivelpiece, S.G. & Volkman, N.J. (1986). Foraging behaviour of gentoo and chinstrap penguins as determined by new radiotelemetry techniques. *The Auk* **103**, 777–81.

Wanless, S., Morris, J.A. & Harris, M.P. (1988). Diving behaviour of guillemot *Uria aalge*, puffin *Fratercula arctica* and razorbill *Alca torda* as shown by radio-telemetry. *Journal of Zoology, London* **216**, 73–81.

Weber, R.E., Hemmingsen, E.A. & Johansen, K. (1974). Functional and biochemical studies of penguin myoglobin. *Comparative Biochemistry and Physiology* **49B**, 197–214.

Weber, R.E., Johansen, K. & Abe, A.S. (1981). Myoglobin from the burrowing reptile *Amphisbaena alba*. Concentrations and functional characteristics. *Comparative Biochemistry and Physiology* **A68**, 159–65.

Williams, T.M. (1983). Locomotion in the north American mink, a semi-aquatic mammal. I. Swimming energetics and body drag. *Journal of Experimental Biology* **103**, 155–68.

Woakes, A.J. & Butler, P.J. (1983). Swimming and diving in tufted ducks, *Aythya fuligula*, with particular reference to heart rate and gas exchange. *Journal of Experimental Biology* **107**, 311–29.

INDEX